Java EE实战精粹

——MyBatis+Spring+Spring MVC

高洪岩 著

人民邮电出版社
北京

图书在版编目（CIP）数据

Java EE实战精粹 : MyBatis+Spring+Spring MVC /
高洪岩著. -- 北京 : 人民邮电出版社, 2020.1（2022.8重印）
ISBN 978-7-115-51902-3

Ⅰ. ①J… Ⅱ. ①高… Ⅲ. ①JAVA语言－程序设计
Ⅳ. ①TP312.8

中国版本图书馆CIP数据核字(2019)第188260号

内容提要

本书主要讲解 Java EE 框架 MyBatis、Spring 和 SpringMVC 的核心开发技术，帮助读者进行“精要”式的学习和项目实战，同时汲取 Java EE 的思想，并最终将其灵活运用到实际工作中。

全书内容共 7 章，分别对 Mybatis、Spring 和 Spring MVC 的基础知识与核心技术实现进行了详细的描述。书中利用大量篇幅介绍了 Spring 中的 DI 与 AOP，这两种技术是 Spring 框架的内部原理，掌握它们是学习 Spring 的重中之重。另外，本书还系统解析了 MyBatis+Spring+SpringMVC 框架的整合，并介绍了如何使用 Spring Boot 开发 Web 软件项目。

本书适用于已具有一定 Java 编程基础的读者，包括具有 Servlet 编程经验，以及在 Java 平台下进行各类软件开发的开发人员、测试人员等。

◆ 著　　　　高洪岩
　责任编辑　陈聪聪
　责任印制　焦志炜

◆ 人民邮电出版社出版发行　　北京市丰台区成寿寺路 11 号
　邮编 100164　　电子邮件 315@ptpress.com.cn
　网址 http://www.ptpress.com.cn
　固安县铭成印刷有限公司印刷

◆ 开本：787×1092 1/16
　印张：28　　　　　　　　2020 年 1 月第 1 版
　字数：675 千字　　　　　2022 年 8 月河北第 2 次印刷

定价：108.00 元

读者服务热线：(010)81055410　印装质量热线：(010)81055316
反盗版热线：(010)81055315
广告经营许可证：京东市监广登字20170147号

前言

关于本书

从用人单位对信息技术（IT）人才招聘的要求来看，越来越趋向于“实战性”，也就是要求员工进入软件公司后能立即融入开发的任务中，快速为软件公司创造巨大的经济利益。本书就本着这个出发点来进行设计。

内容精悍而不失实用价值的主流 Java EE 开源框架图书，应具有包含主流框架中相当重要且核心、常用的内容，而不是那些无用的知识，这样读者就可以快速上手，并沿着这个核心探索出一些方向，自行在工作和学习中不断拓展和挖掘。这就是作者撰写本书的主要目的。

Java EE 的世界非常庞大，以至于世界上没有任何一本书能把它讲得非常完整或详细。要想学好 Java 语言或 Java EE 框架并掌握其中丰富的编码技巧，以及设计模式、代码优化，并将它们综合地熟练应用在软件项目中并没有捷径，只有从零开始练习。

本书章节的编排不但涵盖了学习主流 Java EE 框架所需掌握的核心技术，也涵盖了使用它们进行项目实战的必备知识，主旨就是希望让读者尽快上手，掌握开源 Java EE 框架的核心内容，正确进行项目实战，汲取 Java EE 的思想，并最终将这种思想活用到实际工作中。

现在主流的 Java EE 框架是 SSM 或 Spring Boot，但 Spring Boot 框架仅仅就是一个“盒子”、一个“封装器”。想要实现功能还是需要整合其他第三方的框架，比如 MyBatis、Spring 或 Spring MVC 等，因此，在学习 Spring Boot 之前必须要有 SSM 框架的开发经验，那么能让一位 Java EE 框架初学者从零开始到最终掌握这几个框架，一直是作者的写作目标。有些 Java EE 开源的框架的确能非常大地提高开发效率，但因为使用的人不多，所以覆盖面比较窄。而软件公司在招聘时的技术需求大多数情况下却是“大众化”的，这就要求应聘者在面试前就有主流 Java EE 框架的学习或使用经验。如果找不到合适的教材，读者在学习某一项技术时就根本摸不清哪些知识点是常用的，哪些是不常用的，大大降低学习效率，分散了注意力。

本书面向的读者

首先，本书适合所有 Java 程序员，作为 Java 开源世界的主流框架，Java EE 程序员没有理由不学习它们。其次，本书适合希望学习这些框架编程的在校学生。学校的功课很多，而一本大部头的框架书籍需要花费大量的时间研读，因此在学习效率上，本书可以快速带领读者进入 Java EE 框架开发的殿堂，同时又不会遗漏掉应该掌握的核心技能。

本书的结构

第 1 章将会介绍基于 SQL 映射的 MyBatis 框架，在本章中，读者可以使用此框架操作主流的数据库，并学习 MyBatis 核心 API 的使用，采用自定义封装法来简化 MyBatis 的操作代码，进而提高开发效率。

第 2 章主要讲解了有关 MyBatis 映射的知识，包括<sql>、<resultMap>、<choose>、<set>、<foreach>等常用标签，DB 连接信息存储到 Properties 文件的读取，使用 JDBC 数据源，别名 typeAliases 的配置，CLOB 字段的读取以及分页等必备技术点。

第 3 章和第 4 章开始介绍 Spring 中的 IOC 和 AOP 技术，包含注入、注入原理、动态代理的实现与 AOP 切面的原理。

第 5 章将介绍时下流行的 Spring MVC 框架，读者可体会使用此框架开发一个经典登录功能时使用的技术点，限制 form 提交方式，还要掌握分组分模块开发的技术，重定向/转发的使用，JSON+AJAX+Spring MVC 联合开发，上传/下载的实现，以及数据验证功能的使用，XML 配置文件的处理，业务层 Service 的注入，ModelAndView 对象的使用，以及 HttpSession 在 Spring MVC 中的使用等功能。

第 6 章展现了特别常用的 MyBatis+Spring+Spring MVC 整合，以当前极具实战的组合框架来讲解整合的过程，而不囿于某一个框架本身，整合后的项目代码写法更加统一，便于维护与扩展。

第 7 章完整介绍了使用 Spring Boot 开发 Web 软件项目的过程。通过将 Spring MVC 与 MyBatis、Spring Boot 进行整合，带领读者学以致用。

要学好 Java EE，不但要有较好的 Java 基础，还需要了解多项开发框架的技术，因为 Java EE 框架本身就是采用这些技术开发出来的。

如何使用本书

首先需要声明，本书不是 Java Web 的入门教程，学习本书之前先要对 Java Web 中的 JSP、Servlet 等 Web 技术有所了解，尽量能完整地使用 JSP 或 Servlet 开发一个小型项目，再阅读本书，读者会发现代码的分层更加明确，结构更加清楚。

软件开发实践才是硬道理，设计、排错、拥有更多想法的经验非常重要，因此，请读者使用手中的键盘，练习一下吧！

尽量在读者自己的计算机中动手执行本书的所有代码。仅仅看书和自己动手存在天壤之别，动手运行代码和看书结合起来才能更深地理解框架的各项功能。

如何下载代码

为了环保，本书不提供光盘，读者可以从异步社区下载本书配套资源。

如何与作者联系

由于 Java EE 内容涵盖面广，涉及的知识点非常多，加之作者水平有限，错误之处在所难免，恳请各位读者批评指正。可以通过 QQ：279377921 与作者联系，期待与读者进行技术上的交流。

感谢

在本书出版的过程中，得到公司领导和同事的大力支持，在此表示感谢；感谢家人给予我充足的时间来撰写本书；感谢出生两个多月的儿子高晟京，看到你，我更加有了动力；最后感谢在本书上耗费大量精力的各位编辑，编辑们在工作上仔细、谨慎的工作态度值得我学习。

资源与支持

本书由异步社区出品，社区（https://www.epubit.com/）为您提供相关资源和后续服务。

配套资源

本书提供如下资源：

- 本书配套资源请到异步社区本书购买页处下载。

要获得以上配套资源，请在异步社区本书页面中点击 配套资源 ，跳转到下载界面，按提示进行操作即可。注意：为保证购书读者的权益，该操作会给出相关提示，要求输入提取码进行验证。

提交勘误

作者和编辑尽最大努力来确保书中内容的准确性，但难免会存在疏漏。欢迎您将发现的问题反馈给我们，帮助我们提升图书的质量。

当您发现错误时，请登录异步社区，按书名搜索，进入本书页面，点击“提交勘误”，输入勘误信息，单击“提交”按钮即可。本书的作者和编辑会对您提交的勘误进行审核，确认并接受后，您将获赠异步社区的100积分。积分可用于在异步社区兑换优惠券、样书或奖品。

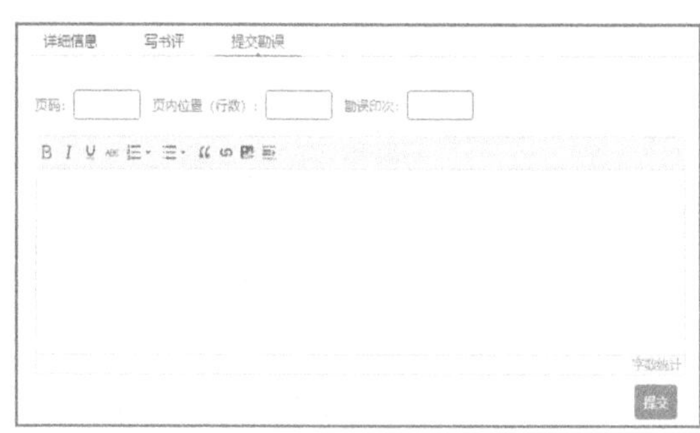

扫码关注本书

扫描下方二维码，您将会在异步社区微信服务号中看到本书信息及相关的服务提示。

与我们联系

我们的联系邮箱是 contact@epubit.com.cn。

如果您对本书有任何疑问或建议，请您发邮件给我们，并请在邮件标题中注明本书书名，以便我们更高效地做出反馈。

如果您有兴趣出版图书、录制教学视频，或者参与图书翻译、技术审校等工作，可以发邮件给我们；有意出版图书的作者也可以到异步社区在线提交投稿（直接访问 www.epubit.com/selfpublish/submission 即可）。

如果您是学校、培训机构或企业，想批量购买本书或异步社区出版的其他图书，也可以发邮件给我们。

如果您在网上发现有针对异步社区出品图书的各种形式的盗版行为，包括对图书全部或部分内容的非授权传播，请您将怀疑有侵权行为的链接发邮件给我们。您的这一举动是对作者权益的保护，也是我们持续为您提供有价值的内容的动力之源。

关于异步社区和异步图书

“**异步社区**”是人民邮电出版社旗下 IT 专业图书社区，致力于出版精品 IT 技术图书和相关学习产品，为作译者提供优质出版服务。异步社区创办于 2015 年 8 月，提供大量精品 IT 技术图书和电子书，以及高品质技术文章和视频课程。更多详情请访问异步社区官网 https://www.epubit.com。

“**异步图书**”是由异步社区编辑团队策划出版的精品 IT 专业图书的品牌，依托于人民邮电出版社近 30 年的计算机图书出版积累和专业编辑团队，相关图书在封面上印有异步图书的 LOGO。异步图书的出版领域包括软件开发、大数据、AI、测试、前端、网络技术等。

异步社区

微信服务号

目录

第 1 章　MyBatis 3 核心技术之必备技能

本章目标：

- 了解 MyBatis 核心对象的生命周期
- 使用 SqlSession 对象或 Mapper 接口操作数据库
- MyBatis 结合 ThreadLocal 类进行 CURD 的封装
- 手动搭建开发环境

1.1　什么是框架

MyBatis 是一个操作数据库的框架，那什么是框架？框架就是软件功能的半成品，框架提供了一个软件项目中通用的功能，将大多数常见的功能进行封装，无须自己重复开发，框架提高了开发及运行效率。在软件公司中，大多数情况是使用框架开发软件项目。

1.2　什么是对象关系映射

MyBatis 是一个基于“ORM”的框架，ORM 的全称是对象关系映射（Object Relational Mapping）。

对象（Object）就是 Java 中的对象，关系（Relational）就是数据库中的数据表，基于“ORM”的框架是把数据在对象和关系之间进行双向转换。

ORM 的细节可以从 3 个方面来介绍。

（1）1 个类对应 1 个表。

（2）1 个类的对象对应表中的 1 行。

（3）1 个类的对象中的属性对应 1 个表中的列。

ORM 映射关系如图 1-1 所示。

MyBatis 框架可以将 Java 类中的数据转化成数据表中的记录，或者将数据表中的记录封装到 Java 类中，过程如图 1-2 所示。

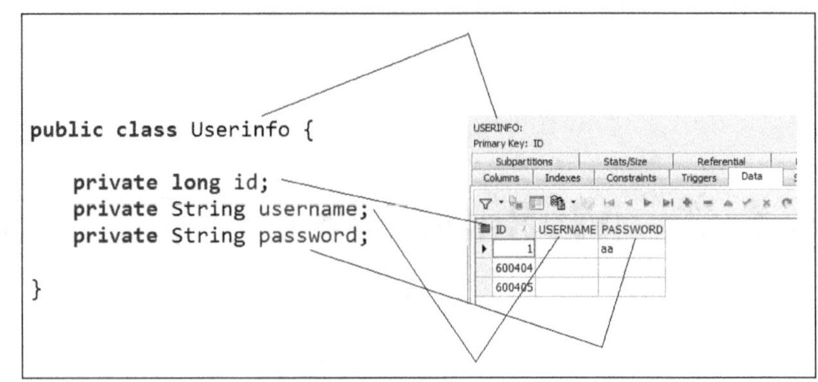

图 1-1 ORM 映射关系

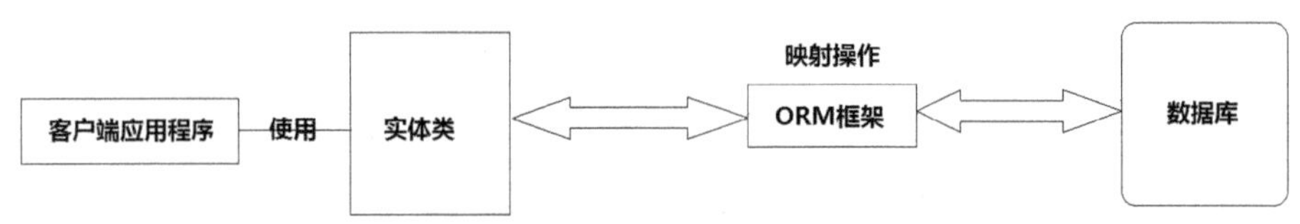

图 1-2 ORM 框架映射的主流程

从图 1-2 的过程来看，程序员不再直接使用 JDBC 对象访问数据库，而是以面向对象的方式来使用实体类，对实体类进行的增加、删除、更新和查询操作都会由 ORM 框架转化成对数据库的增加、删除、更新和查询操作。

ORM 框架内部的核心技术的原理其实就是 JDBC 和反射，这个技术是由 ORM 框架，也就是 MyBatis 来进行封装的。

1.3 MyBatis 的优势

一个事物一定要有优势才不会被淘汰，软件技术也是如此。

MyBatis 是现阶段操作数据库的主流框架，此框架的主要作用就是更加便捷地操作数据库。它具有很多优势，和程序员密切相关的优势主要体现在如下 7 个方面。

（1）ROW 行与 Entity 实体类双向转换：可以将数据表中的 ROW 行与 Entity 实体类进行互相转换，比如将 ResultSet 对象返回的数据自动封装到 Entity 实体类或 List 中，或将 Entity 实体类中的数据转换成数据表中新的一行 ROW。

（2）SQL 语句与 Java 文件分离：可以把 SQL 语句写到 XML 文件中，目的是将 SQL 语句与 Java 文件进行分离，有利于代码的后期维护，也可使代码的分层更加明确。

（3）允许对 SQL 语句进行自定义优化：因为 MyBatis 框架是使用 SQL 语句对数据库进行操作的，所以可以单独地对 SQL 语句进行优化，以提高操作效率。而 Hibernate 框架却做不到这一点，所以 MyBatis 相比 Hibernate 框架就具有很大的优势，这也是现阶段大部分软件公司逐步用 MyBatis 替换掉 Hibernate 框架的主要原因。

（4）减化 DAO 层代码：使用传统的 JDBC 开发方式时，需要写上必要的 DAO 层代码以对数据库进行操作，但这样的代码写法在软件开发的过程中非常不便，因为多个 DAO 类中的大部分 JDBC 代码是冗余的，所以 MyBatis 解决了这个问题。使用 MyBatis 做查询时可以自动将数据表中的数据记录封装到实体类或 Map 中，再将它们放入 List 进行返回。这么常见而且有利于提高开发效率的功能 MyBatis 都可以自由方便地处理，不需要程序员写底层的实现代码，MyBatis 就可以完全进行封装。从此观点来看，使用 MyBatis 框架去开发软件非常方便、快捷，省略了大量冗余的 JDBC 代码，MyBatis 把常用的 JDBC 操作都进行了封装，进而提高开发效率，MyBatis 框架很有使用上的必要性。

（5）半自动化所带来的灵活性：MyBatis 是“半自动化”的“ORM”框架，但它应该算作 SQL 映射框架（SQL Mapper Framework）。将 MyBatis 称为“半自动化的 ORM 框架”是因为 MyBatis 操作数据库时还是使用原始的 SQL 语句，这些 SQL 语句还需要程序员自己来进行设计，这就是半自动化。MyBatis 在使用方式上和全自动化的 ORM 框架 Hibernate 有着非常大的区别，MyBatis 是以 SQL 语句为映射基础，而 Hibernate 是彻底地基于实体类与表进行映射，基本是属于全自动化的 ORM 框架。但正是因为 MyBatis 属于半自动化的 ORM 框架这个特性，所以可以将 SQL 语句灵活多变的特性融入项目开发中。

（6）支持 XML 或 Annotations 注解的方式进行 ORM：MyBatis 可以使用 XML 或 Annotations 注解的方式将数据表中的记录映射成 1 个 Map 或 Java POJO 实体类对象，但推荐使用 XML 方式，该方式也是 MyBatis 官方推荐的。

（7）功能丰富：MyBatis 还可以实现定义 SQL 段落、调用存储过程和进行高级映射等功能。

1.4　ORM 的原理实现

本章将探究 ORM 框架的底层原理，用代码来模拟实现一个微型的 ORM 功能。

1.4.1　使用 JDBC 和反射技术实现泛型 DAO

MyBatis 实现 ROW 行与 Entity 实体类双向转换的原理是基于 JDBC 和反射技术，MyBatis 框架只是对 JDBC 技术进行了轻量级的封装，使程序员更方便地去操作数据库。

为了对 MyBatis 框架核心功能的原理有更加细致的了解，在本节就来实现一下 ROW 行与 Entity 实体类的双向转换。

创建测试用的项目 entity-row-double。

创建获得 Connection 连接对象，代码如下：

```
package dbtools;

import java.sql.Connection;
import java.sql.DriverManager;
import java.sql.SQLException;

public class GetConnection {
    public static Connection getConnection() throws ClassNotFoundException, SQLException {
        String driverName = "oracle.jdbc.OracleDriver";
        String url = "jdbc:oracle:thin:@localhost:1521:orcl";
        String username = "y2";
        String password = "123";
        Class.forName(driverName);
        Connection conn = DriverManager.getConnection(url, username, password);
```

```
            return conn;
        }
    }
```

创建实体类 Userinfo，核心代码如下：

```
public class Userinfo {
    private long id;
    private String username;
    private String password;

    public Userinfo() {
    }

    public Userinfo(long id, String username, String password) {
        super();
        this.id = id;
        this.username = username;
        this.password = password;
    }

    //省略 get 和 set 方法
}
```

创建泛型 DAO 类 BaseDAO，代码如下：

```
package dao;

import java.lang.reflect.Field;
import java.sql.Connection;
import java.sql.PreparedStatement;
import java.sql.ResultSet;
import java.sql.SQLException;
import java.util.ArrayList;
import java.util.List;

import dbtools.GetConnection;

public class BaseDAO<T> {
    // 此方法模拟了 MyBatis 的 save()方法
    // MyBatis 框架内部的核心和本示例基本一样
    // 使用的技术就是 JDBC 和反射
    public void save(T t)
            throws IllegalArgumentException, IllegalAccessException,
ClassNotFoundException, SQLException {
        String sql = "insert into ";
        String colName = "";
        String colParam = "";
        String begin = "(";
        String end = ")";
        Class classRef = t.getClass();
        String tableName = classRef.getSimpleName().toLowerCase();
        sql = sql + tableName;
        List values = new ArrayList();
        Field[] fieldArray = classRef.getDeclaredFields();
        for (int i = 0; i < fieldArray.length; i++) {
            Field eachField = fieldArray[i];
            eachField.setAccessible(true);
```

```
                String eachFieldName = eachField.getName();
                Object eachValue = eachField.get(t);
                colName = colName + "," + eachFieldName;
                colParam = colParam + ",?";
                values.add(eachValue);
            }
            colName = colName.substring(1);
            colParam = colParam.substring(1);
            sql = sql + begin + colName + end;
            sql = sql + " values" + begin + colParam + end;
            System.out.println(sql);
            for (int i = 0; i < values.size(); i++) {
                System.out.println(values.get(i));
            }

            Connection conn = GetConnection.getConnection();
            PreparedStatement ps = conn.prepareStatement(sql);
            for (int i = 0; i < values.size(); i++) {
                ps.setObject(i + 1, values.get(i));
            }
            ps.executeUpdate();
            ps.close();
            conn.close();
        }

        // 此方法模拟了 MyBatis 的 get()方法
        public T get(Class<T> classObject, long id)
                throws InstantiationException, IllegalAccessException, ClassNotFoundException,
SQLException {
            T t = null;
            String sql = "select * from " + classObject.getSimpleName().toLowerCase() + "
where id=?";
            Connection conn = GetConnection.getConnection();
            PreparedStatement ps = conn.prepareStatement(sql);
            ps.setLong(1, id);
            ResultSet rs = ps.executeQuery();
            while (rs.next()) {
                t = classObject.newInstance();
                Field[] fieldArray = classObject.getDeclaredFields();
                for (int i = 0; i < fieldArray.length; i++) {
                    Field eachField = fieldArray[i];
                    eachField.setAccessible(true);
                    String fieldName = eachField.getName();
                    Object value = rs.getObject(fieldName);
                    if (value.getClass().getTypeName().equals("java.math.BigDecimal")) {
                        long longValue = Long.parseLong("" + value);
                        eachField.set(t, longValue);
                    } else {
                        eachField.set(t, value);
                    }
                }
            }
            rs.close();
            ps.close();
            conn.close();
            return t;
        }
```

```
        // 此方法模拟了 MyBatis 的 update()方法
        public void update(T t)
                throws IllegalArgumentException, IllegalAccessException, ClassNotFoundException,
SQLException {
            String sql = "update " + t.getClass().getSimpleName().toLowerCase() + " set ";
            String whereSQL = " where id=?";
            String colName = "";
            Class classRef = t.getClass();
            List values = new ArrayList();
            Field[] fieldArray = classRef.getDeclaredFields();
            long idValue = 0;
            for (int i = 0; i < fieldArray.length; i++) {
                Field eachField = fieldArray[i];
                eachField.setAccessible(true);
                String eachFieldName = eachField.getName();
                if (!eachFieldName.equals("id")) {
                    Object eachValue = eachField.get(t);
                    colName = colName + "," + eachFieldName + "=?";
                    values.add(eachValue);
                } else {
                    Object eachValue = eachField.get(t);
                    idValue = Long.parseLong(eachValue.toString());
                }
            }
            values.add(idValue);

            colName = colName.substring(1);
            sql = sql + colName + whereSQL;
            System.out.println(sql);
            for (int i = 0; i < values.size(); i++) {
                System.out.println(values.get(i));
            }

            Connection conn = GetConnection.getConnection();
            PreparedStatement ps = conn.prepareStatement(sql);
            for (int i = 0; i < values.size(); i++) {
                ps.setObject(i + 1, values.get(i));
            }
            ps.executeUpdate();
            ps.close();
            conn.close();
        }

        // 此方法模拟了 MyBatis 的 delete()方法
        public void delete(T t) throws IllegalArgumentException, IllegalAccessException,
ClassNotFoundException,
                SQLException, NoSuchFieldException, SecurityException {
            String sql = "delete from " + t.getClass().getSimpleName().toLowerCase();
            String whereSQL = " where id=?";
            Class classRef = t.getClass();
            List values = new ArrayList();
            Field idField = classRef.getDeclaredField("id");
            idField.setAccessible(true);
            Object object = idField.get(t);
            values.add(object);

            sql = sql + whereSQL;
            System.out.println(sql);
```

```
        for (int i = 0; i < values.size(); i++) {
            System.out.println(values.get(i));
        }

        Connection conn = GetConnection.getConnection();
        PreparedStatement ps = conn.prepareStatement(sql);
        for (int i = 0; i < values.size(); i++) {
            ps.setObject(i + 1, values.get(i));
        }
        ps.executeUpdate();
        ps.close();
        conn.close();
    }

}
```

增加记录的代码如下：

```
public class Insert {
    public static void main(String[] args)
            throws IllegalArgumentException, IllegalAccessException, ClassNotFoundException,
SQLException {
        Userinfo userinfo = new Userinfo();
        userinfo.setId(1000L);
        userinfo.setUsername("中国");
        userinfo.setPassword("中国人");
        BaseDAO<Userinfo> dao = new BaseDAO<>();
        dao.save(userinfo);
    }
}
```

查询记录的代码如下：

```
public class Select {
    public static void main(String[] args)
            throws InstantiationException, IllegalAccessException, ClassNotFoundException,
SQLException {
        BaseDAO<Userinfo> dao = new BaseDAO<>();
        Userinfo userinfo = dao.get(Userinfo.class, 1000L);
        System.out.println(userinfo.getId() + " " + userinfo.getUsername() + " " +
userinfo.getPassword());
    }
}
```

修改记录的代码如下：

```
public class Update {
    public static void main(String[] args)
            throws InstantiationException, IllegalAccessException, ClassNotFoundException,
SQLException {
        BaseDAO<Userinfo> dao = new BaseDAO<>();
        Userinfo userinfo = dao.get(Userinfo.class, 1000L);
        userinfo.setUsername("xxx");
        userinfo.setPassword("xxxxxx");
        dao.update(userinfo);
    }
}
```

删除记录的代码如下：

```
public class Delete {
    public static void main(String[] args) throws InstantiationException,
IllegalAccessException,
            ClassNotFoundException, SQLException, IllegalArgumentException,
NoSuchFieldException, SecurityException {
        BaseDAO<Userinfo> dao = new BaseDAO<>();
        Userinfo userinfo = dao.get(Userinfo.class, 1000L);
        dao.delete(userinfo);
    }
}
```

以上代码的作用就是使用 JDBC 结合反射技术来将数据表中的 1 行记录和实体类进行双向转换，这也是 ORM 框架的原理。以上代码用到了 JDBC（结合）反射技术，MyBatis 实现 ORM 的原理就是 JDBC 和反射技术。

1.4.2 操作 XML 文件

提到标记语言人们就容易想起 HTML。HTML 提供了很多标签来实现 Web 前端界面的设计，但 HTML 中的标签并不允许自定义。如果想定义一些独有的标签，HTML 就不再可行了，这时可以使用 XML 来实现。

XML 的全称是 eXtensible Markup Language（可扩展标记语言），它可以自定义标记名称与内容，在灵活度上相比 HTML 有大幅提高，经常用在配置以及数据交互领域。

在开发软件项目时，经常会接触 XML 文件，比如 web.xml 文件中就有 XML 代码，XML 代码的主要作用就是配置，那么在 Java 中如何读取 XML 中的内容呢？

创建名称为 xmlTest 的 Java 项目，在项目中引入 dom4j-1.6.1.jar 文件，创建 struts.xml 文件，代码如下：

```
<mymvc>
    <actions>
        <action name="list" class="controller.List">
            <result name="toListJSP">
                /list.jsp
            </result>
            <result name="toShowUserinfoList" type="redirect">
                showUserinfoList.ghy
            </result>
        </action>

        <action name="showUserinfoList" class="controller.ShowUserinfoList">
            <result name="toShowUserinfoListJSP">
                /showUserinfoList.jsp
            </result>
        </action>
    </actions>
</mymvc>
```

创建 Reader 类，代码如下：

```
package test;

import java.util.List;
```

```java
import org.dom4j.Attribute;
import org.dom4j.Document;
import org.dom4j.DocumentException;
import org.dom4j.Element;
import org.dom4j.io.SAXReader;

public class Reader {

    public static void main(String[] args) {
        try {
            SAXReader reader = new SAXReader();
            Document document = reader.read(reader.getClass()
                    .getResourceAsStream("/struts.xml"));

            Element mymvcElement = document.getRootElement();
            System.out.println(mymvcElement.getName());
            Element actionsElement = mymvcElement.element("actions");
            System.out.println(actionsElement.getName());
            System.out.println("");
            List<Element> actionList = actionsElement.elements("action");
            for (int i = 0; i < actionList.size(); i++) {
                Element actionElement = actionList.get(i);
                System.out.println(actionElement.getName());
                System.out.print("name="
                        + actionElement.attribute("name").getValue());
                System.out.println("action class="
                        + actionElement.attribute("class").getValue());

                List<Element> resultList = actionElement.elements("result");
                for (int j = 0; j < resultList.size(); j++) {
                    Element resultElement = resultList.get(j);
                    System.out.print("  result name="
                            + resultElement.attribute("name").getValue());
                    Attribute typeAttribute = resultElement.attribute("type");
                    if (typeAttribute != null) {
                        System.out.println(" type=" + typeAttribute.getValue());
                    } else {
                        System.out.println("");
                    }
                    System.out.println("   " + resultElement.getText().trim());
                    System.out.println("");
                }

                System.out.println("");
            }

        } catch (DocumentException e) {
            // TODO Auto-generated catch block
            e.printStackTrace();
        }
    }
}
```

程序运行后的结果如图 1-3 所示。

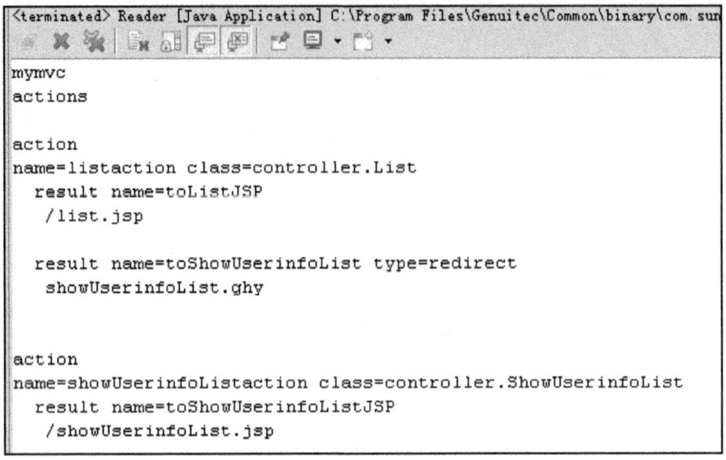

```
<terminated> Reader [Java Application] C:\Program Files\Genuitec\Common\binary\com.sun
mymvc
actions

action
name=listaction class=controller.List
  result name=toListJSP
   /list.jsp

  result name=toShowUserinfoList type=redirect
   showUserinfoList.ghy


action
name=showUserinfoListaction class=controller.ShowUserinfoList
  result name=toShowUserinfoListJSP
   /showUserinfoList.jsp
```

图 1-3　用 dom4j 解析的 XML 文件内容

上面的示例是读取解析 XML 文件，那么如何创建 XML 文件呢？继续创建一个名称为 createXML 的 Java 项目，并创建 Java 类 Writer，代码如下：

```
package test;

import java.io.FileWriter;
import java.io.IOException;

import org.dom4j.Document;
import org.dom4j.DocumentHelper;
import org.dom4j.Element;
import org.dom4j.io.OutputFormat;
import org.dom4j.io.XMLWriter;

public class Writer {

    public static void main(String[] args) {
        try {
            Document document = DocumentHelper.createDocument();
            Element mymvcElement = document.addElement("mymvc");
            Element actionsElement = mymvcElement.addElement("actions");
            // /
            Element listActionElement = actionsElement.addElement("action");
            listActionElement.addAttribute("name", "list");
            listActionElement.addAttribute("class", "controller.List");

            Element toListJSPResultElement = listActionElement
                    .addElement("result");
            toListJSPResultElement.addAttribute("name", "toListJSP");
            toListJSPResultElement.setText("/list.jsp");

            Element toShowUserinfoListResultElement = listActionElement
                    .addElement("result");
            toShowUserinfoListResultElement.addAttribute("name",
                    "toShowUserinfoList");
            toShowUserinfoListResultElement.addAttribute("type", "redirect");
            toShowUserinfoListResultElement.setText("showUserinfoList.ghy");
            // /
```

```
            Element showUserinfoListActionElement = actionsElement
                    .addElement("action");
            showUserinfoListActionElement.addAttribute("name",
                    "showUserinfoList");
            showUserinfoListActionElement.addAttribute("class",
                    "controller.ShowUserinfoList");

            Element toShowUserinfoListJSPResultElement = showUserinfoListActionElement
                    .addElement("result");
            toShowUserinfoListJSPResultElement.addAttribute("name",
                    "toShowUserinfoListJSP");
            toShowUserinfoListResultElement.setText("/showUserinfoList.jsp");
            // /

            OutputFormat format = OutputFormat.createPrettyPrint();
            XMLWriter writer = new XMLWriter(new FileWriter("ghy.xml"), format);
            writer.write(document);
            writer.close();

        } catch (IOException e) {
            // TODO Auto-generated catch block
            e.printStackTrace();
        }

    }

}
```

程序运行后在当前项目中创建名称为 ghy.xml 的文件，文件内容如图 1-4 所示。

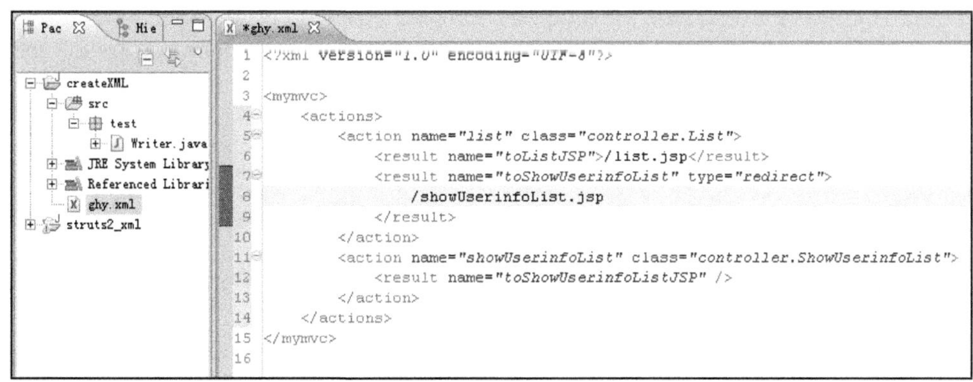

图 1-4　创建的 ghy.xml 中的内容

上面的示例是创建 XML 文件，那么如何修改 XML 文件呢？继续创建一个名称为 Update 的 Java 类，代码如下：

```
package dom4jTest;

import java.io.FileWriter;
import java.io.IOException;
import java.util.List;

import org.dom4j.Attribute;
import org.dom4j.Document;
import org.dom4j.DocumentException;
```

```
import org.dom4j.Element;
import org.dom4j.io.OutputFormat;
import org.dom4j.io.SAXReader;
import org.dom4j.io.XMLWriter;

public class Update {

    public static void main(String[] args) throws IOException {
        try {
            SAXReader reader = new SAXReader();
            Document document = reader.read(reader.getClass().getResourceAsStream("/test.xml"));
            Element mymvcElement = document.getRootElement();
            Element actionsElement = mymvcElement.element("actions");
            List<Element> actionList = actionsElement.elements("action");
            for (int i = 0; i < actionList.size(); i++) {
                Element actionElement = actionList.get(i);
                List<Element> resultList = actionElement.elements("result");
                for (int j = 0; j < resultList.size(); j++) {
                    Element resultElement = resultList.get(j);
                    String resultName = resultElement.attribute("name").getValue();
                    if (resultName.equals("toShowUserinfoList")) {
                        Attribute typeAttribute = resultElement.attribute("type");
                        if (typeAttribute != null) {
                            typeAttribute.setValue("zzzzzzzzzzzzzzzzzzzzzz");
                            resultElement.setText("xxxxxxxxxxxxxxxxxx");
                        }
                    }
                }
            }
            OutputFormat format = OutputFormat.createPrettyPrint();
            XMLWriter writer = new XMLWriter(new FileWriter("src\\ghy.xml"), format);
            writer.write(document);
            writer.close();
        } catch (DocumentException e) {
            e.printStackTrace();
        }
    }

}
```

产生的 XML 文件内容如下：

```
<?xml version="1.0" encoding="UTF-8"?>

<mymvc>
  <actions>
    <action name="list" class="controller.List">
      <result name="toListJSP">/list.jsp</result>
      <result name="toShowUserinfoList" type="zzzzzzzzzzzzzzzzzzzzzz">xxxxxxxxxxxxxxxxxx</result>
    </action>
    <action name="showUserinfoList" class="controller.ShowUserinfoList">
      <result name="toShowUserinfoListJSP">/showUserinfoList.jsp</result>
    </action>
  </actions>
</mymvc>
```

成功更改 XML 文件中的属性值与文本内容。

那么如何删除 XML 中的 Node 节点呢？创建 Delete.java，代码如下：

```
package dom4jTest;

import java.io.FileWriter;
import java.io.IOException;
import java.util.List;

import org.dom4j.Document;
import org.dom4j.DocumentException;
import org.dom4j.Element;
import org.dom4j.io.OutputFormat;
import org.dom4j.io.SAXReader;
import org.dom4j.io.XMLWriter;

public class Delete {

    public static void main(String[] args) throws IOException {
        try {
            SAXReader reader = new SAXReader();
            Document document = reader.read(reader.getClass().getResourceAsStream("/test.xml"));
            Element mymvcElement = document.getRootElement();
            Element actionsElement = mymvcElement.element("actions");
            List<Element> actionList = actionsElement.elements("action");
            for (int i = 0; i < actionList.size(); i++) {
                Element actionElement = actionList.get(i);
                List<Element> resultList = actionElement.elements("result");
                Element resultElement = null;
                boolean isFindNode = false;
                for (int j = 0; j < resultList.size(); j++) {
                    resultElement = resultList.get(j);
                    String resultName = resultElement.attribute("name").getValue();
                    if (resultName.equals("toShowUserinfoList")) {
                        isFindNode = true;
                        break;
                    }
                }
                if (isFindNode == true) {
                    actionElement.remove(resultElement);
                }
            }
            OutputFormat format = OutputFormat.createPrettyPrint();
            XMLWriter writer = new XMLWriter(new FileWriter("src\\ghy.xml"), format);
            writer.write(document);
            writer.close();
        } catch (DocumentException e) {
            e.printStackTrace();
        }
    }

}
```

产生的 XML 文件内容如下：

```
<?xml version="1.0" encoding="UTF-8"?>

<mymvc>
    <actions>
```

```
        <action name="list" class="controller.List">
            <result name="toListJSP">/list.jsp</result>
        </action>
        <action name="showUserinfoList" class="controller.ShowUserinfoList">
            <result name="toShowUserinfoListJSP">/showUserinfoList.jsp</result>
        </action>
    </actions>
</mymvc>
```

成功删除 Node 节点。

那么如何删除属性 Attr 呢？创建 DeleteAttr 类，代码如下：

```
package dom4jTest;

import java.io.FileWriter;
import java.io.IOException;
import java.util.List;

import org.dom4j.Attribute;
import org.dom4j.Document;
import org.dom4j.DocumentException;
import org.dom4j.Element;
import org.dom4j.io.OutputFormat;
import org.dom4j.io.SAXReader;
import org.dom4j.io.XMLWriter;

public class DeleteAttr {

    public static void main(String[] args) throws IOException {
        try {
            SAXReader reader = new SAXReader();
            Document document = reader.read(reader.getClass().getResourceAsStream("/test.xml"));
            Element mymvcElement = document.getRootElement();
            Element actionsElement = mymvcElement.element("actions");
            List<Element> actionList = actionsElement.elements("action");
            for (int i = 0; i < actionList.size(); i++) {
                Element actionElement = actionList.get(i);
                List<Element> resultList = actionElement.elements("result");
                for (int j = 0; j < resultList.size(); j++) {
                    Element resultElement = resultList.get(j);
                    String resultName = resultElement.attribute("name").getValue();
                    if (resultName.equals("toShowUserinfoList")) {
                        Attribute typeAttribute = resultElement.attribute("type");
                        if (typeAttribute != null) {
                            resultElement.remove(typeAttribute);
                        }
                    }
                }
            }
            OutputFormat format = OutputFormat.createPrettyPrint();
            XMLWriter writer = new XMLWriter(new FileWriter("src\\ghy.xml"), format);
            writer.write(document);
            writer.close();
        } catch (DocumentException e) {
            e.printStackTrace();
        }
    }

}
```

产生的 XML 文件代码如下：

```
<?xml version="1.0" encoding="UTF-8"?>

<mymvc>
    <actions>
        <action name="list" class="controller.List">
            <result name="toListJSP">/list.jsp</result>
            <result name="toShowUserinfoList">showUserinfoList.ghy</result>
        </action>
        <action name="showUserinfoList" class="controller.ShowUserinfoList">
            <result name="toShowUserinfoListJSP">/showUserinfoList.jsp</result>
        </action>
    </actions>
</mymvc>
```

成功删除 XML 文件中的 type 属性。

1.5　准备 MyBatis 的开发环境

本节将搭建 MyBatis 的开发环境，为后面的学习做好准备工作。

1.5.1　下载 Eclipse

进入开发工具 Eclipse 的官方网址并进入下载页面，单击“Download Packages”链接下载“Eclipse IDE for Java EE Developers”，也就是 64 位版本的 Eclipse，如图 1-5 所示。

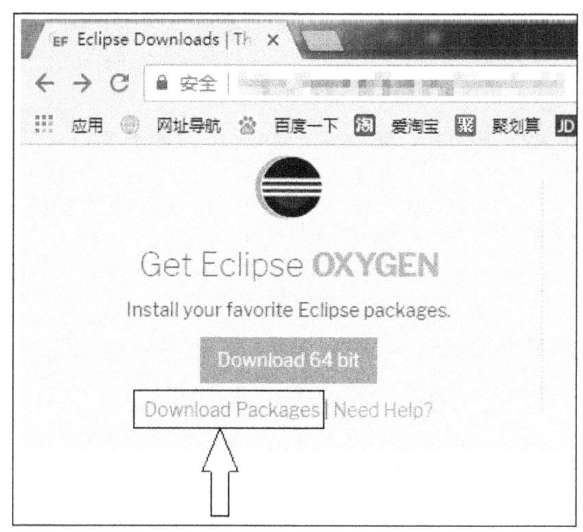

图 1-5　单击“Download Packages”链接下载 Eclipse

下载成功后，将 Eclipse 解压到 C:\MyBatisEclipse 路径中。

1.5.2　下载 MyBatis

由于 MyBatis 是第三方的框架，javaee.jar 中并不包含它的 API，因此需要单独进行下载。

进入 GitHub 官网后找到 MyBatis 项目，打开网页后看到如图 1-6 所示的界面。

图 1-6 MyBatis 项目页面

单击“Download Latest”链接后打开下载页面，如图 1-7 所示。

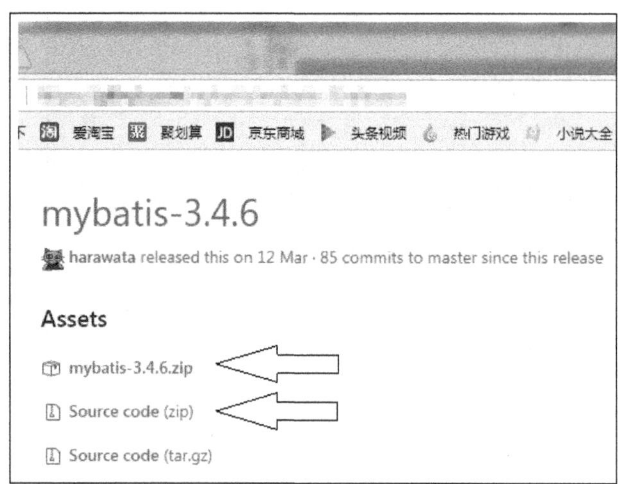

图 1-7 准备下载 MyBatis 框架

下载 MyBatis 框架压缩包 mybatis-3.4.6.zip 以及源代码 Source code（zip）文件。

解压 mybatis-3.4.6.zip 文件可以看到 MyBatis 相关的资料，比如 MyBatis 的核心 JAR 包 mybatis-3.4.6.jar 和 PDF 格式的帮助文档 mybatis-3.4.6.pdf，文件夹 lib 中存放的是 mybatis-3.4.6.jar 依赖的其他 JAR 包文件，内容如图 1-8 所示。

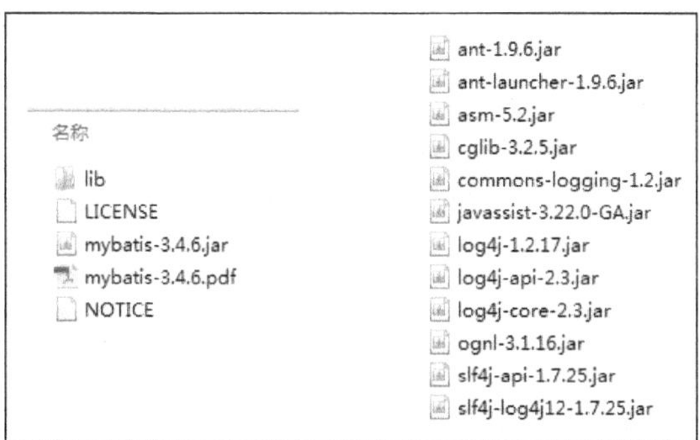

图 1-8 MyBatis 压缩包中的资料

1.5.3 在 Eclipse 中创建 Library 库

把mybatis-3.4.6.jar和ojdbc8.jar驱动文件复制到lib 文件夹中，其中一共有 14 个 JAR 包文件，总大小约为 10MB，如图 1-9 所示。

图 1-9 所有的 JAR 包文件

在后面的学习中需要在 Eclipse 里创建很多 Java 项目，如果在每个 Java 项目中都复制约 10MB 大小的 JAR 包文件，则非常占用硬盘空间。解决的办法就是在 Eclipse 中创建一个 Library 库，向这个 Library 库中添加 14 个 JAR 包文件，然后让每个 Java 项目引用这个 Library 库即可，这样做就大大降低了硬盘占用率。

进入 Eclipse，在"Preferences"界面中依次选择"Java"→"Build Path"→"User Libraries"界面，如图 1-10 所示。

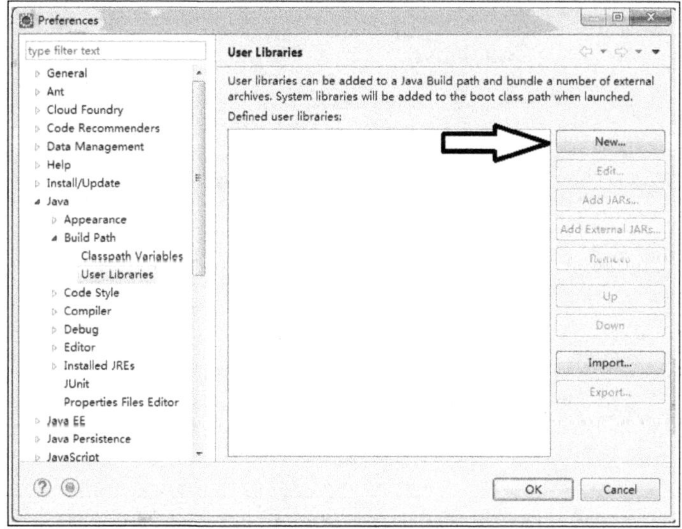

图 1-10 进入 User Libraries 界面

单击“New…”按钮创建 Library 库，弹出界面如图 1-11 所示。

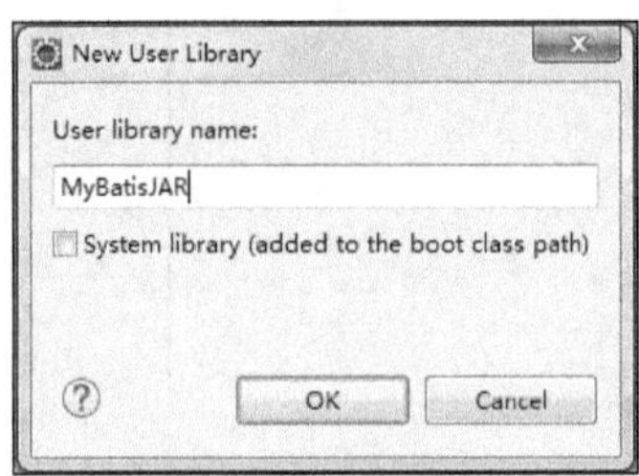

图 1-11　设置 Library 库的名称

输入 Library 库的名称为“MyBatisJAR”，单击“OK”按钮，再单击“Add External JARs…”按钮添加扩展的 JAR 包，如图 1-12 所示。

扩展 JAR 包文件添加成功后再单击右下角的“Apply and Close”按钮完成创建 Library 库，如图 1-13 所示。

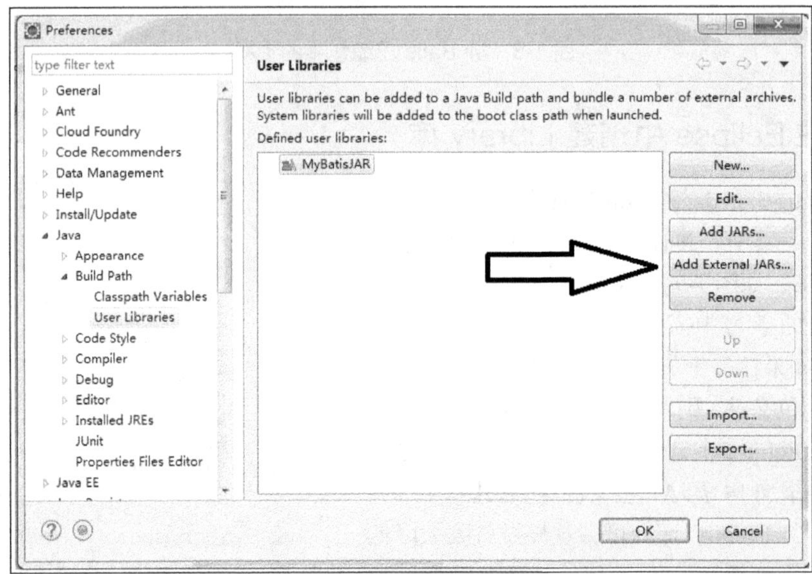

图 1-12　单击“Add External JARs…”按钮添加扩展 JAR 包

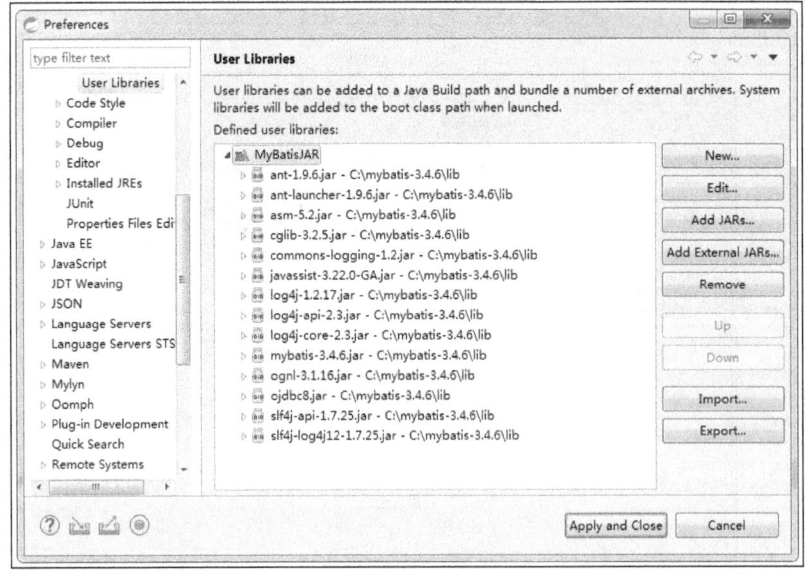

图 1-13　自定义 Library 库 MyBatisJAR 创建完毕

1.5.4 创建 Java 项目并引用 Library 库

创建名称为 libraryTest 的 Java 项目，在创建向导中单击“Add Library...”按钮添加 Library 库，效果如图 1-14 所示。

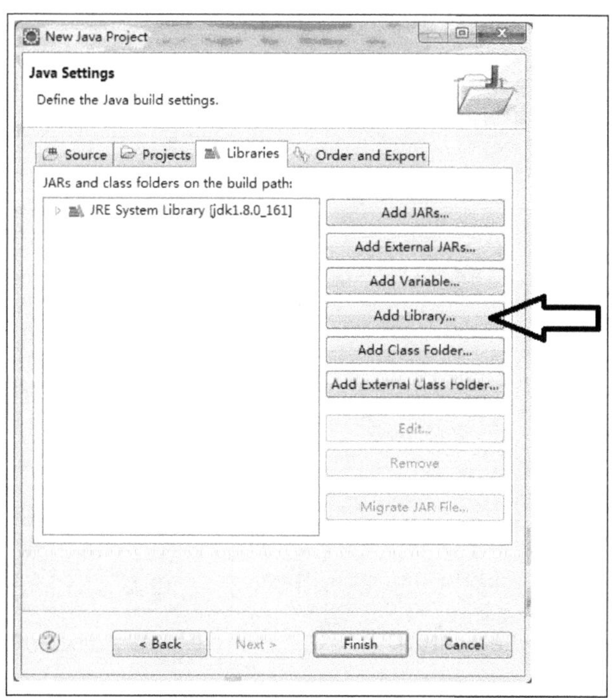

图 1-14 单击“Add Library...”按钮添加 Library 库

在弹出的界面中选择“User Library”选项，如图 1-15 所示。

选择名称为“MyBatisJAR”的 Library 库，如图 1-16 所示。

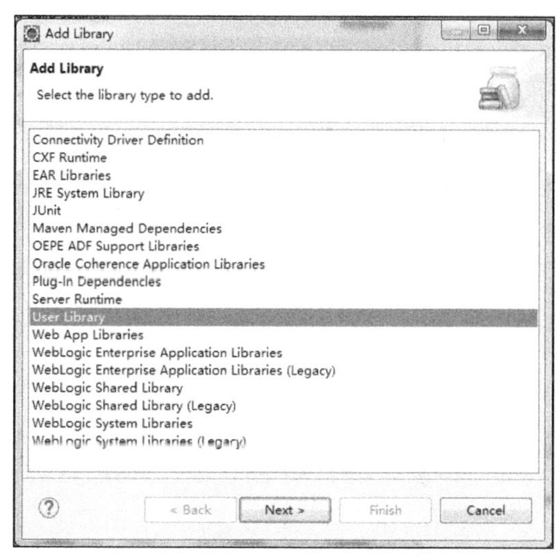

图 1-15 选择“User Library”选项

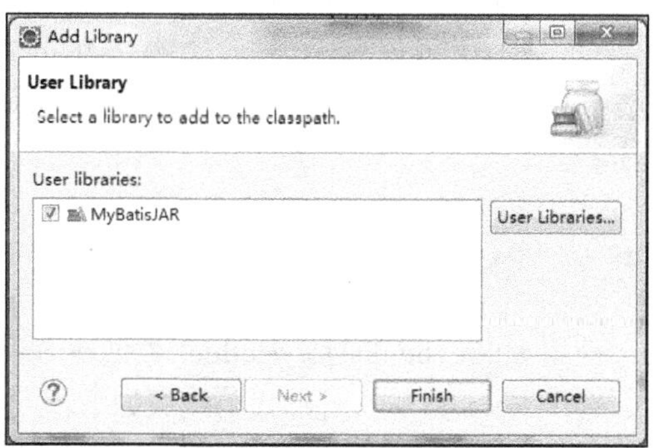

图 1-16 选择名称为“MyBatisJAR”的 Library 库

单击“Finish”按钮后可以发现项目引用了 MyBatisJAR 库，如图 1-17 所示。

再单击“Finish”按钮完成 Java 项目的创建。

Java 项目结构如图 1-18 所示。

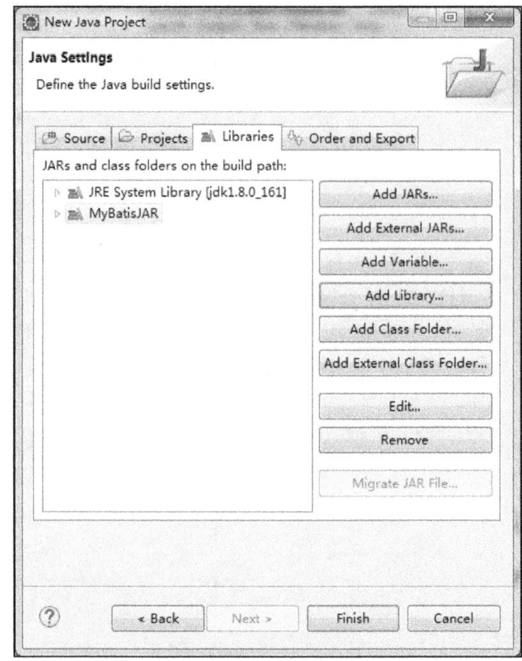

图 1-17　成功引入 MyBatisJAR 库

图 1-18　创建完成的 Java 项目结构

在新建的 Java 类中就可以使用 mybatis.jar 包中的类来操作数据库了。

1.6　创建 SqlSessionFactory 和 SqlSession 对象

开门见山，是快速学习一门技术的优选方式。

MyBatis 框架的核心是 SqlSessionFactoryBuilder、SqlSessionFactory 和 SqlSession 对象，这三者之间的创建关系如下。

SqlSessionFactoryBuilder 创建出 SqlSessionFactory，SqlSessionFactory 创建出 SqlSession。

使用 SqlSessionFactoryBuilder 类创建 SqlSessionFactory 对象的方式可以来自于一个 XML 配置文件，还可以来自于一个实例化的 Configuration 对象，因为使用 XML 方式创建 SqlSessionFactory 对象在使用上比较广泛，而且也是官方所推荐的，所以在下面的小节就进行此实验。

1.6.1　XML 配置文件模板

使用 SqlSessionFactoryBuilder 类创建 SqlSessionFactory 对象可以来自于一个 XML 配置文件，这个 XML 配置文件的代码模板如下：

```
<?xml version="1.0" encoding="UTF-8" ?>
<!DOCTYPE configuration
```

```
PUBLIC "-//mybatis.org//DTD Config 3.0//EN"
"mybatis-3-config.dtd">
<configuration>
    <properties>
        <property name="" value="" />
    </properties>
    <settings>
        <setting name="" value="" />
    </settings>
    <typeAliases>
        <typeAlias type="" />
        <package name="" />
    </typeAliases>
    <typeHandlers>
        <typeHandler handler="" />
        <package name="" />
    </typeHandlers>
    <objectFactory type="">
    </objectFactory>
    <objectWrapperFactory type=""></objectWrapperFactory>
    <reflectorFactory type="" />
    <plugins>
        <plugin interceptor=""></plugin>
    </plugins>
    <environments default="">
        <environment id="">
            <transactionManager type="" />
            <dataSource type="">
                <property name="" value="" />
            </dataSource>
        </environment>
    </environments>
    <databaseIdProvider type="">
        <property name="" value="" />
    </databaseIdProvider>
    <mappers>
        <mapper />
        <package name="" />
    </mappers>
</configuration>
```

上面的代码就是 MyBatis 配置文件的模板，但不要对上面模板中冗长的代码产生不安，因为在初期学习的过程中并不会使用上面全部的配置标记。另外，也不要死记硬背模板中的代码，因为在 PDF 帮助文档中提供了一个极简版模板，复制那个极简版的 XML 模板代码就可以搭建 MyBatis 的开发环境。

注意： 模板中标记的顺序不能改变，不然会出现异常。

1.6.2　使用 XML 配置文件创建 SqlSessionFactory 对象

SqlSessionFactory 对象存储 MyBatis 环境的全局信息。

创建名称为 mybatis1 的 Java 项目。

根据 XML 配置文件来创建 SqlSessionFactory 对象的核心代码，具体如下：

```
public class Test1 {
    public static void main(String[] args) {
        try {
            String resource = "mybatis-config.xml";
            InputStream inputStream = Resources.getResourceAsStream(resource);
            SqlSessionFactory sqlSessionFactory = new SqlSessionFactoryBuilder().build
(inputStream);
            System.out.println(sqlSessionFactory);
        } catch (IOException e) {
            e.printStackTrace();
        }
    }
}
```

在代码中使用 Resources 类将 mybatis-config.xml 文件转换成 InputStream 输入流，再把 InputStream 输入流传入 SqlSessionFactoryBuilder 类的 build()方法里来创建 SqlSessionFactory 对象。

类 SqlSessionFactoryBuilder 的主要作用就是根据 mybatis-config.xml 配置文件中的信息来创建 SqlSessionFactory 对象。

配置文件 mybatis-config.xml 的代码及解释如下：

```
<?xml version="1.0" encoding="UTF-8" ?>
<!DOCTYPE configuration
PUBLIC "-//mybatis.org//DTD Config 3.0//EN"
"http://mybatis.org/dtd/mybatis-3-config.dtd">
<configuration>
    <!--在配置文件中可以有多个<environment id="development">配置 -->
    <!--目的就是在配置文件中保存多个数据库的连接信息 -->
    <!--而使用<environments default="development">代码的作用就是 -->
    <!--默认使用 id 为 development 的数据库连接 -->
    <environments default="development">
        <!--定义 id 为 development 的数据库连接 -->
        <environment id="development">
            <!-- 定义数据库的事务要由程序员的代码进行控制 -->
            <transactionManager type="JDBC" />
            <!-- 使用 MyBatis 自己提供的连接池来处理 Connection 对象 -->
            <dataSource type="POOLED">
                <!-- 连接数据库的 4 大变量 -->
                <property name="driver" value="${driver}" />
                <property name="url" value="${url}" />
                <property name="username" value="${username}" />
                <property name="password" value="${password}" />
            </dataSource>
        </environment>
    </environments>
</configuration>
```

前面小节介绍过，MyBatis 配置文件中的代码是不需要死记硬背的，因此，上面 mybatis-config.xml 配置文件中的代码来自于 PDF 帮助文档并少量更改，PDF 帮助文档中的 XML 配置文件模板代码示例如图 1-19 所示。

配置文件 mybatis-config.xml 的主要作用就是定义如何连接数据库，包含连接数据库所用到的 username、password 及 url 等参数，但在本实验中，文件 mybatis-config.xml 里并没有实质的属性值，而是使用${xxxx}作为替代。这是因为在获取 SqlSessionFactory 工厂对象时，不需要提供这些具体的参数值。

```
<?xml version="1.0" encoding="UTF-8" ?>
<!DOCTYPE configuration
  PUBLIC "-//[illegible]//DTD Config 3.0//EN"
  "http://[illegible]/dtd/mybatis-3-config.dtd">
<configuration>
  <environments default="development">
    <environment id="development">
      <transactionManager type="JDBC"/>
      <dataSource type="POOLED">
        <property name="driver" value="${driver}"/>
        <property name="url" value="${url}"/>
        <property name="username" value="${username}"/>
        <property name="password" value="${password}"/>
      </dataSource>
    </environment>
  </environments>
  <mappers>
    <mapper resource="org/mybatis/example/BlogMapper.xml"/>
  </mappers>
</configuration>
```

While there is a lot more to the XML configuration file, the above example points out the most critical parts. Notice the XML header, required to validate the XML document. The body of the environment element contains the environment configuration for transaction management and connection pooling. The mappers element contains a list of mappers – the XML files and/or annotated Java interface classes that contain the SQL code and mapping definitions.

图 1-19 帮助文档提供的 XML 配置文件的极简代码

程序运行后并没有出现异常，控制台打印的信息如下：

```
org.apache.ibatis.session.defaults.DefaultSqlSessionFactory@1554909b
```

DefaultSqlSessionFactory.java 是 SqlSessionFactory.java 接口的实现类，具有实现关系，效果如图 1-20 所示。

SqlSessionFactory - org.apache.ibatis.session
DefaultSqlSessionFactory - org.apache.ibatis.session.defaults
SqlSessionManager - org.apache.ibatis.session

图 1-20 实现关系

到此，使用 XML 配置文件创建 SqlSessionFactory 对象是成功的。

1.6.3 创建 SqlSession 对象

对数据库执行增加、删除、更新和查询操作是需要使用 SqlSession 对象的。使用 SqlSessionFactory 可以创建 SqlSession 对象。

示例核心代码如下：

```
public class Test2 {
    public static void main(String[] args) {
        try {
            String resource = "mybatis-config.xml";
            InputStream inputStream = Resources.getResourceAsStream(resource);
            SqlSessionFactory sqlSessionFactory = new SqlSessionFactoryBuilder().
build(inputStream);
```

```
                System.out.println(sqlSessionFactory);
                SqlSession sqlSession = sqlSessionFactory.openSession();
                System.out.println(sqlSession);
                sqlSession.commit();
                sqlSession.close();
            } catch (IOException e) {
                e.printStackTrace();
            }
        }
}
```

调用 SqlSessionFactory 对象的 openSession()方法来创建 SqlSession 对象。

程序运行后，在控制台输出的结果如下：

```
org.apache.ibatis.session.defaults.DefaultSqlSessionFactory@1554909b
org.apache.ibatis.session.defaults.DefaultSqlSession@42d3bd8b
```

获得 SqlSession 对象并调用下面的 4 种方法可以对数据库进行操作：

```
    // session.insert(arg0); //增加操作
    // session.delete(arg0); //删除操作
    // session.update(arg0); //更新操作
    // session.selectList(arg0); //查询操作
```

DefaultSqlSession.java 是 SqlSession.java 接口的实现类，具有实现关系，效果如图 1-21 所示。

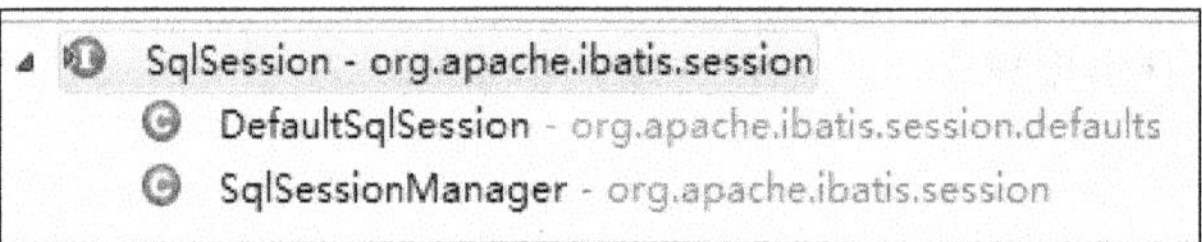

图 1-21　实现关系

到此，使用 SqlSessionFactory 创建 SqlSession 对象是成功的。

1.6.4　SqlSessionFactoryBuilder 和 SqlSessionFactory 的 API

虽然现在正在学习新的知识，包括要熟悉新的类名、新的方法名称、新的包名等，但是 MyBatis 在 API 的设计结构上是相当简洁的，大部分是重载的方法。下面来看一看 SqlSessionFactoryBuilder 类的结构，效果如图 1-22 所示。

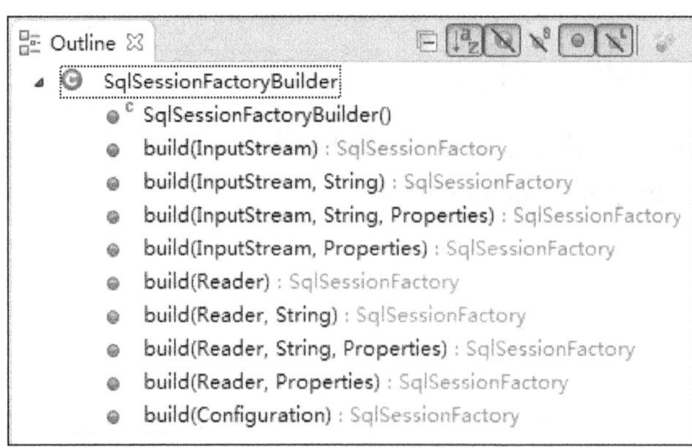

图 1-22　SqlSessionFactoryBuilder 类结构

再来看一下 SqlSessionFactory 类的结构，如图 1-23 所示。

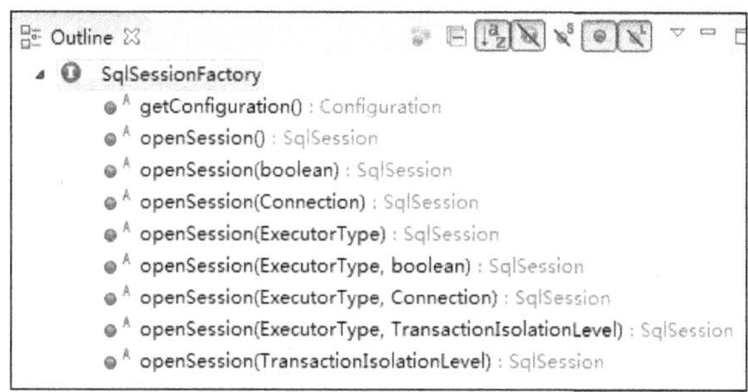

图 1-23　SqlSessionFactory 类的结构

从图 1-22 和图 1-23 中可以看到，基本全是重载的方法，主要就是通过 SqlSessionFactoryBuilder.build()方法取得 SqlSessionFactory 对象，再使用 SqlSessionFactory.openSession()方法取得 SqlSession 对象。

SqlSession 主要的作用是对数据库进行 CURD 操作。

1.7 在 Eclipse 中安装 MyBatis Generator 插件

使用 MyBatis 操作数据库要使用 SQL 映射文件和实体类，但这两个文件的代码比较繁杂，尤其是如果数据表的字段很多，则手写实体类将会成为噩梦。此种情况也存在于 Hibernate 框架中。因此，为了提高开发效率，MyBatis 官方提供 1 个 Eclipse 插件，名称为 MyBatis Generator。该插件主要的功能就是根据数据表结构生成对应的 SQL 映射文件和实体类，此插件需要在 Eclipse 中在线安装，安装过程如下。

进入 Eclipse 的 Marketplace，效果如图 1-24 所示。

图 1-24　进入 Eclipse 的 Marketplace

搜索关键字“mybatis”后出现了“MyBatis Generator”插件，效果如图 1-25 所示。

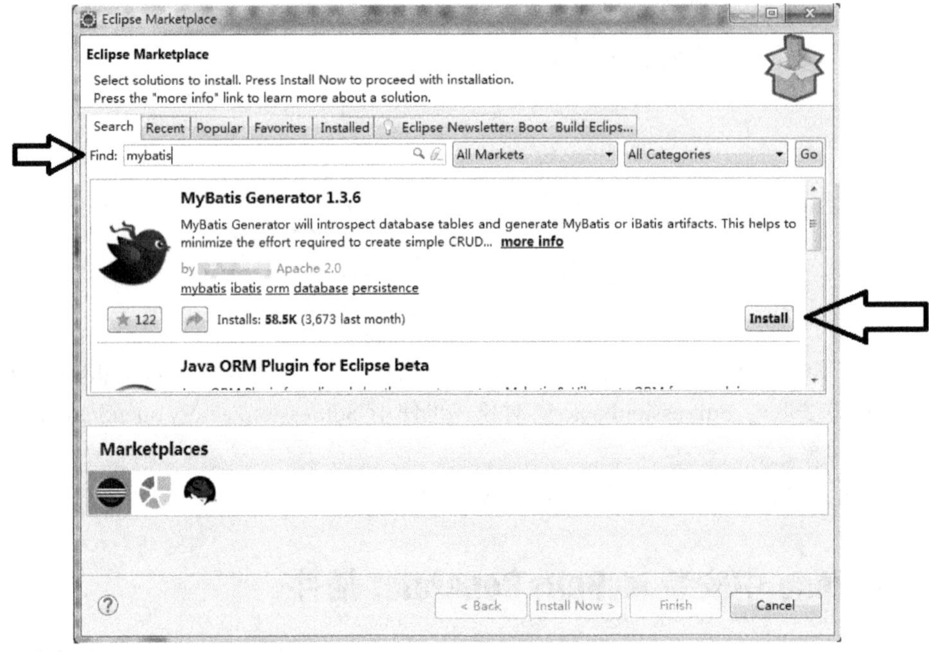

图 1-25 搜索到“MyBatis Generator”插件

单击“Install”按钮在线安装，弹出如图 1-26 所示的界面。

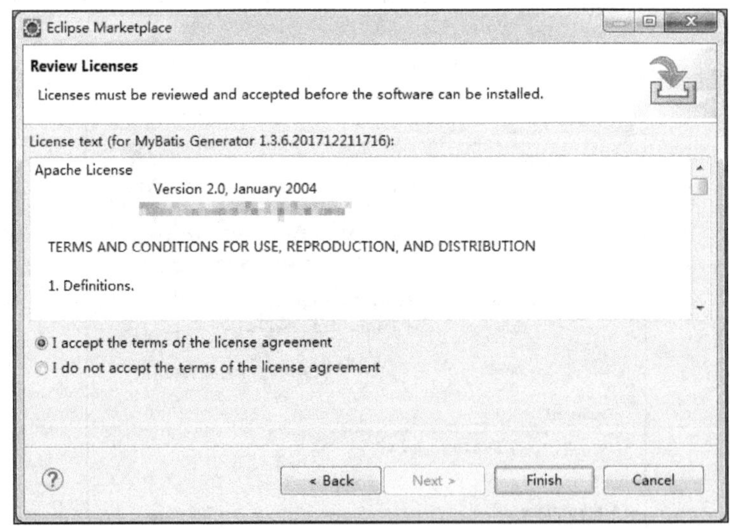

图 1-26 接受许可

单击“Finish”按钮开始安装，安装过程如图 1-27 所示。

在安装过程中会弹出提示，如图 1-28 所示，信息内容是安装的软件有未签名的内容，询问是否继续，单击“Install anyway”按钮继续安装。

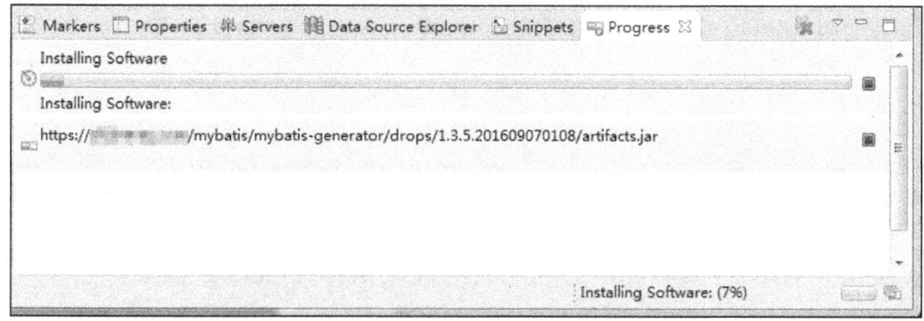

图 1-27 安装过程

插件安装完毕后询问是否重新启动 Eclipse，如图 1-29 所示，单击“Restart Now”按钮重新启动 Eclipse。

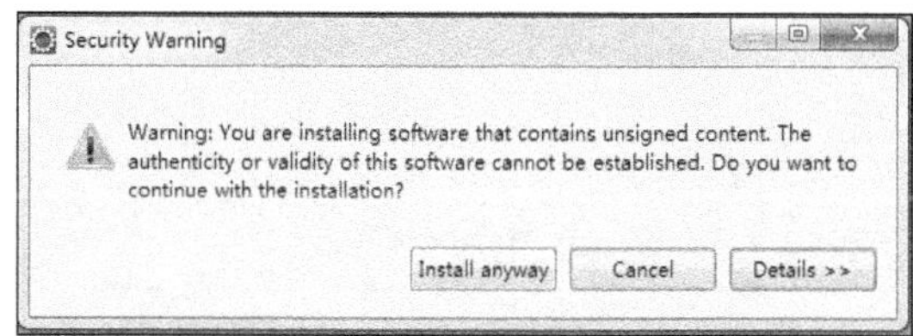

图 1-28 询问是否继续安装

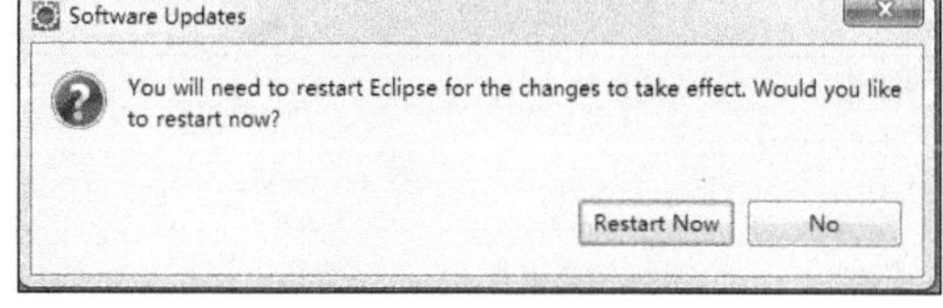

图 1-29 是否重启 Eclipse

重新启动 Eclipse 后再次进入 Eclipse 中的 Marketplace，进入“Installed”标签页，发现插件 MyBatis Generator 已经安装成功，效果如图 1-30 所示。

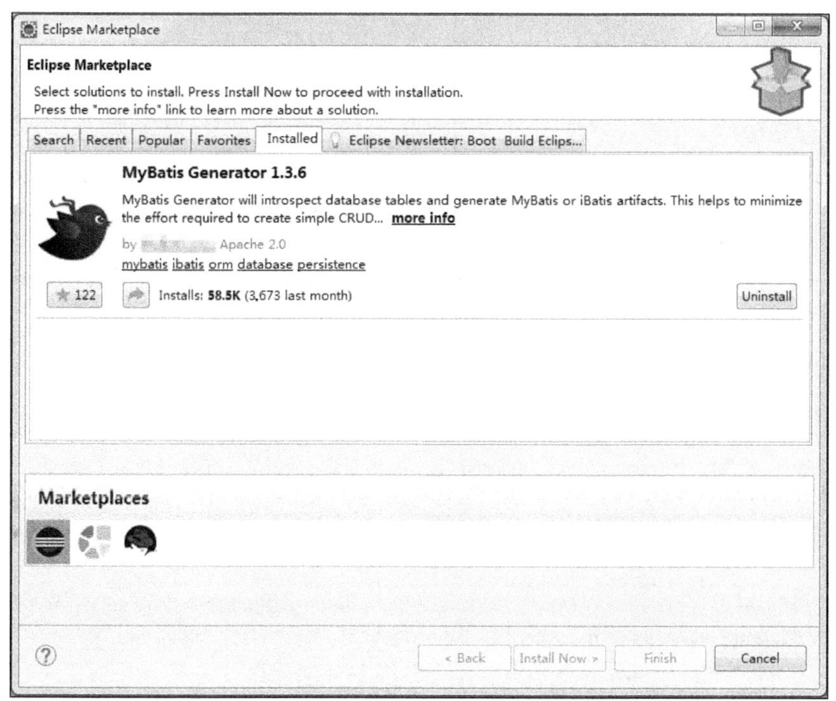

图 1-30 插件 MyBatis Generator 已经安装成功

插件安装成功后就可以在新建 Java 项目时创建 MyBatis Generator Configuration File.xml 文

件了，通过这个文件就可以将数据表的信息逆向成 SQL 映射文件以及实体类，不需要手写代码了。

1.8 使用 MyBatis Generator 工具逆向的代码操作 Oracle 数据库

本节就要使用 MyBatis Generator 工具逆向的代码操作 Oracle 数据库。

1.8.1 进行逆向操作

新建名称为 GeneratorOracle 的 Java 项目，并关联 MyBatisJAR 库。

然后在 src 节点下单击鼠标右键，新建 1 个 MyBatis 的 Generator 配置文件，如图 1-31 所示。

单击“Next”按钮出现如图 1-32 所示界面。

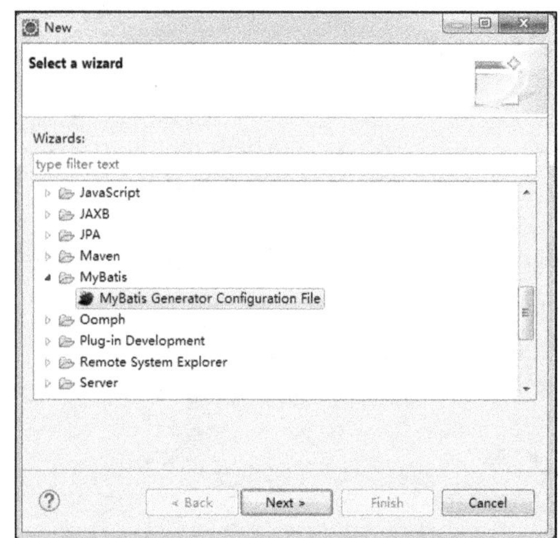

图 1-31 创建生成 ORM 的配置 XML 文件

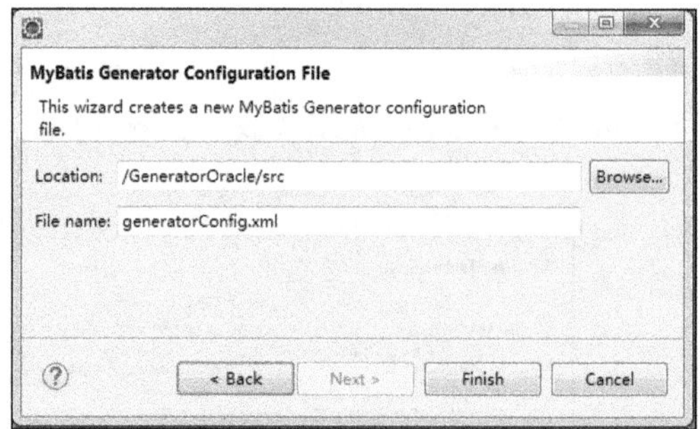

图 1-32 将 generatorConfig.xml 文件放入 src 节点下即可

在图 1-32 界面中不需要更改配置，默认即可，单击“Finish”按钮完成 Generator 配置文件的创建。

更改 generatorConfig.xml 的配置文件，代码如下：

```
<?xml version="1.0" encoding="UTF-8"?>
<!DOCTYPE generatorConfiguration PUBLIC "-//mybatis.org//DTD MyBatis Generator Configuration 1.0//EN" "http://mybatis.org/dtd/mybatis-generator-config_1_0.dtd">
<generatorConfiguration>
    <context id="context1">
        <jdbcConnection
            connectionURL="jdbc:oracle:thin:@localhost:1521:orcl"
            driverClass="oracle.jdbc.OracleDriver" password="123" userId="y2" />
        <javaModelGenerator targetPackage="entity"
            targetProject="GeneratorOracle" />
        <sqlMapGenerator targetPackage="sqlmapping"
            targetProject="GeneratorOracle" />
        <javaClientGenerator targetPackage="client"
            targetProject="GeneratorOracle" type="XMLMAPPER" />
        <table schema="y2" tableName="userinfo">
```

```
        </table>
    </context>
</generatorConfiguration>
```

注意：在连接 Oracle 数据库时，<table>标记的 schema 属性值 y2 是登录 Oracle 数据库的用户名。

配置文件 generatorConfig.xml 是 MyBatis Generator 插件中必备的文件，通过此文件可以将数据表的结构逆向出对应的实体类、SQL 映射文件以及客户端代码，使用这些代码就可以对数据表进行增加、删除、更新和查询操作。

数据表 userinfo 的表结构如图 1-33 所示。

Columns | Indexes | Constraints | Triggers | Data | Script | Grants | Synonyms | Partitions | Sub

Column Name	ID	PK	Index Pos	Null?	Data Type
ID	1	1	1	N	NUMBER (18)
USERNAME	2			Y	VARCHAR2 (50 Byte)
PASSWORD	3			Y	VARCHAR2 (50 Byte)

图 1-33 userinfo 数据表结构

配置文件 generatorConfig.xml 准备就绪后，单击图 1-34 中的菜单项“Run MyBatis Generator”。

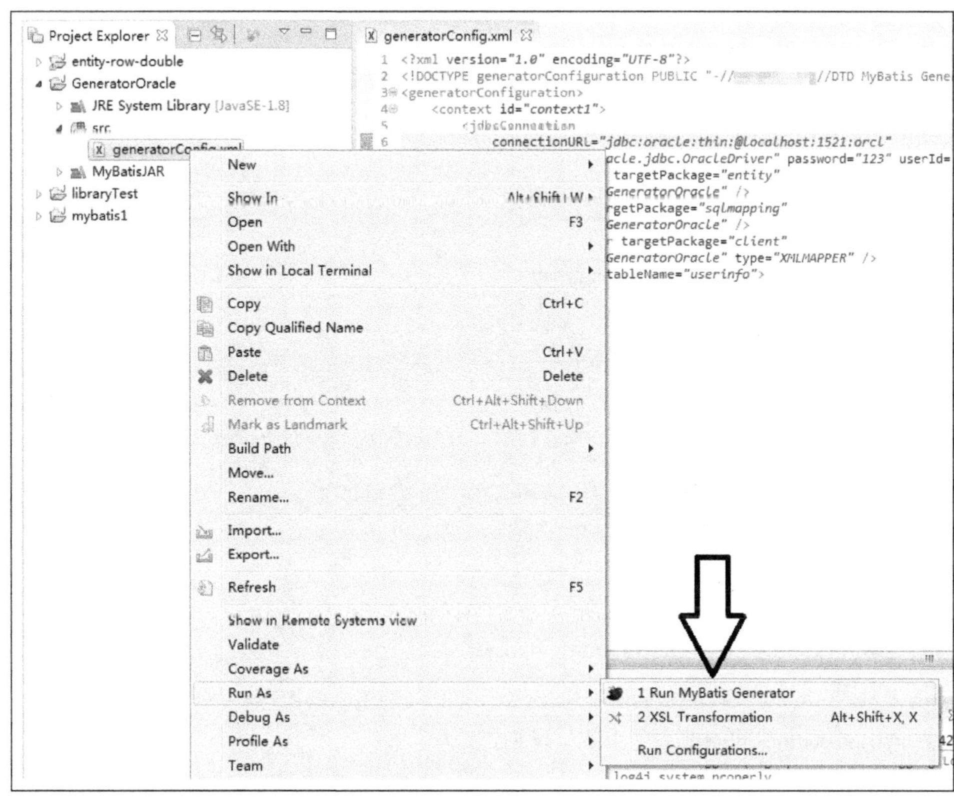

图 1-34 根据 XML 配置文件生成 SQL 映射文件

在控制台输出成功逆向的信息，如图 1-35 所示。

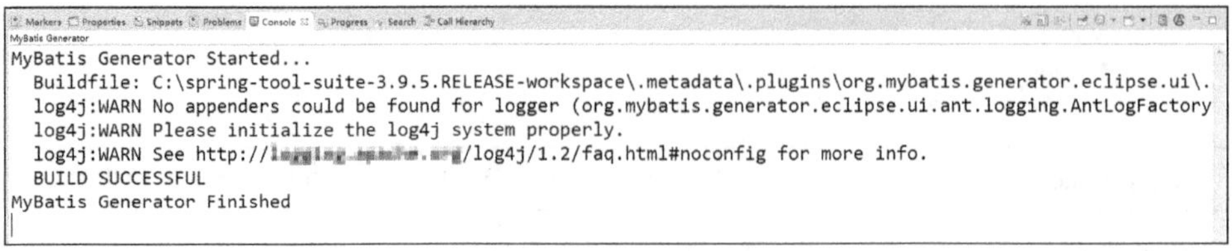

```
MyBatis Generator Started...
  Buildfile: C:\spring-tool-suite-3.9.5.RELEASE-workspace\.metadata\.plugins\org.mybatis.generator.eclipse.ui\.
  log4j:WARN No appenders could be found for logger (org.mybatis.generator.eclipse.ui.ant.logging.AntLogFactory
  log4j:WARN Please initialize the log4j system properly.
  log4j:WARN See http://logging.apache.org/log4j/1.2/faq.html#noconfig for more info.
  BUILD SUCCESSFUL
MyBatis Generator Finished
```

图 1-35 逆向操作成功

注意： 如果在逆向的过程中出现如下异常：

```
    C:\spring-tool-suite-3.9.5.RELEASE-workspace\.metadata\.plugins\org.mybatis.generator
.eclipse.ui\.generatedAntScripts\GeneratorOracle-generatorConfig.xml.xml:4:
java.lang.RuntimeException: Exception getting JDBC Driver
    java.lang.RuntimeException: Exception getting JDBC Driver
```

那么说明并没有找到 JDBC 驱动，可以先选择右键菜单项“Run As”→“Run Configurations...”，然后在“Classpath”标签页中添加 JDBC 驱动。

成功逆向后的 Java 项目结构如图 1-36 所示。

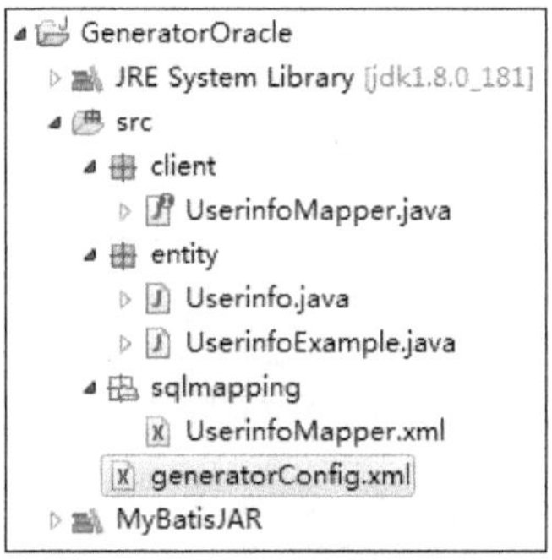

至此，操作数据库的基础文件已经准备完毕，下面开始在 Oracle 数据库的 userinfo 数据表中添加 1 条记录。

```
GeneratorOracle
  JRE System Library [jdk1.8.0_181]
  src
    client
      UserinfoMapper.java
    entity
      Userinfo.java
      UserinfoExample.java
    sqlmapping
      UserinfoMapper.xml
    generatorConfig.xml
  MyBatisJAR
```

图 1-36 项目结构

1.8.2 操作数据库

在 src 中创建 mybatis-config.xml 文件，代码如下：

```xml
<?xml version="1.0" encoding="UTF-8" ?>
<!DOCTYPE configuration
PUBLIC "-//mybatis.org//DTD Config 3.0//EN"
"http://mybatis.org/dtd/mybatis-3-config.dtd">
<configuration>
    <environments default="development">
        <environment id="development">
            <transactionManager type="JDBC" />
            <dataSource type="POOLED">
                <property name="driver" value="oracle.jdbc.OracleDriver" />
                <property name="url"
                    value="jdbc:oracle:thin:@localhost:1521:orcl" />
                <property name="username" value="y2" />
                <property name="password" value="123" />
            </dataSource>
        </environment>
    </environments>
    <mappers>
        <mapper resource="sqlmapping/UserinfoMapper.xml" />
    </mappers>
</configuration>
```

注意： 在 mybatis-config.xml 配置文件中添加如下配置：

```xml
<mappers>
    <mapper resource="sqlmapping/UserinfoMapper.xml" />
</mappers>
```

因为包 sqlmapping 中的文件 UserinfoMapper.xml 里存储了 SQL 语句，所以在配置文件中需要关联。

类核心代码如下：

```
public class Test {
    public static void main(String[] args) throws IOException {
        Userinfo userinfo = new Userinfo();
        userinfo.setUsername("中国");
        userinfo.setPassword("中国人");

        String configFile = "mybatis-config.xml";
        InputStream configStream = Resources.getResourceAsStream(configFile);
        SqlSessionFactoryBuilder builder = new SqlSessionFactoryBuilder();
        SqlSessionFactory factory = builder.build(configStream);
        SqlSession session = factory.openSession();
        session.insert("insert", userinfo);
        session.commit();
        session.close();
    }
}
```

方法 main()中的代码是一个经典的 insert 数据表的功能，从代码中可以看到 MyBatis 用相当精简的 API 就可以完全地控制数据表中的记录。可见 MyBatis 在学习、开发等方面成本都是比较低的。

程序运行后在控制台输出的异常信息如下：

```
Exception in thread "main" org.apache.ibatis.exceptions.PersistenceException:
### Error updating database.  Cause: java.sql.SQLIntegrityConstraintViolationException:
ORA-01400: 无法将 NULL 插入 ("Y2"."USERINFO"."ID")

### The error may involve client.UserinfoMapper.insert-Inline
### The error occurred while setting parameters
### SQL: insert into Y2.USERINFO (ID, USERNAME, PASSWORD      )     values (?, ?, ?      )
### Cause: java.sql.SQLIntegrityConstraintViolationException: ORA-01400: 无法将 NULL 插入 ("Y2"."USERINFO"."ID")

 at org.apache.ibatis.exceptions.ExceptionFactory.wrapException(ExceptionFactory.java:30)
 at org.apache.ibatis.session.defaults.DefaultSqlSession.update(DefaultSqlSession.java:200)
 at org.apache.ibatis.session.defaults.DefaultSqlSession.insert(DefaultSqlSession.java:185)
 at test.Test.main(Test.java:24)
Caused by: java.sql.SQLIntegrityConstraintViolationException: ORA-01400: 无法将 NULL 插入 ("Y2"."USERINFO"."ID")

 at oracle.jdbc.driver.T4CTTIoer11.processError(T4CTTIoer11.java:494)
```

出错信息提示不能对 id 主键列赋 null 空值，由于 Oracle 数据库的主键并不是自增的，在 SQL 语句中需要对 id 列进行传值，但本示例中并未对 id 属性进行显式赋值，并且在 SQL 语句中也没有使用序列，因此出现了异常。在正常情况下，在 UserinfoMapper.xml 文件的 insert 语句中需要结合序列来实现添加记录的功能，这样可以免去对 id 属性传值的 Java 代码。逆向生成错误的 insert 语句如下所示：

```
<insert id="insert" parameterType="entity.Userinfo">
  <!--
    WARNING - @mbg.generated
```

```
    This element is automatically generated by MyBatis Generator, do not modify.
    This element was generated on Mon Sep 10 16:59:40 CST 2018.
  -->
  insert into Y2.USERINFO (ID, USERNAME, PASSWORD
    )
  values (#{id,jdbcType=NUMERIC}, #{username,jdbcType=VARCHAR}, #{password,jdbcType=VARCHAR}
    )
</insert>
```

上面生成的 SQL 语句并未使用序列，也并未对 id 属性传值，因此出现了异常。

程序代码 session.insert（"insert", userinfo）；的第 1 个参数"insert"就是 SQL 映射文件 UserinfoMapper.xml 中的配置<insert id=*"insert"* parameterType=*"entity.Userinfo"*>的 id 值，代表要执行 id 值为 insert 的 SQL 语句，insert()方法的第 2 个参数的数据类型由以下配置代码来决定：

```
parameterType="entity.Userinfo"
```

上面程序在运行时出现异常，说明 SQL 映射文件 UserinfoMapper.xml 中的代码都是错误的，此时要删除 client、entity 和 sqlmapping 包以及包中的所有文件，因为要重新进行逆向。

为了在 insert 语句中使用 Oracle 数据库中的序列来对 id 列进行传值，在逆向之前需要更改 generatorConfig.xml 配置文件中的配置，添加<generatedKey>标签后的配置代码如下：

```
<table schema="y2" tableName="userinfo">
    <generatedKey column="id"
        sqlStatement="select idauto.nextval from dual" identity="false" />
</table>
```

标签<generatedKey>的主要作用就是在生成 insert 的 SQL 语句时使用名称为 idauto 序列的 nextval 值来作为主键的 id 值，identity="false"属性说明主键不是自增的，而是由序列生成的。

如果想同时逆向多个表，那么在 generatorConfig.xml 文件中写入多个<table>标签即可，示例代码如下：

```
<table schema="y2" tableName="userinfo">
    <generatedKey column="id"
        sqlStatement="select idauto.nextval from dual" identity="false" />
</table>
<table schema="y2" tableName="A">
    <generatedKey column="id"
        sqlStatement="select idauto.nextval from dual" identity="false" />
</table>
<table schema="y2" tableName="B">
    <generatedKey column="id"
        sqlStatement="select idauto.nextval from dual" identity="false" />
</table>
```

配置文件 generatorConfig.xml 准备结束后重新进行逆向操作，在 sqlmapping 包中生成最新版正确的 SQL 映射文件。最新版正确的 insert 语句如下：

```
<insert id="insert" parameterType="entity.Userinfo">
  <!--
```

```
    WARNING - @mbg.generated
    This element is automatically generated by MyBatis Generator, do not modify.
    This element was generated on Mon Sep 10 17:11:11 CST 2018.
  -->
  <selectKey keyProperty="id" order="BEFORE" resultType="java.lang.Long">
    select idauto.nextval from dual
  </selectKey>
  insert into Y2.USERINFO (ID, USERNAME, PASSWORD
    )
  values (#{id,jdbcType=NUMERIC}, #{username,jdbcType=VARCHAR}, #{password,jdbcType=VARCHAR}
    )
</insert>
```

再次运行 Test.java 类，成功在数据表中添加了 1 条记录。

1.9 使用 MyBatis Generator 工具逆向的代码操作 MySQL 数据库

本节就要使用 MyBatis Generator 工具逆向的代码操作 MySQL 数据库。

MySQL 在默认情况下是不允许远程连接到数据库的，连接时会出现异常信息：

"Host '某个 IP' is not allowed to connect to this MySQL server"

因此，要进行一下配置，步骤如下。

（1）在 CMD 中输入命令：mysql -uroot -p123，登录 MySQL 控制台。

（2）执行切换数据库命令：use mysql，切换到操作 mysql 数据库。

（3）执行更新 SQL 语句：update user set host = '%' where user = 'root';。

（4）重新启动 MySQL 服务后，远程就可以连接了。注意要关闭操作系统的防火墙。

1.9.1 进行逆向操作

MySQL 数据库中的 userinfo 数据表结构如图 1-37 所示。

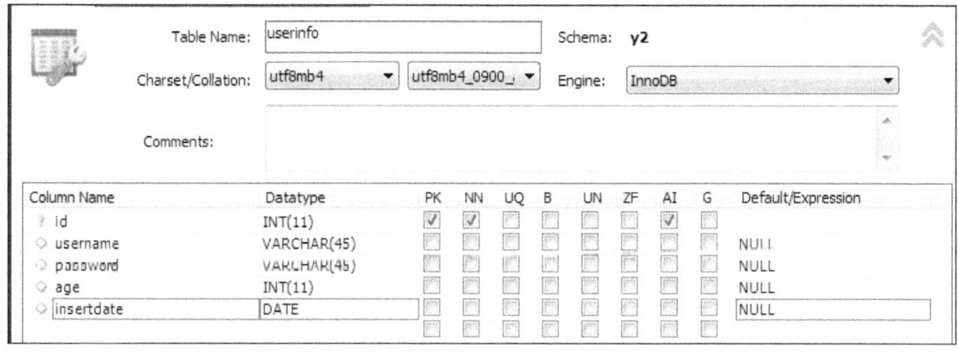

图 1-37 数据表 userinfo 结构

创建名称为 GeneratorMySQL 的项目，更改配置文件 generatorConfig.xml，代码如下：

```
<?xml version="1.0" encoding="UTF-8"?>
<!DOCTYPE generatorConfiguration PUBLIC "-//mybatis.org//DTD MyBatis Generator
Configuration 1.0//EN" "http://mybatis.org/dtd/mybatis-generator-config_1_0.dtd">
```

```
<generatorConfiguration>
    <context id="context1">
        <jdbcConnection
            connectionURL="jdbc:mysql://localhost:3306/y2"
            driverClass="com.mysql.jdbc.Driver" password="123123" userId="root" />
        <javaModelGenerator targetPackage="entity"
            targetProject="GeneratorMySQL" />
        <sqlMapGenerator targetPackage="sqlmapping"
            targetProject="GeneratorMySQL" />
        <javaClientGenerator targetPackage="client"
            targetProject="GeneratorMySQL" type="XMLMAPPER" />
        <table schema="y2" tableName="userinfo">
        </table>
    </context>
</generatorConfiguration>
```

注意：在连接 MySQL 数据库时，<table>标记的 schema 属性值 y2 是操作 MySQL 目标数据库的名字。

对 generatorConfig.xml 文件进行逆向时，会出现错误，效果如图 1-38 所示。

异常信息提示找不到 MySQL 的 JDBC 驱动，解决异常的办法是在 MyBatisJAR 库中添加 MySQL 的 JDBC 驱动，效果如图 1-39 所示。

MySQL 的 JDBC 驱动 JAR 包添加成功后再次执行逆向操作，又出现异常效果，如图 1-40 所示。

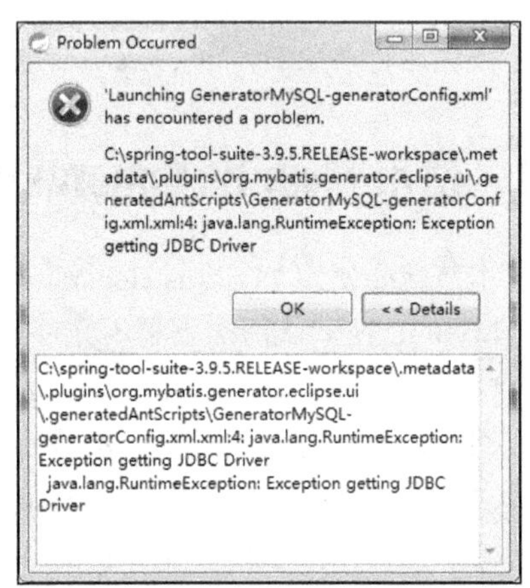

图 1-38 缺少 MySQL 的 JDBC 驱动程序

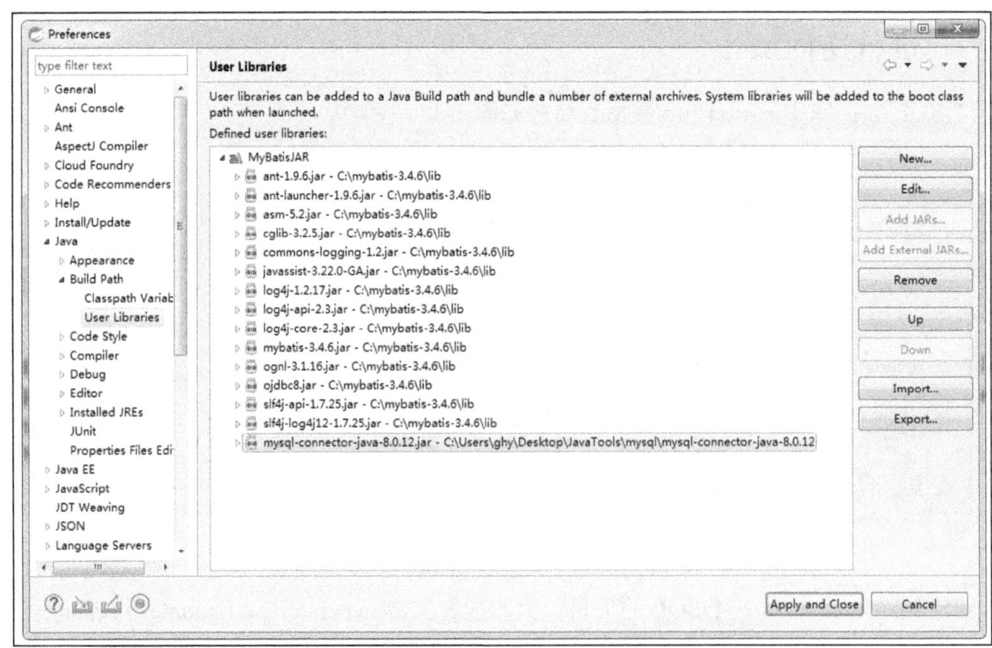

图 1-39 在 MyBatisJAR 库中添加 MySQL 驱动

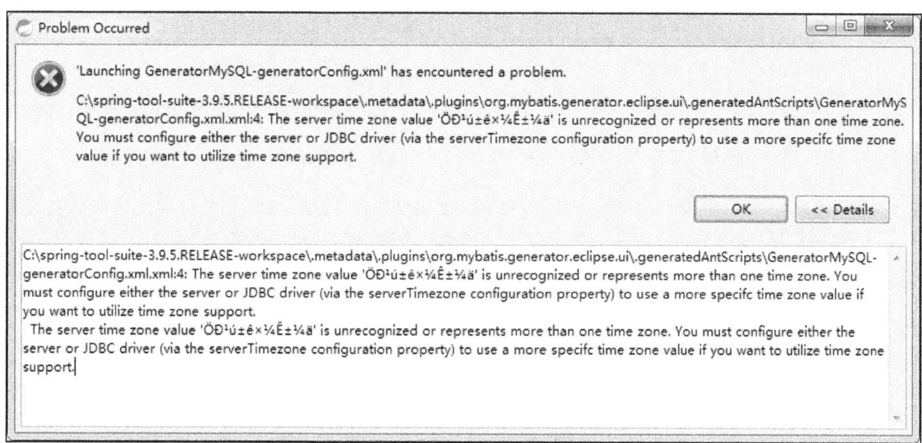

图 1-40 设置时区有异常

解决时区的问题需要进入文件夹 C:\ProgramData\MySQL\MySQL Server 8.0 并编辑 my.ini 文件，在[mysqld]节点下添加配置代码：

```
[mysqld]
default-time-zone='+08:00'
```

重启 MySQL 服务后再次进行逆向没有出现异常，表明成功逆向，Java 项目结构如图 1-41 所示。

GeneratorMySQL
JRE System Library [jdk1.8.0_181]
src
client
UserinfoMapper.java
entity
Userinfo.java
UserinfoExample.java
sqlmapping
UserinfoMapper.xml
generatorConfig.xml
MyBatisJAR

图 1-41 成功逆向后的项目结构

1.9.2 操作数据库

在 src 中创建 mybatis-config.xml 文件，代码如下：

```
<?xml version="1.0" encoding="UTF-8" ?>
<!DOCTYPE configuration
PUBLIC "-//mybatis.org//DTD Config 3.0//EN"
"http://mybatis.org/dtd/mybatis-3-config.dtd">
<configuration>
    <environments default="development">
        <environment id="development">
            <transactionManager type="JDBC" />
            <dataSource type="POOLED">
                <property name="driver" value="com.mysql.jdbc.Driver" />
                <property name="url" value="jdbc:mysql://localhost:3306/y2" />
                <property name="username" value="root" />
                <property name="password" value="123123" />
            </dataSource>
        </environment>
    </environments>
    <mappers>
        <mapper resource="sqlmapping/UserinfoMapper.xml" />
    </mappers>
</configuration>
```

运行类代码如下：

```java
public class Test {
    public static void main(String[] args) throws IOException {
        Userinfo userinfo = new Userinfo();
        userinfo.setUsername("中国");
        userinfo.setPassword("中国人");
        userinfo.setAge(123);
        userinfo.setInsertdate(new Date());

        String configFile = "mybatis-config.xml";
        InputStream configStream = Resources.getResourceAsStream(configFile);
        SqlSessionFactoryBuilder builder = new SqlSessionFactoryBuilder();
        SqlSessionFactory factory = builder.build(configStream);
        SqlSession session = factory.openSession();
        session.insert("insert", userinfo);
        session.commit();
        session.close();
    }
}
```

程序运行后在 userinfo 数据表中添加了新的记录。

1.10 自建环境使用 SqlSession 操作 Oracle 和 MySQL 数据库

前面章节都是使用 MyBatis Generator 插件生成的实体类和 SQL 映射文件来操作数据库，并不能从基础上掌握 MyBatis 框架的使用，本小节将从零起步，以自搭建开发环境开始，再到使用 SqlSession 对象实现经典功能 CURD，并且是针对 Oracle 和 MySQL 这两种主流数据库的。

1.10.1 针对 Oracle 的 CURD

本节将演示使用 SqlSession 对 Oracle 数据库进行 CURD 操作。

1. 准备开发环境

（1）创建数据表。创建 userinfo 数据表，表结构如图 1-42 所示。

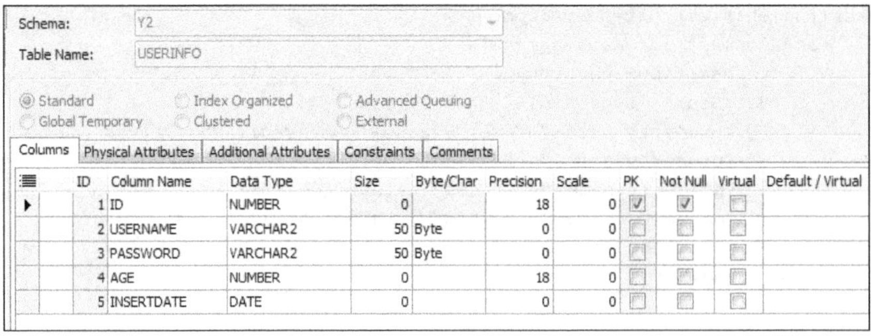

Schema: Y2
Table Name: USERINFO

Standard　Index Organized　Advanced Queuing
Global Temporary　Clustered　External

Columns | Physical Attributes | Additional Attributes | Constraints | Comments

ID	Column Name	Data Type	Size	Byte/Char	Precision	Scale	PK	Not Null	Virtual	Default / Virtual
1	ID	NUMBER	0		18	0	✓	✓		
2	USERNAME	VARCHAR2	50	Byte	0	0				
3	PASSWORD	VARCHAR2	50	Byte	0	0				
4	AGE	NUMBER	0		18	0				
5	INSERTDATE	DATE	0		0	0				

图 1-42 userinfo 数据表结构

（2）创建名称为 mybatis_sqlsession_curd_oracle 的 Java 项目。

（3）准备 generatorConfig.xml 逆向配置文件，主要作用是根据数据表结构只逆向出实体类，不包含 SQL 映射文件以及客户端代码，配置代码如下：

```
<?xml version="1.0" encoding="UTF-8"?>
<!DOCTYPE generatorConfiguration PUBLIC "-//mybatis.org//DTD MyBatis Generator
Configuration 1.0//EN" "http://mybatis.org/dtd/mybatis-generator-config_1_0.dtd">
<generatorConfiguration>
    <context id="context1">
        <jdbcConnection
            connectionURL="jdbc:oracle:thin:@localhost:1521:orcl"
            driverClass="oracle.jdbc.OracleDriver" password="123" userId="y2" />
        <javaModelGenerator targetPackage="entity"
            targetProject="mybatis_sqlsession_curd_oracle" />
        <table schema="y2" tableName="userinfo">
            <generatedKey column="id"
                sqlStatement="select idauto.nextval from dual" identity="false" />
        </table>
    </context>
</generatorConfiguration>
```

开始进行逆向操作，逆向出实体类。

（4）在 src 路径下创建连接数据库的配置文件 mybatis-config.xml，代码如下：

```
<?xml version="1.0" encoding="UTF-8" ?>
<!DOCTYPE configuration
PUBLIC "-//mybatis.org//DTD Config 3.0//EN"
"mybatis-3-config.dtd">
<configuration>
    <environments default="development">
        <environment id="development">
            <transactionManager type="JDBC" />
            <dataSource type="POOLED">
                <property name="driver" value="oracle.jdbc.OracleDriver" />
                <property name="url"
                    value="jdbc:oracle:thin:@localhost:1521:orcl" />
                <property name="username" value="y2" />
                <property name="password" value="123" />
            </dataSource>
        </environment>
    </environments>
    <mappers>
        <mapper resource="sqlmapping/userinfoMapping.xml" />
    </mappers>
</configuration>
```

sqlmapping 包中的 userinfoMapping.xml 是 SQL 映射文件。因为本节是从零起步开始搭建开发环境，所以与 SQL 映射有关的配置代码需要自己手写，这也是本节的重点，sqlmapping 文件并不通过 MyBatis Generator 插件获得。

（5）SQL 映射文件 userinfoMapping.xml 的初始内容如下：

```
<?xml version="1.0" encoding="UTF-8" ?>
<!DOCTYPE mapper PUBLIC "-//mybatis.org//DTD Mapper 3.0//EN" "mybatis-3-mapper.dtd">
<mapper namespace="mybatis.testcurd">
</mapper>
```

（6）创建获取 SqlSession 对象的工具类，代码如下：

```
public class DBTools {
    public static SqlSession getSqlSession() throws IOException {
        String configFile = "mybatis-config.xml";
        InputStream configStream = Resources.getResourceAsStream(configFile);
        SqlSessionFactoryBuilder builder = new SqlSessionFactoryBuilder();
        SqlSessionFactory factory = builder.build(configStream);
        return factory.openSession();
    }
}
```

（7）在项目中添加两个 dtd 文件，分别是 src 路径中的 mybatis-3-config.dtd、sqlmapping 包中的 mybatis-3-mapper.dtd。添加这两个 dtd 文件的目的是在开发 XML 配置文件或 SQL 映射文件时实现自动提示功能。这两个 dtd 文件复制自 mybatis.jar 文件中的 org\apache\ibatis\builder\xml 路径下。

2. 插入记录并返回主键值

在 SQL 映射文件 userinfoMapping.xml 中添加如下配置代码：

```
<insert id="insertUserinfo" parameterType="entity.Userinfo">
    <selectKey resultType="java.lang.Long" keyProperty="id"
        order="BEFORE">
        select idauto.nextval from dual
    </selectKey>
    insert into
    userinfo(id,username,password,age,insertDate)
    values(#{id},#{username},#{password},#{age},#{insertdate})
</insert>
```

其中<selectKey>的 order="BEFORE"属性的含义是 select 语句比 insert 语句先执行，resultType 的属性值 java.lang.Long 表示将序列返回的数字转成 Long 类型，keyProperty="id"的作用是将这个 Long 值放入 parameterType 的 Userinfo 的 id 属性中。

```
<selectKey resultType="java.lang.Long" keyProperty="id"
    order="BEFORE">
    select idauto.nextval from dual
</selectKey>
```

此段配置代码的主要功能是根据序列对象生成一个主键 id 值，将 id 值放入 Userinfo 对象的 id 属性中，然后再执行 insert 语句插入到数据表里。使用序列生成的 id 值还可以在代码中获取，也就是插入一条记录后使用程序代码可以从 Userinfo 对象的 id 属性中获取刚才插入记录的 id 值。

属性 parameterType 定义参数类型，属性 resultType 定义返回值的类型。

创建 Java 类，核心代码如下：

```
public class Insert1 {
    public static void main(String[] args) throws IOException {
        Userinfo userinfo = new Userinfo();
        userinfo.setUsername("中国");
        userinfo.setPassword("中国人");
        userinfo.setAge(100L);
        userinfo.setInsertdate(new Date());
```

```
            SqlSession session = DBTools.getSqlSession();
            session.insert("insertUserinfo", userinfo);
            session.commit();
            session.close();

            System.out.println(userinfo.getId());
        }
    }
```

方法 insert() 的第 1 个参数是 SQL 映射文件配置<insert id="insertUserinfo">的 id 值，代表要执行哪个 SQL 语句。

程序执行后在控制台输出刚才新添加记录的 id 主键值，并且在 userinfo 数据表中可以看到新的记录。

如果方法 insert() 的第 1 个参数值在 SQL 映射文件中并不存在，则运行程序出现异常，创建测试，代码如下：

```
public class Insert2 {
    public static void main(String[] args) throws IOException {
        Userinfo userinfo = new Userinfo();
        userinfo.setUsername("中国");
        userinfo.setPassword("中国人");
        userinfo.setAge(100L);
        userinfo.setInsertdate(new Date());

        SqlSession session = DBTools.getSqlSession();
        session.insert("insert2", userinfo);
        session.commit();
        session.close();

        System.out.println(userinfo.getId());
    }
}
```

程序运行后出现异常：

```
    Caused by: java.lang.IllegalArgumentException: Mapped Statements collection does not
contain value for insert2
```

异常信息提示映射集合中不存在 insert2 这个映射名。

3. 根据 id 值查询记录

在 SQL 映射文件 userinfoMapping.xml 中添加如下配置代码：

```
<select id="getUserinfoById" parameterType="long"
    resultType="entity.Userinfo">
    select * from
    userinfo where id=#{id}
</select>
```

创建 Java 类，核心代码如下：

```
public class SelectById {
    public static void main(String[] args) throws IOException {
        SqlSession session = DBTools.getSqlSession();
        Userinfo userinfo = session.selectOne("getUserinfoById", 600410L);
```

```
            System.out.println(userinfo.getId() + " " + userinfo.getUsername() + " " +
userinfo.getPassword() + " "
                    + userinfo.getAge() + " " + userinfo.getInsertdate());
            session.commit();
            session.close();
        }
    }
```

程序运行后在控制台输出单条记录的信息。

4. 查询所有记录

在 SQL 映射文件 userinfoMapping.xml 中添加如下配置代码：

```
<select id="getAllUserinfo" resultType="entity.Userinfo">
    select * from userinfo
</select>
```

创建 Java 类，核心代码如下：

```
public class SelectAll {
    public static void main(String[] args) throws IOException {
        SqlSession session = DBTools.getSqlSession();
        List<Userinfo> listUserinfo = session.selectList("getAllUserinfo");
        for (int i = 0; i < listUserinfo.size(); i++) {
            Userinfo userinfo = listUserinfo.get(i);
            System.out.println(userinfo.getId() + " " + userinfo.getUsername() + " "
+ userinfo.getPassword() + " "
                    + userinfo.getAge() + " " + userinfo.getInsertdate());
        }
        session.commit();
        session.close();
    }
}
```

程序运行后在控制台输出全部记录的信息。

5. 更新记录

在 SQL 映射文件 userinfoMapping.xml 中添加如下配置代码：

```
<update id="updateUserinfoById" parameterType="entity.Userinfo">
    update userinfo
    set
    username=#{username},password=#{password},age=#{age},insertDate=#{insertdate}
    where id=#{id}
</update>
```

创建 Java 类，核心代码如下：

```
public class UpdateById {
    public static void main(String[] args) throws IOException {
        SqlSession session = DBTools.getSqlSession();
        Userinfo userinfo = session.selectOne("getUserinfoById", 600410L);
        userinfo.setUsername("x");
        userinfo.setPassword("xx");
        userinfo.setAge(200L);
        session.update("updateUserinfoById", userinfo);
```

```
        session.commit();
        session.close();
    }
}
```

程序运行后成功执行更新操作。

6. 删除记录

在 SQL 映射文件 userinfoMapping.xml 中添加如下配置代码：

```
<delete id="deleteUserinfoById" parameterType="long">
    delete from
    userinfo where id=#{id}
</delete>
```

创建 Java 类，核心代码如下：

```
public class DeleteById {
    public static void main(String[] args) throws IOException {
        SqlSession session = DBTools.getSqlSession();
        session.delete("deleteUserinfoById", 600410L);
        session.commit();
        session.close();
    }
}
```

程序运行后成功执行删除操作。

到此，自搭建开发环境并使用 SqlSession 针对 Oracle 数据库的 CURD 操作结束。

1.10.2 针对 MySQL 的 CURD

本节将演示使用 SqlSession 对 MySQL 数据库进行 CURD 操作。

1. 准备开发环境

（1）使用原有的 userinfo 数据表。

（2）创建名称为 mybatis_sqlsession_curd_mysql 的 Java 项目。

（3）准备 generatorConfig.xml 逆向配置文件，主要作用是根据数据表结构只逆向出实体类，不包含 SQL 映射文件以及客户端代码，代码如下：

```
<?xml version="1.0" encoding="UTF-8"?>
<!DOCTYPE generatorConfiguration PUBLIC "-//mybatis.org//DTD MyBatis Generator
Configuration 1.0//EN" "http://mybatis.org/dtd/mybatis-generator-config_1_0.dtd">
<generatorConfiguration>
    <context id="context1">
        <jdbcConnection
            connectionURL="jdbc:mysql://localhost:3306/y2"
            driverClass="com.mysql.jdbc.Driver" password="123123" userId="root" />
        <javaModelGenerator targetPackage="entity"
            targetProject="mybatis_sqlsession_curd_mysql" />
        <table schema="y2" tableName="userinfo">
        </table>
    </context>
</generatorConfiguration>
```

开始进行逆向操作，逆向出实体类。

（4）在 src 路径下创建连接数据库的配置文件 mybatis-config.xml，代码如下：

```
<?xml version="1.0" encoding="UTF-8" ?>
<!DOCTYPE configuration
PUBLIC "-//mybatis.org//DTD Config 3.0//EN"
"mybatis-3-config.dtd">
<configuration>
    <environments default="development">
        <environment id="development">
            <transactionManager type="JDBC" />
            <dataSource type="POOLED">
                <property name="driver" value="com.mysql.jdbc.Driver" />
                <property name="url" value="jdbc:mysql://localhost:3306/y2" />
                <property name="username" value="root" />
                <property name="password" value="123123" />
            </dataSource>
        </environment>
    </environments>
    <mappers>
        <mapper resource="sqlmapping/userinfoMapping.xml" />
    </mappers>
</configuration>
```

sqlmapping 包中的 userinfoMapping.xml 是 SQL 映射文件。因为本节是从零起步开始搭建开发环境，所以与 SQL 映射有关的配置代码需要自己手写，并不能通过 MyBatis Generator 插件获得。

（5）SQL 映射文件 userinfoMapping.xml 的代码如下：

```
<?xml version="1.0" encoding="UTF-8" ?>
<!DOCTYPE mapper PUBLIC "-//mybatis.org//DTD Mapper 3.0//EN" "mybatis-3-mapper.dtd">
<mapper namespace="mybatis.testcurd">
    <insert id="insertUserinfo" parameterType="entity.Userinfo"
        useGeneratedKeys="true" keyProperty="id">
        insert into
        userinfo(username,password,age,insertdate)
        values(#{username},#{password},#{age},#{insertdate})
    </insert>

    <select id="getUserinfoById" parameterType="long"
        resultType="entity.Userinfo">
        select *
        from userinfo where id=#{userId}
    </select>

    <select id="getAllUserinfo" resultType="entity.Userinfo">
        select * from userinfo
        order
        by id asc
    </select>

    <delete id="deleteUserinfoById" parameterType="long">
        delete from
        userinfo where
        id=#{userId}
    </delete>
```

```
    <update id="updateUserinfoById" parameterType="entity.Userinfo">
        update userinfo
        set
        username=#{username},
        password=#{password},
        age=#{age},
        insertdate=#{insertdate}
        where id=#{id}
    </update>

</mapper>
```

下列配置代码中的 useGeneratedKeys="true"代表使用数据库的主键自增机制：

```
<insert id="insertUserinfo" parameterType="entity.Userinfo"
    useGeneratedKeys="true" keyProperty="id">
    insert into
    userinfo(username,password,age,insertdate)
    values(#{username},#{password},#{age},#{insertdate})
</insert>
```

由于 MySQL 数据库的主键具有自增机制，因此在这里不使用类似于 Oracle 数据库的序列来产生主键值。属性 keyProperty="id"代表把最新的 id 值放入 entity.Userinfo 的 id 属性里，以方便在程序代码中获得刚刚插入的记录 id 值。

（6）创建获取 SqlSession 对象的工具类，代码如下：

```
public class DBTools {
    public static SqlSession getSqlSession() throws IOException {
        String configFile = "mybatis-config.xml";
        InputStream configStream = Resources.getResourceAsStream(configFile);
        SqlSessionFactoryBuilder builder = new SqlSessionFactoryBuilder();
        SqlSessionFactory factory = builder.build(configStream);
        return factory.openSession();
    }
}
```

（7）在项目中添加两个 dtd 文件，分别是 src 路径中的 mybatis-3-config.dtd、sqlmapping 包中的 mybatis-3-mapper.dtd，添加这两个 dtd 文件是为了在开发 XML 配置文件或 SQL 映射文件时实现自动提示功能，这两个 dtd 文件复制自 mybatis.jar 文件中的 org\apache\ibatis\builder\xml 路径下。

2．插入记录并返回主键值

创建 Java 类，代码如下：

```
public class Insert1 {
    public static void main(String[] args) throws IOException {
        Userinfo userinfo = new Userinfo();
        userinfo.setUsername("中国");
        userinfo.setPassword("中国人");
        userinfo.setAge(100);
        userinfo.setInsertdate(new Date());

        SqlSession session = DBTools.getSqlSession();
        session.insert("insertUserinfo", userinfo);
```

```
        session.commit();
        session.close();

        System.out.println(userinfo.getId());
    }
}
```

核心代码和操作 Oracle 数据库的代码基本一致，运行后在控制台输出刚刚插入记录的 id 主键值，并且在 MySQL 数据表中添加新的记录。

3．其他业务方法的测试

其他业务方法的代码和操作 Oracle 数据库大体一致，并且已经成功运行，详细程序可查阅随书下载的源代码。

到此，自搭建开发环境并使用 SqlSession 针对 MySQL 数据库的 CURD 操作到结束。

1.10.3 向 SQL 映射传入参数类型

常见的向 SQL 映射传入参数类型有如下 5 种。

（1）传入简单数据类型。

（2）传入复杂数据类型。

（3）传入 Map 数据类型。

（4）传入简单数组/复杂数组数据类型。

（5）传入 List<Long/Entity/Map>数据类型。

创建新的项目 mybatis_sqlsession_parameterType 来测试这 5 种情况。

1．传入简单数据类型

先来测试第 1 种：传入简单数据类型。

SQL 映射代码如下：

```
<select id="test1" parameterType="long"
    resultType="entity.Userinfo">
    select * from userinfo
    where id=#{id}
</select>
```

创建 Java 类，代码如下：

```
public class Test1 {
    public static void main(String[] args) throws IOException {
        SqlSession session = DBTools.getSqlSession();
        Userinfo userinfo = session.selectOne("test1", 100L);
        System.out.println(userinfo.getId() + " " + userinfo.getUsername() + " " +
userinfo.getPassword() + " "
                + userinfo.getAge() + " " + userinfo.getInsertdate());
        session.commit();
        session.close();
    }
}
```

2. 传入复杂数据类型

继续测试第 2 种：传入复杂数据类型。

SQL 映射代码如下：

```
<select id="test2" parameterType="entity.Userinfo"
    resultType="entity.Userinfo">
    select * from userinfo
    where id=#{id}
</select>
```

创建 Java 类，代码如下：

```
public class Test2 {
    public static void main(String[] args) throws IOException {
        Userinfo queryUserinfo = new Userinfo();
        queryUserinfo.setId(100L);
        SqlSession session = DBTools.getSqlSession();
        Userinfo userinfo = session.selectOne("test2", queryUserinfo);
        System.out.println(userinfo.getId() + " " + userinfo.getUsername() + " " +
userinfo.getPassword() + " "
                + userinfo.getAge() + " " + userinfo.getInsertdate());
        session.commit();
        session.close();
    }
}
```

3. 传入 Map 数据类型

继续测试第 3 种：传入 Map 数据类型。

SQL 映射代码如下：

```
<select id="test3" parameterType="map"
    resultType="entity.Userinfo">
    select * from userinfo
    where id=#{id}
</select>
```

创建 Java 类，代码如下：

```
public class Test3 {
    public static void main(String[] args) throws IOException {
        Map map = new HashMap();
        map.put("id", 100);
        SqlSession session = DBTools.getSqlSession();
        Userinfo userinfo = session.selectOne("test3", map);
        System.out.println(userinfo.getId() + " " + userinfo.getUsername() + " " +
userinfo.getPassword() + " "
                + userinfo.getAge() + " " + userinfo.getInsertdate());
        session.commit();
        session.close();
    }
}
```

4. 传入简单数组/复杂数组数据类型

继续测试第 4 种：传入简单数组/复杂数组数据类型。

先来看一下以简单数据类型的数组作为参数进行传递的情况。

SQL 映射代码如下：

```
<select id="test41" parameterType="int[]"
    resultType="entity.Userinfo">
    select * from userinfo
    where id=#{array[0]} or
    id=#{array[1]} or
    id=#{array[2]}
</select>
```

创建 Java 类，代码如下：

```
public class Test41 {
    public static void main(String[] args) throws IOException {
        long[] idArray = new long[] { 88, 89, 90 };
        SqlSession session = DBTools.getSqlSession();
        List<Userinfo> listUserinfo = session.selectList("test41", idArray);
        for (int i = 0; i < listUserinfo.size(); i++) {
            Userinfo userinfo = listUserinfo.get(i);
            System.out.println(userinfo.getId() + " " + userinfo.getUsername() + " "
+ userinfo.getPassword() + " "
                    + userinfo.getAge() + " " + userinfo.getInsertdate());
        }
        session.commit();
        session.close();
    }
}
```

再来看一下以复杂数据类型的数组作为参数进行传递的情况。

SQL 映射代码如下：

```
<select id="test42" parameterType="Object[]"
    resultType="entity.Userinfo">
    select * from userinfo
    where id=#{array[0].id} or
    id=#{array[1].id} or
    id=#{array[2].id}
</select>
```

创建 Java 类，代码如下：

```
public class Test42 {
    public static void main(String[] args) throws IOException {
        Userinfo userinfo1 = new Userinfo();
        userinfo1.setId(88L);
        Userinfo userinfo2 = new Userinfo();
        userinfo2.setId(89L);
        Userinfo userinfo3 = new Userinfo();
        userinfo3.setId(90L);

        Userinfo[] userinfoArray = new Userinfo[3];
        userinfoArray[0] = userinfo1;
        userinfoArray[1] = userinfo2;
        userinfoArray[2] = userinfo3;

        SqlSession session = DBTools.getSqlSession();
        List<Userinfo> listUserinfo = session.selectList("test42", userinfoArray);
```

```
            for (int i = 0; i < listUserinfo.size(); i++) {
                Userinfo userinfo = listUserinfo.get(i);
                System.out.println(userinfo.getId() + " " + userinfo.getUsername() + " "
+ userinfo.getPassword() + " "
                        + userinfo.getAge() + " " + userinfo.getInsertdate());
            }
            session.commit();
            session.close();
        }
    }
```

5. 传入 List<Long/Entity/Map>数据类型

最后测试第 5 种：传入 List<Long/Entity/Map> 数据类型。

（1）List 中存储简单数据类型的 SQL 映射，代码如下：

```
<select id="test51" parameterType="list"
    resultType="entity.Userinfo">
    select * from userinfo
    where id=#{list[0]} or
    id=#{list[1]}
    or
    id=#{list[2]}
</select>
```

创建 Java 类，代码如下：

```
public class Test51 {
    public static void main(String[] args) throws IOException {
        List idList = new ArrayList();
        idList.add(88);
        idList.add(89);
        idList.add(90);

        SqlSession session = DBTools.getSqlSession();
        List<Userinfo> listUserinfo = session.selectList("test51", idList);
        for (int i = 0; i < listUserinfo.size(); i++) {
            Userinfo userinfo = listUserinfo.get(i);
            System.out.println(userinfo.getId() + " " + userinfo.getUsername() + " "
+ userinfo.getPassword() + " "
                    + userinfo.getAge() + " " + userinfo.getInsertdate());
        }
        session.commit();
        session.close();
    }
}
```

（2）List 中存储复杂数据类型的 SQL 映射，代码如下：

```
<select id="test52" parameterType="list"
    resultType="entity.Userinfo">
    select * from userinfo
    where id=#{list[0].id} or
    id=#{list[1].id}
</select>
```

创建 Java 类，代码如下：

```
public class Test52 {
    public static void main(String[] args) throws IOException {
        Userinfo userinfo1 = new Userinfo();
        userinfo1.setId(88L);
        Userinfo userinfo2 = new Userinfo();
        userinfo2.setId(89L);

        List idList = new ArrayList();
        idList.add(userinfo1);
        idList.add(userinfo2);

        SqlSession session = DBTools.getSqlSession();
        List<Userinfo> listUserinfo = session.selectList("test52", idList);
        for (int i = 0; i < listUserinfo.size(); i++) {
            Userinfo userinfo = listUserinfo.get(i);
            System.out.println(userinfo.getId() + " " + userinfo.getUsername() + " " +
userinfo.getPassword() + " "
                    + userinfo.getAge() + " " + userinfo.getInsertdate());
        }
        session.commit();
        session.close();
    }
}
```

（3）List 中存储 Map 数据类型的 SQL 映射，代码如下：

```
<select id="test53" parameterType="list"
    resultType="entity.Userinfo">
    select * from userinfo
    where id=#{list[0].myKey1} or
    id=#{list[1].myKey2} or
    id=#{list[2].myKey3}
</select>
```

创建 Java 类，代码如下：

```
public class Test53 {
    public static void main(String[] args) throws IOException {
        Map map1 = new HashMap();
        map1.put("myKey1", 88L);
        Map map2 = new HashMap();
        map2.put("myKey2", 89L);
        Map map3 = new HashMap();
        map3.put("myKey3", 90L);

        List list = new ArrayList();
        list.add(map1);
        list.add(map2);
        list.add(map3);

        SqlSession session = DBTools.getSqlSession();
        List<Userinfo> listUserinfo = session.selectList("test53", list);
        for (int i = 0; i < listUserinfo.size(); i++) {
            Userinfo userinfo = listUserinfo.get(i);
            System.out.println(userinfo.getId() + " " + userinfo.getUsername() + " " +
userinfo.getPassword() + " "
                    + userinfo.getAge() + " " + userinfo.getInsertdate());
        }
```

```
        session.commit();
        session.close();
    }
}
```

1.10.4 从 SQL 映射取得返回值类型

常见的从 SQL 映射取得返回值类型有如下 3 种类型。

（1）返回简单数据类型。

（2）返回复杂数据类型。

（3）返回 Map 数据类型。

在项目 mybatis_sqlsession_resultType 中测试这 3 种情况。

1. 返回简单数据类型

先来测试第 1 种：返回简单数据类型。

SQL 映射代码如下：

```
<select id="test1" resultType="int">
    select count(*) from
    userinfo
</select>
```

创建 Java 类，代码如下：

```
public class Test1 {
    public static void main(String[] args) throws IOException {
        SqlSession session = DBTools.getSqlSession();
        int count = session.selectOne("test1");
        System.out.println("count=" + count);
        session.commit();
        session.close();
    }
}
```

2. 返回复杂数据类型

继续测试第 2 种：返回复杂数据类型。

SQL 映射代码如下：

```
<select id="test2" resultType="entity.Userinfo">
    select * from userinfo
    where id=100
</select>
```

创建 Java 类，代码如下：

```
public class Test2 {
    public static void main(String[] args) throws IOException {
        SqlSession session = DBTools.getSqlSession();
        Userinfo userinfo = session.selectOne("test2");
        System.out.println(userinfo.getId() + " " + userinfo.getUsername() + " " +
userinfo.getPassword() + " "
                + userinfo.getAge() + " " + userinfo.getInsertdate());
```

```
        session.commit();
        session.close();
    }
}
```

3. 返回 Map 数据类型

继续测试第 3 种：返回 Map 数据类型。

SQL 映射代码如下：

```
<select id="test31" resultType="map">
    select * from userinfo
    where
    id=90
</select>
```

创建 Java 类，代码如下：

```
public class Test31 {
    public static void main(String[] args) throws IOException {
        SqlSession session = DBTools.getSqlSession();
        Map map = session.selectOne("test31");
        System.out.println(map.get("ID"));
        System.out.println(map.get("USERNAME"));
        System.out.println(map.get("PASSWORD"));
        System.out.println(map.get("AGE"));
        System.out.println(map.get("INSERTDATE"));
        session.commit();
        session.close();
    }
}
```

从 map 中取值时，传入的 key 必须是大写的形式，因为本实验操作的数据库是 Oracle，该数据库所有的列名都是大写的形式。

如果查询的结果是动态计算出来的，那么可以使用以下两种方式来获得列中的值。

（1）使用别名。

SQL 映射代码如下：

```
<select id="test32" resultType="map">
    select sum(age) "sumAge" from
    userinfo
</select>
```

创建 Java 类，代码如下：

```
public class Test32 {
    public static void main(String[] args) throws IOException {
        SqlSession session = DBTools.getSqlSession();
        Map map = session.selectOne("test32");
        System.out.println(map.get("sumAge"));
        session.commit();
        session.close();
    }
}
```

（2）使用默认列名。

SQL 映射代码如下：

```
<select id="test33" resultType="map">
    select sum(age) from userinfo
</select>
```

创建 Java 类，代码如下：

```
public class Test33 {
    public static void main(String[] args) throws IOException {
        SqlSession session = DBTools.getSqlSession();
        Map map = session.selectOne("test33");
        Iterator iterator = map.keySet().iterator();
        while (iterator.hasNext()) {
            System.out.println("列名: " + iterator.next());
        }
        System.out.println(map.get("SUM(AGE)"));
        session.commit();
        session.close();
    }
}
```

4. 返回多个记录的情况

selectList()方法返回的 List 中可以存储简单、复杂以及 Map 数据类型。

（1）List 中存储简单数据类型。

如果 resultType 属性值是 long，SQL 映射示例代码如下：

```
<select id="test6" resultType="long">
    select id from userinfo
    order by id
    asc
</select>
```

selectList()方法返回的数据类型是 List<Long>，运行类代码如下：

```
public class Test6 {
    public static void main(String[] args) throws IOException {
        SqlSession session = DBTools.getSqlSession();
        List<Long> listMap = session.selectList("test6");
        for (int i = 0; i < listMap.size(); i++) {
            System.out.println(listMap.get(i));
        }
        session.commit();
        session.close();
    }
}
```

（2）List 中存储复杂数据类型。

如果执行如下的 SQL 映射代码：

```
<select id="test4" resultType="entity.Userinfo">
    select * from userinfo
    order by id
    asc
</select>
```

后查询的结果大于 1 条记录，则 selectList()方法返回的数据类型是 List<Userinfo>，因为 resultType 属性值是 entity.Userinfo 实体类数据类型，运行类代码如下：

```
public class Test4 {
    public static void main(String[] args) throws IOException {
        SqlSession session = DBTools.getSqlSession();
        List<Userinfo> listUserinfo = session.selectList("test4");
        System.out.println(listUserinfo.size());
        session.commit();
        session.close();
    }
}
```

（3）List 中存储 Map。

同理，如果 resultType 属性值是 map，SQL 映射示例代码如下：

```
<select id="test5" resultType="map">
    select * from userinfo
    order by id
    asc
</select>
```

selectList()方法返回的数据类型是 List<Map>，运行类代码如下：

```
public class Test5 {
    public static void main(String[] args) throws IOException {
        SqlSession session = DBTools.getSqlSession();
        List<Map> listMap = session.selectList("test5");
        System.out.println(listMap.size());
        session.commit();
        session.close();
    }
}
```

1.10.5 SQL 映射文件中命名空间的作用

当多个 SQL 映射文件中的 id 值一样时，使用 SqlSession 操作数据库时会出现异常。

创建测试用的 namespaceError 项目，并创建 userinfoMappingA.xml 映射文件，代码如下：

```
<mapper namespace="AAAAA">
    <insert id="insertUserinfo" parameterType="entity.Userinfo">
        insert into
        userinfo(id,username)
        values(idauto.nextval,'A')
    </insert>
</mapper>
```

创建名称为 userinfoMappingB.xml 的映射文件，代码如下：

```
<?xml version="1.0" encoding="UTF-8" ?>
<!DOCTYPE mapper
PUBLIC "-//mybatis.org//DTD Mapper 3.0//EN"
"http://mybatis.org/dtd/mybatis-3-mapper.dtd">
<mapper namespace="BBBBB">
    <insert id="insertUserinfo" parameterType="entity.Userinfo">
        insert into
```

```
        userinfo(id,username)
        values(idauto.nextval,'B')
    </insert>
</mapper>
```

再将这两个 SQL 映射文件使用<mapper>标签注册到 mybatis-config.xml 配置文件中，代码如下：

```
<mappers>
    <mapper resource="mapping/userinfoMappingA.xml" />
    <mapper resource="mapping/userinfoMappingB.xml" />
</mappers>
```

在这两个 SQL 映射文件中都有<insert>的 id 属性值为 insertUserinfo 的配置代码，那么在执行如下 Java 代码后就出现了异常：

```
public class Insert1 {
    public static void main(String[] args) throws IOException {
        SqlSession session = DBTools.getSqlSession();
        session.insert("insertUserinfo");
        session.commit();
        session.close();
    }
}
```

异常信息如下：

```
    Caused by: java.lang.IllegalArgumentException: insertUserinfo is ambiguous in Mapped
Statements collection (try using the full name including the namespace, or rename one of
the entries)
```

异常信息提示 insertUserinfo 并不是确定的名字，而是模糊的。这时可以尝试使用全名称的方式来调用此 SQL 映射。所谓的“全名称”方式就是指 sqlId 之前写上命名空间，对命名空间的命名可以写上表的名称，或者是业务的名称，这样有助于区分重复的 sqlId，更改后的 Java 代码如下：

```
public class Insert2 {
    public static void main(String[] args) throws IOException {
        SqlSession session = DBTools.getSqlSession();
        session.insert("BBBBB.insertUserinfo");
        session.commit();
        session.close();
    }
}
```

执行 insert()方法时在第 1 个参数加上了命名空间 BBBBB，使用.小数点作为间隔，再执行程序就不会出现异常了。

命名空间的主要作用是防止 SQL 映射中的 id 值一样导致出现异常。

1.11 自建环境使用 Mapper 接口操作 Oracle 和 MySQL 数据库

本节将从零起步，以自搭建开发环境开始，再到使用 Mapper 接口实现经典功能 CURD，并且针对 Oracle 和 MySQL 这两种主流数据库。

前面知识点中的代码都是使用如下程序对数据库进行操作：

```
sqlSession.insert("sqlId");
sqlSession.delete("sqlId");
sqlSession.update("sqlId");
sqlSession.selectList("sqlId");
```

对方法 insert()、delete()、update()和 selectList()传入 String sqlId 来达到调用 SQL 语句进而对数据库操作的目的，完全是面向 String 字符串类型的 sqlId 编程。虽然能达到实现操作数据库的目的，但这种代码的写法却是不规范的。理想中规范的写法不是面向 String sqlId 编程，而是面向接口编程。新版的 MyBatis 提供了“接口-SQL 映射”功能，使程序员完全面向 Mapper 接口进行编程，相比 String sqlId 的方式，使用 Mapper 在代码规范方面上了一个台阶。

1.11.1 接口-SQL 映射的对应关系

“接口-SQL 映射”的对应关系如图 1-43 所示。

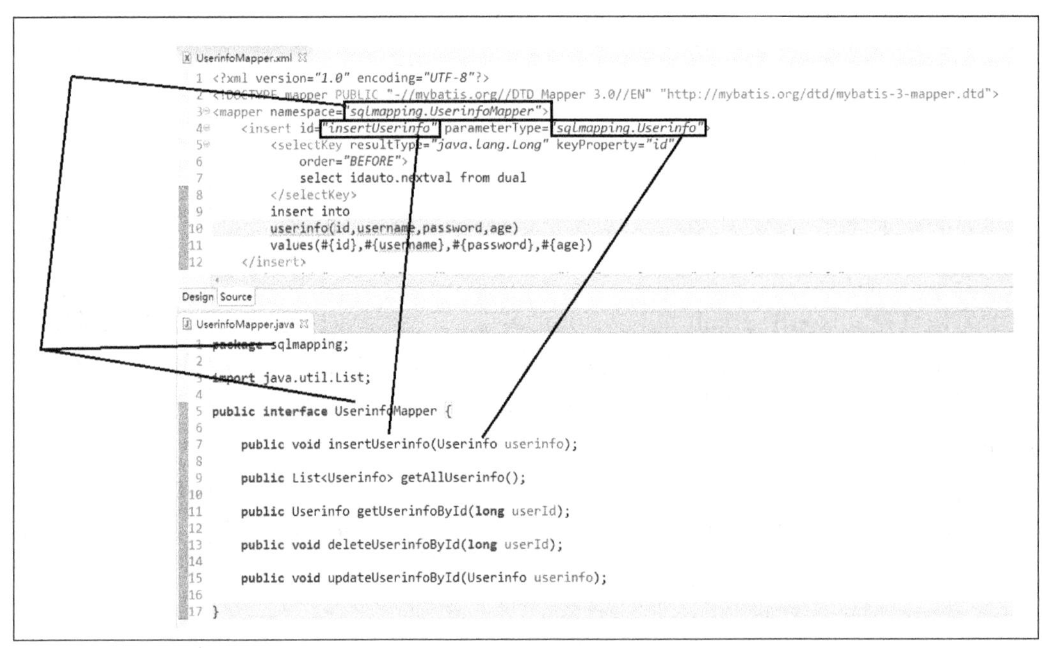

图 1-43 接口-SQL 映射原理

“接口-SQL 映射”的原理如下。

SQL 映射文件 UserinfoMapper.xml 中的 namespace 属性值 sqlmapping.UserinfoMapper 代表该映射对应的就是 sqlmapping 包中的 UserinfoMapper.java 接口，而<insert>标签的 id 属性值 insertUserinfo 就是 UserinfoMapper.java 接口中的 public void insertUserinfo（Userinfo userinfo）方法，<insert>标签的 parameterType 属性值 sqlmapping.Userinfo 就是 public void insertUserinfo (Userinfo userinfo)方法参数类型，只要它们一一对应，就能实现“接口-SQL 映射”，程序员完全以面向接口的方式设计软件。

1.11.2 针对 Oracle 的 CURD

本节将演示使用 Mapper 接口对 Oracle 数据库进行 CURD 操作。

1. 准备开发环境

（1）使用原有 userinfo 数据表。

（2）创建名称为 mybatis_mapper_curd_oracle 的 Java 项目。

（3）准备 generatorConfig.xml 逆向配置文件，主要作用是根据数据表结构只逆向出实体类，不包含 SQL 映射文件以及客户端代码，配置代码如下：

```
<?xml version="1.0" encoding="UTF-8"?>
<!DOCTYPE generatorConfiguration PUBLIC "-//mybatis.org//DTD MyBatis Generator Configuration 1.0//EN" "http://mybatis.org/dtd/mybatis-generator-config_1_0.dtd">
<generatorConfiguration>
    <context id="context1">
        <jdbcConnection
            connectionURL="jdbc:oracle:thin:@localhost:1521:orcl"
            driverClass="oracle.jdbc.OracleDriver" password="123" userId="y2" />
        <javaModelGenerator targetPackage="entity"
            targetProject="mybatis_mapper_curd_oracle" />
        <table schema="y2" tableName="userinfo">
            <generatedKey column="id"
                sqlStatement="select idauto.nextval from dual" identity="false" />
        </table>
    </context>
</generatorConfiguration>
```

开始进行逆向操作，逆向出实体类。

（4）在 src 路径下创建连接数据库的配置文件 mybatis-config.xml，代码如下：

```
<?xml version="1.0" encoding="UTF-8" ?>
<!DOCTYPE configuration
PUBLIC "-//mybatis.org//DTD Config 3.0//EN"
"http://mybatis.org/dtd/mybatis-3-config.dtd">
<configuration>
    <environments default="development">
        <environment id="development">
            <transactionManager type="JDBC" />
            <dataSource type="POOLED">
                <property name="driver" value="oracle.jdbc.OracleDriver" />
                <property name="url"
                    value="jdbc:oracle:thin:@localhost:1521:orcl" />
                <property name="username" value="y2" />
                <property name="password" value="123" />
            </dataSource>
        </environment>
    </environments>
    <mappers>
        <mapper resource="sqlmapping/UserinfoMapper.xml" />
    </mappers>
</configuration>
```

sqlmapping 包中的 UserinfoMapper.xml 是 SQL 映射文件。因为本节是从零起步搭建开发环境，所以与 SQL 映射有关的配置代码需要自己手写，并不能通过 MyBatis Generator 插件获得。

（5）SQL 映射文件 UserinfoMapper.xml 的初始代码如下：

```
<?xml version="1.0" encoding="UTF-8" ?>
<!DOCTYPE mapper PUBLIC "-//mybatis.org//DTD Mapper 3.0//EN" "mybatis-3-mapper.dtd">
<mapper namespace="sqlmapping.UserinfoMapper">
</mapper>
```

SQL 映射文件 UserinfoMapper.xml 的名称可以和接口名称不一样，比如 SQL 映射文件名称为 UserinfoMapper.xml，接口名称是 XXUserinfoMapper.java，可以在 SQL 映射文件中配置代码如下：

```
<mapper namespace="sqlmapping.XXUserinfoMapper">
```

这样的写法也是正确的。

（6）创建与 UserinfoMapper.xml 文件匹配的 Mapper 映射接口 UserinfoMapper.java，接口 UserinfoMapper.java 的初始代码如下：

```
public interface UserinfoMapper {
}
```

（7）创建获取 SqlSession 对象的工具类，代码如下：

```
public class DBTools {
    public static SqlSession getSqlSession() throws IOException {
        String configFile = "mybatis-config.xml";
        InputStream configStream = Resources.getResourceAsStream(configFile);
        SqlSessionFactoryBuilder builder = new SqlSessionFactoryBuilder();
        SqlSessionFactory factory = builder.build(configStream);
        return factory.openSession();
    }
}
```

（8）在项目中添加两个 dtd 文件，分别是 src 路径中的 mybatis-3-config.dtd、sqlmapping 包中的 mybatis-3-mapper.dtd，添加这两个 dtd 文件是为了在开发 XML 配置文件或 SQL 映射文件时实现自动提示功能。这两个 dtd 文件复制自 mybatis.jar 文件中的 org\apache\ibatis\builder\xml。

2．插入记录并返回主键值

在 SQL 映射文件 UserinfoMapper.xml 中添加如下配置代码：

```
<insert id="insertUserinfo" parameterType="entity.Userinfo">
    <selectKey order="BEFORE" resultType="java.lang.Long"
        keyProperty="id">
        select idauto.nextval from dual
    </selectKey>
    insert into userinfo(id,username,password,age,insertdate)
    values(#{id},#{username},#{password},#{age},#{insertdate})
</insert>
```

在 Mapper 接口文件 UserinfoMapper.java 中添加如下代码：

```
public interface UserinfoMapper {
    public void insertUserinfo(Userinfo userinfo);
}
```

创建 Java 类，核心代码如下：

```
public class Insert {
    public static void main(String[] args) throws IOException {

        Userinfo userinfo = new Userinfo();
        userinfo.setUsername("中国");
        userinfo.setPassword("中国人");
        userinfo.setAge(100L);
        userinfo.setInsertdate(new Date());

        SqlSession session = DBTools.getSqlSession();
        UserinfoMapper mapper = session.getMapper(UserinfoMapper.class);
        mapper.insertUserinfo(userinfo);
        session.commit();
        session.close();

        System.out.println("id=" + userinfo.getId());
    }
}
```

程序执行后在控制台输出刚才新添加记录的 id 主键值，并且在 userinfo 数据表中可以看到新的记录。

3. 根据 id 值查询记录

在 SQL 映射文件 UserinfoMapper.xml 中添加如下配置代码：

```
<select id="selectById" parameterType="long"
    resultType="entity.Userinfo">
    select *
    from userinfo where id=#{userId}
</select>
```

在 Mapper 接口文件 UserinfoMapper.java 中添加如下代码：

```
public interface UserinfoMapper {
    public Userinfo selectById(long userId);
}
```

创建 Java 类，核心代码如下：

```
public class SelectById {
    public static void main(String[] args) throws IOException {
        SqlSession session = DBTools.getSqlSession();
        UserinfoMapper mapper = session.getMapper(UserinfoMapper.class);
        Userinfo userinfo = mapper.selectById(600413L);
        System.out.println(userinfo.getId() + " " + userinfo.getUsername() + " " +
userinfo.getPassword() + " "
                + userinfo.getAge() + " " + userinfo.getInsertdate());
        session.commit();
        session.close();
    }
}
```

程序运行后在控制台输出单条记录的信息。

4. 查询所有记录

在 SQL 映射文件 UserinfoMapper.xml 中添加如下配置代码：

```
<select id="selectAll" resultType="entity.Userinfo">
    select * from userinfo order
    by id asc
</select>
```

在 Mapper 接口文件 UserinfoMapper.java 中添加如下代码：

```
public interface UserinfoMapper {
    public List<Userinfo> selectAll();
}
```

创建 Java 类，核心代码如下：

```
public class SelectAll {
    public static void main(String[] args) throws IOException {
        SqlSession session = DBTools.getSqlSession();
        UserinfoMapper mapper = session.getMapper(UserinfoMapper.class);
        List<Userinfo> listUserinfo = mapper.selectAll();
        for (int i = 0; i < listUserinfo.size(); i++) {
            Userinfo userinfo = listUserinfo.get(i);
            System.out.println(userinfo.getId() + " " + userinfo.getUsername() + " "
+ userinfo.getPassword() + " "
                    + userinfo.getAge() + " " + userinfo.getInsertdate());
        }
        session.commit();
        session.close();
    }
}
```

程序运行后在控制台输出全部记录的信息。

5. 更新记录

在 SQL 映射文件 UserinfoMapper.xml 中添加如下配置代码：

```
<update id="updateById" parameterType="entity.Userinfo">
    update userinfo set
    username=#{username},
    password=#{password},
    age=#{age},
    insertdate=#{insertdate}
    where id=#{id}
</update>
```

在 Mapper 接口文件 UserinfoMapper.java 中添加如下代码：

```
public interface UserinfoMapper {
    public Userinfo selectById(long userId);
    public void updateById(Userinfo userinfo);
}
```

创建 Java 类，核心代码如下：

```
public class UpdateById {
    public static void main(String[] args) throws IOException {
        SqlSession session = DBTools.getSqlSession();
        UserinfoMapper mapper = session.getMapper(UserinfoMapper.class);
        Userinfo userinfo = mapper.selectById(2142502L);

        userinfo.setUsername("xxxxxxxxxxxx");

        mapper.updateById(userinfo);

        session.commit();
        session.close();
    }
}
```

程序运行后成功执行更新操作。

6. 删除记录

在 SQL 映射文件 UserinfoMapper.xml 中添加如下配置代码：

```
<delete id="deleteById" parameterType="long">
    delete
    from userinfo where
    id=#{userId}
</delete>
```

在 Mapper 接口文件 UserinfoMapper.java 中添加如下代码：

```
public interface UserinfoMapper {
    public void deleteById(long userId);
}
```

创建 Java 类，核心代码如下：

```
public class DeleteById {
    public static void main(String[] args) throws IOException {
        SqlSession session = DBTools.getSqlSession();
        UserinfoMapper mapper = session.getMapper(UserinfoMapper.class);
        mapper.deleteById(2152493L);
        session.commit();
        session.close();
    }
}
```

程序运行后成功执行删除操作。

到此，自搭建开发环境并使用 Mapper 针对 Oracle 数据库的 CURD 操作结束。

1.11.3 针对 MySQL 的 CURD

本节将演示使用 Mapper 对 MySQL 数据库进行 CURD 操作。

1. 准备开发环境

（1）使用原有 userinfo 数据表。

（2）创建名称为 mybatis_mapper_curd_mysql 的 Java 项目。

（3）准备 generatorConfig.xml 逆向配置文件，主要作用是根据数据表结构只逆向出实体类，不包含 SQL 映射文件以及客户端代码，配置代码如下：

```
<?xml version="1.0" encoding="UTF-8"?>
<!DOCTYPE generatorConfiguration PUBLIC "-//mybatis.org//DTD MyBatis Generator
Configuration 1.0//EN" "http://mybatis.org/dtd/mybatis-generator-config_1_0.dtd">
<generatorConfiguration>
    <context id="context1">
        <jdbcConnection
            connectionURL="jdbc:mysql://localhost:3306/y2"
            driverClass="com.mysql.jdbc.Driver" password="123123" userId="root" />
        <javaModelGenerator targetPackage="entity"
            targetProject="mybatis_mapper_curd_mysql" />
        <table schema="y2" tableName="userinfo">
        </table>
    </context>
</generatorConfiguration>
```

开始进行逆向操作，逆向出实体类。

（4）在 src 路径下创建连接数据库的配置文件 mybatis-config.xml，代码如下：

```
<?xml version="1.0" encoding="UTF-8" ?>
<!DOCTYPE configuration
PUBLIC "-//mybatis.org//DTD Config 3.0//EN"
"http://mybatis.org/dtd/mybatis-3-config.dtd">
<configuration>
    <environments default="development">
        <environment id="development">
            <transactionManager type="JDBC" />
            <dataSource type="POOLED">
                <property name="driver" value="com.mysql.jdbc.Driver" />
                <property name="url" value="jdbc:mysql://localhost:3306/y2" />
                <property name="username" value="root" />
                <property name="password" value="123123" />
            </dataSource>
        </environment>
    </environments>
    <mappers>
        <mapper resource="sqlmapping/UserinfoMapper.xml" />
    </mappers>
</configuration>
```

sqlmapping 包中的 UserinfoMapper.xml 是 SQL 映射文件。因为本节是从零起步搭建开发环境，所以与 SQL 映射有关的配置代码需要自己手写，并不能通过 MyBatis Generator 插件获得。

（5）SQL 映射文件 UserinfoMapper.xml 的代码如下：

```
<?xml version="1.0" encoding="UTF-8"?>
<!DOCTYPE mapper PUBLIC "-//mybatis.org//DTD Mapper 3.0//EN" "mybatis-3-mapper.dtd">
<mapper namespace="sqlmapping.UserinfoMapper">
    <insert id="insertUserinfo" parameterType="entity.Userinfo"
        useGeneratedKeys="true" keyProperty="id">
        insert into
        userinfo(id,username,password,age,insertdate)
        values(#{id},#{username},#{password},#{age},#{insertdate})
    </insert>
```

```
    <select id="selectAll" resultType="entity.Userinfo">
        select * from userinfo order
        by id asc
    </select>

    <select id="selectById" parameterType="long"
        resultType="entity.Userinfo">
        select *
        from userinfo where id=#{userId}
    </select>

    <delete id="deleteById" parameterType="long">
        delete
        from userinfo where
        id=#{userId}
    </delete>

    <update id="updateById" parameterType="entity.Userinfo">
        update userinfo set
        username=#{username},
        password=#{password},
        age=#{age},
        insertdate=#{insertdate}
        where id=#{id}
    </update>
</mapper>
```

（6）创建与 UserinfoMapper.xml 文件匹配的 Mapper 映射接口 UserinfoMapper.java，接口 UserinfoMapper 的代码如下：

```
public interface UserinfoMapper {
    public void insertUserinfo(Userinfo userinfo);
    public List<Userinfo> selectAll();
    public Userinfo selectById(long userId);
    public void deleteById(long userId);
    public void updateById(Userinfo userinfo);
}
```

（7）创建获取 SqlSession 对象的工具类，代码如下：

```
public class DBTools {
    public static SqlSession getSqlSession() throws IOException {
        String configFile = "mybatis-config.xml";
        InputStream configStream = Resources.getResourceAsStream(configFile);
        SqlSessionFactoryBuilder builder = new SqlSessionFactoryBuilder();
        SqlSessionFactory factory = builder.build(configStream);
        return factory.openSession();
    }
}
```

（8）在项目中添加两个 dtd 文件，分别是 src 路径中的 mybatis-3-config.dtd、sqlmapping 包中的 mybatis-3-mapper.dtd，添加这两个 dtd 文件是为了在开发 XML 配置文件或 SQL 映射文件时实现自动提示功能。这两个 dtd 文件复制自 mybatis.jar 文件中的 org\apache\ibatis\builder\xml。

2．插入记录并返回主键值

创建 Java 类，代码如下：

```
public class Insert {
    public static void main(String[] args) throws IOException {
        Userinfo userinfo = new Userinfo();
        userinfo.setUsername("中国");
        userinfo.setPassword("中国人");
        userinfo.setAge(100);
        userinfo.setInsertdate(new Date());

        SqlSession session = DBTools.getSqlSession();
        UserinfoMapper mapper = session.getMapper(UserinfoMapper.class);
        mapper.insertUserinfo(userinfo);
        session.commit();
        session.close();

        System.out.println("id=" + userinfo.getId());
    }
}
```

核心代码和操作 Oracle 数据库的代码基本一致，运行后在控制台输出刚刚插入记录的 id 主键值，并且在 MySQL 数据表中添加了新的记录。

3．其他业务方法的测试

其他业务方法的代码和操作 Oracle 数据库大体一致，并且已经成功运行，详细程序可查阅随书下载的源代码。

到此，自搭建开发环境并使用 Mapper 针对 MySQL 数据库的 CURD 操作结束。

1.11.4　向 Mapper 接口传入参数类型

常见的向 Mapper 接口传入参数类型有如下 5 种。

（1）传入简单数据类型。

（2）传入复杂数据类型。

（3）传入 Map 数据类型。

（4）传入简单数组/复杂数组数据类型。

（5）传入 List<Long/Entity/Map>数据类型。

创建新的项目 mybatis_mapper_parameterType 来测试这 5 种情况。

1．传入简单数据类型

先来测试第 1 种：传入简单数据类型。

参数个数为 1 个的 SQL 映射代码如下：

```
<select id="test1" resultType="entity.Userinfo">
    select *
    from userinfo where
```

```
    id=#{id}
</select>
```

在 Mapper 接口文件 UserinfoMapper.java 中添加如下代码：

```
public interface UserinfoMapper {
    public Userinfo test1(long id);
}
```

创建 Java 类，代码如下：

```
public class Test1 {
    public static void main(String[] args) throws IOException {
        SqlSession session = DBTools.getSqlSession();
        UserinfoMapper mapper = session.getMapper(UserinfoMapper.class);
        Userinfo userinfo = mapper.test1(89L);
        System.out.println(userinfo.getId() + " " + userinfo.getUsername() + " " +
userinfo.getPassword() + " "
                + userinfo.getAge() + " " + userinfo.getInsertdate());
        session.commit();
        session.close();
    }
}
```

参数个数为多个的 SQL 映射代码如下：

```
<select id="test11" resultType="entity.Userinfo">
    select *
    from userinfo where
    id=#{arg0} or id=#{arg1}
</select>
```

在 Mapper 接口文件 UserinfoMapper.java 中添加如下代码：

```
public interface UserinfoMapper {
    public List<Userinfo> test11(long id1, long id2);
}
```

创建 Java 类，代码如下：

```
public class Test11 {
    public static void main(String[] args) throws IOException {
        SqlSession session = DBTools.getSqlSession();
        UserinfoMapper mapper = session.getMapper(UserinfoMapper.class);
        List<Userinfo> listUserinfo = mapper.test11(88L, 89L);
        for (int i = 0; i < listUserinfo.size(); i++) {
            Userinfo userinfo = listUserinfo.get(i);
            System.out.println(userinfo.getId() + " " + userinfo.getUsername() + " "
+ userinfo.getPassword() + " "
                    + userinfo.getAge() + " " + userinfo.getInsertdate());
        }
        session.commit();
        session.close();
    }
}
```

2. 传入复杂数据类型

继续测试第 2 种：传入复杂数据类型。

参数个数为 1 个的 SQL 映射代码如下：

```
<select id="test2" resultType="entity.Userinfo">
    select *
    from userinfo where
    id=#{id}
</select>
```

在 Mapper 接口文件 UserinfoMapper.java 中添加如下代码：

```
public interface UserinfoMapper {
    public Userinfo test2(Userinfo userinfo);
}
```

创建 Java 类，代码如下：

```
public class Test2 {
    public static void main(String[] args) throws IOException {
        Userinfo queryUserinfo = new Userinfo();
        queryUserinfo.setId(89L);

        SqlSession session = DBTools.getSqlSession();
        UserinfoMapper mapper = session.getMapper(UserinfoMapper.class);
        Userinfo userinfo = mapper.test2(queryUserinfo);
        System.out.println(userinfo.getId() + " " + userinfo.getUsername() + " " +
userinfo.getPassword() + " "
                + userinfo.getAge() + " " + userinfo.getInsertdate());
        session.commit();
        session.close();
    }
}
```

参数个数为多个的 SQL 映射代码如下：

```
<select id="test22" resultType="entity.Userinfo">
    select *
    from userinfo where
    id=#{arg0.id} or id=#{arg1.id}
</select>
```

在 Mapper 接口文件 UserinfoMapper.java 中添加如下代码：

```
public interface UserinfoMapper {
    public List<Userinfo> test22(Userinfo userinfo1, Userinfo userinfo2);
}
```

创建 Java 类，代码如下：

```
public class Test22 {
    public static void main(String[] args) throws IOException {
        Userinfo queryUserinfo1 = new Userinfo();
        queryUserinfo1.setId(89L);
        Userinfo queryUserinfo2 = new Userinfo();
        queryUserinfo2.setId(90L);

        SqlSession session = DBTools.getSqlSession();
        UserinfoMapper mapper = session.getMapper(UserinfoMapper.class);
        List<Userinfo> listUserinfo = mapper.test22(queryUserinfo1, queryUserinfo2);
```

```
        for (int i = 0; i < listUserinfo.size(); i++) {
            Userinfo userinfo = listUserinfo.get(i);
            System.out.println(userinfo.getId() + " " + userinfo.getUsername() + " "
+ userinfo.getPassword() + " "
                    + userinfo.getAge() + " " + userinfo.getInsertdate());
        }
        session.commit();
        session.close();
    }
}
```

3. 传入 Map 数据类型

继续测试第 3 种：传入 Map 数据类型。

SQL 映射代码如下：

```
<select id="test3" resultType="entity.Userinfo">
    select *
    from userinfo where
    id=#{findId}
</select>
```

在 Mapper 接口文件 UserinfoMapper.java 中添加如下代码：

```
public interface UserinfoMapper {
    public Userinfo test3(Map map);
}
```

创建 Java 类，代码如下：

```
public class Test3 {
    public static void main(String[] args) throws IOException {
        Map map = new HashMap();
        map.put("findId", 89L);

        SqlSession session = DBTools.getSqlSession();
        UserinfoMapper mapper = session.getMapper(UserinfoMapper.class);
        Userinfo userinfo = mapper.test3(map);
        System.out.println(userinfo.getId() + " " + userinfo.getUsername() + " " +
userinfo.getPassword() + " "
                + userinfo.getAge() + " " + userinfo.getInsertdate());
        session.commit();
        session.close();
    }
}
```

4. 传入简单数组/复杂数组数据类型

继续测试第 4 种：传入简单数组/复杂数组数据类型。

先来看一下简单数组数据类型作为参数进行传递的情况。

SQL 映射代码如下：

```
<select id="test41" resultType="entity.Userinfo">
    select *
    from userinfo where
    id=#{array[0]} or
```

```
    id=#{array[1]} or
    id=#{array[2]}
</select>
```

在 Mapper 接口文件 UserinfoMapper.java 中添加如下代码：

```
public interface UserinfoMapper {
    public List<Userinfo> test41(long[] idArray);
}
```

创建 Java 类，代码如下：

```
public class Test41 {
    public static void main(String[] args) throws IOException {
        long[] idArray = new long[] { 88, 89, 90 };

        SqlSession session = DBTools.getSqlSession();
        UserinfoMapper mapper = session.getMapper(UserinfoMapper.class);
        List<Userinfo> listUserinfo = mapper.test41(idArray);
        for (int i = 0; i < listUserinfo.size(); i++) {
            Userinfo userinfo = listUserinfo.get(i);
            System.out.println(userinfo.getId() + " " + userinfo.getUsername() + " "
+ userinfo.getPassword() + " "
                    + userinfo.getAge() + " " + userinfo.getInsertdate());
        }
        session.commit();
        session.close();
    }
}
```

再来看一下复杂数组数据类型作为参数进行传递的情况。

SQL 映射代码如下：

```
<select id="test42" resultType="entity.Userinfo">
    select *
    from userinfo where
    id=#{array[0].id} or
    id=#{array[1].id} or
    id=#{array[2].id}
</select>
```

在 Mapper 接口文件 UserinfoMapper.java 中添加如下代码：

```
public interface UserinfoMapper {
    public List<Userinfo> test42(Userinfo[] userinfoArray);
}
```

创建 Java 类，代码如下：

```
public class Test42 {
    public static void main(String[] args) throws IOException {
        Userinfo userinfo1 = new Userinfo();
        userinfo1.setId(88L);
        Userinfo userinfo2 = new Userinfo();
        userinfo2.setId(89L);
        Userinfo userinfo3 = new Userinfo();
        userinfo3.setId(90L);

        Userinfo[] userinfoArray = { userinfo1, userinfo2, userinfo3 };
```

```
        SqlSession session = DBTools.getSqlSession();
        UserinfoMapper mapper = session.getMapper(UserinfoMapper.class);
        List<Userinfo> listUserinfo = mapper.test42(userinfoArray);
        for (int i = 0; i < listUserinfo.size(); i++) {
            Userinfo userinfo = listUserinfo.get(i);
            System.out.println(userinfo.getId() + " " + userinfo.getUsername() + " "
+ userinfo.getPassword() + " "
                    + userinfo.getAge() + " " + userinfo.getInsertdate());
        }
        session.commit();
        session.close();
    }
}
```

最后来看一下简单数组与复杂数据联合使用作为参数进行传递的情况。

SQL 映射代码如下：

```
<select id="test43" resultType="entity.Userinfo">
    select *
    from userinfo where
    id=#{arg0[0]} or
    id=#{arg0[1]} or
    id=#{arg1[0].id} or id=#{arg1[1].id}
</select>
```

在 Mapper 接口文件 UserinfoMapper.java 中添加如下代码：

```
public interface UserinfoMapper {
    public List<Userinfo> test43(long[] idArray, Userinfo[] userinfoArray);
}
```

创建 Java 类，代码如下：

```
public class Test43 {
    public static void main(String[] args) throws IOException {
        long[] idArray = new long[] { 88, 89 };

        Userinfo userinfo1 = new Userinfo();
        userinfo1.setId(90L);
        Userinfo userinfo2 = new Userinfo();
        userinfo2.setId(91L);
        Userinfo[] userinfoArray = { userinfo1, userinfo2 };

        SqlSession session = DBTools.getSqlSession();
        UserinfoMapper mapper = session.getMapper(UserinfoMapper.class);
        List<Userinfo> listUserinfo = mapper.test43(idArray, userinfoArray);
        for (int i = 0; i < listUserinfo.size(); i++) {
            Userinfo userinfo = listUserinfo.get(i);
            System.out.println(userinfo.getId() + " " + userinfo.getUsername() + " "
+ userinfo.getPassword() + " "
                    + userinfo.getAge() + " " + userinfo.getInsertdate());
        }
        session.commit();
        session.close();
    }
}
```

5. 传入 List<Long/Entity/Map>数据类型

最后测试第 5 种：传入 List<Long/Entity/Map>数据类型。

（1）List 中存储简单数据类型的 SQL 映射代码如下：

```
<select id="test51" resultType="entity.Userinfo">
    select *
    from userinfo where
    id=#{list[0]} or
    id=#{list[1]}
</select>
```

在 Mapper 接口文件 UserinfoMapper.java 中添加如下代码：

```
public interface UserinfoMapper {
    public List<Userinfo> test51(List<Long> idList);
}
```

创建 Java 类，代码如下：

```
public class Test51 {
    public static void main(String[] args) throws IOException {
        List idList = new ArrayList();
        idList.add(88);
        idList.add(89);

        SqlSession session = DBTools.getSqlSession();
        UserinfoMapper mapper = session.getMapper(UserinfoMapper.class);
        List<Userinfo> listUserinfo = mapper.test51(idList);
        for (int i = 0; i < listUserinfo.size(); i++) {
            Userinfo userinfo = listUserinfo.get(i);
            System.out.println(userinfo.getId() + " " + userinfo.getUsername() + " "
+ userinfo.getPassword() + " "
                    + userinfo.getAge() + " " + userinfo.getInsertdate());
        }
        session.commit();
        session.close();
    }
}
```

（2）List 中存储复杂数据类型的 SQL 映射代码如下：

```
<select id="test52" resultType="entity.Userinfo">
    select *
    from userinfo where
    id=#{list[0].id} or
    id=#{list[1].id}
</select>
```

在 Mapper 接口文件 UserinfoMapper.java 中添加如下代码：

```
public interface UserinfoMapper {
    public List<Userinfo> test52(List<Userinfo> idList);
}
```

创建 Java 类，代码如下：

```
public class Test52 {
    public static void main(String[] args) throws IOException {
        Userinfo userinfo1 = new Userinfo();
        userinfo1.setId(88L);
        Userinfo userinfo2 = new Userinfo();
        userinfo2.setId(89L);

        List listParam = new ArrayList();
        listParam.add(userinfo1);
        listParam.add(userinfo2);

        SqlSession session = DBTools.getSqlSession();
        UserinfoMapper mapper = session.getMapper(UserinfoMapper.class);
        List<Userinfo> listUserinfo = mapper.test52(listParam);
        for (int i = 0; i < listUserinfo.size(); i++) {
            Userinfo userinfo = listUserinfo.get(i);
            System.out.println(userinfo.getId() + " " + userinfo.getUsername() + " "
+ userinfo.getPassword() + " "
                        + userinfo.getAge() + " " + userinfo.getInsertdate());
        }
        session.commit();
        session.close();
    }
}
```

（3）List 中存储 Map 数据类型的 SQL 映射代码如下：

```
<select id="test53" resultType="entity.Userinfo">
    select *
    from userinfo where
    id=#{list[0].myKey1} or
    id=#{list[1].myKey2}
</select>
```

在 Mapper 接口文件 UserinfoMapper.java 中添加如下代码：

```
public interface UserinfoMapper {
    public List<Userinfo> test53(List<Map> idList);
}
```

创建 Java 类，代码如下：

```
public class Test53 {
    public static void main(String[] args) throws IOException {
        Map map1 = new HashMap();
        map1.put("myKey1", 89L);
        Map map2 = new HashMap();
        map2.put("myKey2", 90L);

        List listParam = new ArrayList();
        listParam.add(map1);
        listParam.add(map2);

        SqlSession session = DBTools.getSqlSession();
        UserinfoMapper mapper = session.getMapper(UserinfoMapper.class);
        List<Userinfo> listUserinfo = mapper.test53(listParam);
        for (int i = 0; i < listUserinfo.size(); i++) {
            Userinfo userinfo = listUserinfo.get(i);
```

```
                System.out.println(userinfo.getId() + " " + userinfo.getUsername() + " "
+ userinfo.getPassword() + " "
                        + userinfo.getAge() + " " + userinfo.getInsertdate());
            }
            session.commit();
            session.close();
        }
    }
```

1.11.5 从 SQL 映射取得返回值类型

常见的从 SQL 映射取得返回值类型有如下 3 种。

（1）返回简单数据类型。

（2）返回复杂数据类型。

（3）返回 Map 数据类型。

在项目 mybatis_mapper_resultType 中测试这 3 种情况。

1．返回简单数据类型

先来测试第 1 种：返回简单数据类型。

SQL 映射代码如下：

```
<select id="test1" resultType="int">
    select count(*) from
userinfo
</select>
```

在 Mapper 接口文件 UserinfoMapper.java 中添加如下代码：

```
public interface UserinfoMapper {
    public int test1();
}
```

创建 Java 类，代码如下：

```
public class Test1 {
    public static void main(String[] args) throws IOException {
        SqlSession session = DBTools.getSqlSession();
        UserinfoMapper mapper = session.getMapper(UserinfoMapper.class);
        int count = mapper.test1();
        System.out.println("count=" + count);
        session.commit();
        session.close();
    }
}
```

2．返回复杂数据类型

继续测试第 2 种：返回复杂数据类型。

SQL 映射代码如下：

```
<select id="test2" resultType="entity.Userinfo">
    select * from userinfo
    where id=90
</select>
```

在 Mapper 接口文件 UserinfoMapper.java 中添加如下代码：

```
public interface UserinfoMapper {
    public Userinfo test2();
}
```

创建 Java 类，代码如下：

```
public class Test2 {
    public static void main(String[] args) throws IOException {
        SqlSession session = DBTools.getSqlSession();
        UserinfoMapper mapper = session.getMapper(UserinfoMapper.class);
        Userinfo userinfo = mapper.test2();
        System.out.println(userinfo.getId() + " " + userinfo.getUsername() + " " +
userinfo.getPassword() + " "
                + userinfo.getAge() + " " + userinfo.getInsertdate());
        session.commit();
        session.close();
    }
}
```

3. 返回 Map 数据类型

继续测试第 3 种：返回 Map 数据类型。

SQL 映射代码如下：

```
<select id="test31" resultType="map">
    select * from userinfo
    where id=90
</select>
```

在 Mapper 接口文件 UserinfoMapper.java 中添加如下代码：

```
public interface UserinfoMapper {
    public Map test31();
}
```

创建 Java 类，代码如下：

```
public class Test31 {
    public static void main(String[] args) throws IOException {
        SqlSession session = DBTools.getSqlSession();
        UserinfoMapper mapper = session.getMapper(UserinfoMapper.class);
        Map map = mapper.test31();
        System.out.println(map.get("ID"));
        System.out.println(map.get("USERNAME"));
        System.out.println(map.get("PASSWORD"));
        System.out.println(map.get("AGE"));
        System.out.println(map.get("INSERTDATE"));
        session.commit();
        session.close();
    }
}
```

如果查询的结果是动态计算出来的，那么可以使用以下两种方式来获得列中的值。

（1）使用别名。

SQL 映射代码如下：

```
<select id="test32" resultType="map">
    select sum(age) "sumAge" from
    userinfo
</select>
```

在 Mapper 接口文件 UserinfoMapper.java 中添加如下代码：

```
public interface UserinfoMapper {
    public Map test32();
}
```

创建 Java 类，代码如下：

```
public class Test32 {
    public static void main(String[] args) throws IOException {
        SqlSession session = DBTools.getSqlSession();
        UserinfoMapper mapper = session.getMapper(UserinfoMapper.class);
        Map map = mapper.test32();
        System.out.println(map.get("sumAge"));
        session.commit();
        session.close();
    }
}
```

（2）使用默认列名。

SQL 映射代码如下：

```
<select id="test33" resultType="map">
    select sum(age) from userinfo
</select>
```

在 Mapper 接口文件 UserinfoMapper.java 中添加如下代码：

```
public interface UserinfoMapper {
    public Map test33();
}
```

创建 Java 类，代码如下：

```
public class Test33 {
    public static void main(String[] args) throws IOException {
        SqlSession session = DBTools.getSqlSession();
        UserinfoMapper mapper = session.getMapper(UserinfoMapper.class);
        Map map = mapper.test33();
        Iterator iterator = map.keySet().iterator();
        while (iterator.hasNext()) {
            System.out.println("列名：" + iterator.next());
        }
        System.out.println(map.get("SUM(AGE)"));
        session.commit();
        session.close();
    }
}
```

4．返回多个记录的情况

方法 selectList()返回的 List 中可以存储简单、复杂和 Map 数据类型。

（1）List 中存储简单数据类型。

SQL 映射示例代码如下：

```
<select id="test6" resultType="long">
    select id from userinfo
    order by id
    asc
</select>
```

在 Mapper 接口文件 UserinfoMapper.java 中添加如下代码：

```
public interface UserinfoMapper {
    public List<Long> test6();
}
```

运行类代码如下：

```
public class Test6 {
    public static void main(String[] args) throws IOException {
        SqlSession session = DBTools.getSqlSession();
        UserinfoMapper mapper = session.getMapper(UserinfoMapper.class);
        List<Long> listLong = mapper.test6();
        for (int i = 0; i < listLong.size(); i++) {
            Long eachId = listLong.get(i);
            System.out.println(eachId);
        }
        session.commit();
        session.close();
    }
}
```

（2）List 中存储复杂数据类型。

SQL 映射示例代码如下：

```
<select id="test4" resultType="entity.Userinfo">
    select * from userinfo
    order by id
    asc
</select>
```

在 Mapper 接口文件 UserinfoMapper.java 中添加如下代码：

```
public interface UserinfoMapper {
    public List<Userinfo> test4();
}
```

运行类代码如下：

```
public class Test4 {
    public static void main(String[] args) throws IOException {
        SqlSession session = DBTools.getSqlSession();
        UserinfoMapper mapper = session.getMapper(UserinfoMapper.class);
        List<Userinfo> listUserinfo = mapper.test4();
        for (int i = 0; i < listUserinfo.size(); i++) {
            Userinfo userinfo = listUserinfo.get(i);
            System.out.println(userinfo.getId() + " " + userinfo.getUsername() + " "
+ userinfo.getPassword() + " "
                            + userinfo.getAge() + " " + userinfo.getInsertdate());
        }
        session.commit();
```

```
        session.close();
    }
}
```

（3）List 中存储 Map 数据类型。

SQL 映射示例代码如下：

```
<select id="test5" resultType="map">
    select * from userinfo
    order by id
    asc
</select>
```

在 Mapper 接口文件 UserinfoMapper.java 中添加如下代码：

```
public interface UserinfoMapper {
    public List<Map> test5();
}
```

运行类代码如下：

```
public class Test5 {
    public static void main(String[] args) throws IOException {
        SqlSession session = DBTools.getSqlSession();
        UserinfoMapper mapper = session.getMapper(UserinfoMapper.class);
        List<Map> listMap = mapper.test5();
        for (int i = 0; i < listMap.size(); i++) {
            Map map = listMap.get(i);
            System.out.println(map.get("ID") + " " + map.get("USERNAME") + " " + map.
get("PASSWORD") + " "
                    + map.get("AGE") + " " + map.get("INSERTDATE"));
        }
        session.commit();
        session.close();
    }
}
```

1.12　MyBatis 核心对象的生命周期与封装

在前面对两种主流数据库实现基本的 CURD 后，对 MyBatis 核心对象在使用上应该不再陌生，本节将会继续介绍这些核心对象的生命周期。

对象的生命周期也就是对象从创建到销毁的过程，但在此过程中，如果实现的代码质量不太优质，那么很容易造成程序上的错误或效率的降低。

（1）SqlSessionFactoryBuilder 对象可以被 JVM 所实例化，使用或者销毁。一旦使用 SqlSessionFactoryBuilder 对象创建了 SqlSessionFactory 后，SqlSessionFactoryBuilder 类就不需要存在了，也就是不需要保持此对象的状态，可以任由 JVM 销毁，因此 SqlSessionFactoryBuilder 对象的最佳使用范围是在方法之内，也就是说，可以在方法内部声明 SqlSessionFactoryBuilder 对象来创建 SqlSessionFactory 对象。

（2）SqlSessionFactory 对象是由 SqlSessionFactoryBuilder 对象创建而来。一旦 SqlSessionFactory 类的实例被创建，该实例在应用程序执行期间应该始终存在，根本不需要每一次操作数据库时都重新创建它。因为创建 SqlSessionFactory 对象比较耗时，所以应用它的最佳方式就是写一个单例模式，或使用 Spring 框架来实现单例模式以对 SqlSessionFactory 对象进行有效的管理。SqlSessionFactory 对象是线程安全的。

（3）SqlSession 对象是由 SqlSessionFactory 类创建而来，需要注意的是，每个线程都应该有它自己的 SqlSession 实例。SqlSession 的实例不能共享使用，因为它是线程不安全的，所以千万不要在 Servlet 中声明该对象的 1 个实例变量。因为 Servlet 是单例的，声明成实例变量会造成线程安全问题，也绝不能将 SqlSession 对象放在一个类的静态字段甚至是实例字段中，还不可以将 SqlSession 对象放在 HttpSession 会话或 ServletContext 上下文中。在接收到 HTTP 请求后可以打开 1 个 SqlSession 对象操作数据库，在响应之前需要关闭 SqlSession。关闭 SqlSession 很重要，读者应该确保使用 finally 块来关闭它。下面的示例就是一个确保 SqlSession 对象正常关闭的基本模式代码：

```
public class insertUserinfo extends HttpServlet {
    public void doGet(HttpServletRequest request, HttpServletResponse response)
            throws ServletException, IOException {
        SqlSession sqlSession = GetSqlSession.getSqlSession();
        try {
            // sqlSession curd code
            sqlSession.commit();
        } catch (Exception e) {
            sqlSession.rollback();
            e.printStackTrace();
        } finally {
            sqlSession.close();
        }

    }
}
```

1.12.1 创建 GetSqlSessionFactory 类

根据前面学习到的生命周期的知识，后面将对 MyBatis 核心代码进行封装，这样更有助于提高数据 CURD 的方便性。

创建测试项目，名称为 mybatis_threadlocal。

创建 GetSqlSessionFactory 类，完整代码如下：

```
public class GetSqlSessionFactory {
    private static SqlSessionFactory factory;

    private GetSqlSessionFactory() {
    }

    synchronized public static SqlSessionFactory getSqlSessionFactory() throws IOException {
        if (factory == null) {
            String configFile = "mybatis-config.xml";
            InputStream configStream = Resources.getResourceAsStream(configFile);
            SqlSessionFactoryBuilder builder = new SqlSessionFactoryBuilder();
            factory = builder.build(configStream);
        }
        return factory;
    }

}
```

在 GetSqlSessionFactory.java 类中使用单例设计模式来取得 SqlSessionFactory 对象。

1.12.2 创建 GetSqlSession 类

核心代码如下：

```
public class GetSqlSession {

    private static ThreadLocal<SqlSession> tl = new ThreadLocal<>();

    public static SqlSession getSqlSession() throws IOException {
        SqlSession session = tl.get();
        if (session == null) {
            session = GetSqlSessionFactory.getSqlSessionFactory().openSession();
            tl.set(session);
        } else {
        }
        return session;
    }

    public static void commit() {
        if (tl.get() != null) {
            tl.get().commit();
            tl.get().close();
            tl.set(null);
        }
    }

    public static void rollback() {
        if (tl.get() != null) {
            tl.get().rollback();
            tl.get().close();
            tl.set(null);
        }
    }
}
```

1.12.3 创建 SQL 映射接口

核心代码如下：

```
public interface UserinfoMapper {
    public void insert(Userinfo userinfo);
    public void deleteById(long id);
    public void updateById(Userinfo userinfo);
    public List<Userinfo> selectAll();
    public Userinfo selectById(long id);
}
```

1.12.4 创建 SQL 映射文件

代码如下：

```
<?xml version="1.0" encoding="UTF-8" ?>
<!DOCTYPE mapper
```

```
PUBLIC "-//mybatis.org//DTD Mapper 3.0//EN"
"mybatis-3-mapper.dtd">
<mapper namespace="sqlmapping.UserinfoMapper">
    <insert id="insert" parameterType="entity.Userinfo">
        <selectKey order="BEFORE" resultType="java.lang.Long"
            keyProperty="id">
            select idauto.nextval from dual
        </selectKey>
        insert into userinfo(id,username,password,age,insertdate)
        values(#{id},#{username},#{password},#{age},#{insertdate})
    </insert>

    <select id="selectAll" resultType="entity.Userinfo">
        select * from userinfo order
        by id asc
    </select>

    <select id="selectById" parameterType="long"
        resultType="entity.Userinfo">
        select *
        from userinfo where id=#{id}
    </select>

    <update id="updateById" parameterType="entity.Userinfo">
        update userinfo set
        username=#{username},
        password=#{password},
        age=#{age},
        insertdate=#{insertdate}
        where id=#{id}
    </update>

    <delete id="deleteById" parameterType="long">
        delete
        from userinfo where
        id=#{id}
    </delete>
</mapper>
```

1.12.5 测试多次获取的 SqlSession 对象是否为同一个

首先测试多次获取的 SqlSession 对象是不是同一个，核心代码如下：

```
public class Test1 {
    public static void main(String[] args) throws IOException {
        System.out.println(GetSqlSession.getSqlSession());
        System.out.println(GetSqlSession.getSqlSession());
        System.out.println(GetSqlSession.getSqlSession());
        System.out.println(GetSqlSession.getSqlSession());
        System.out.println(GetSqlSession.getSqlSession());
        System.out.println(GetSqlSession.getSqlSession());
    }
}
```

程序运行后在控制台输出的信息如下：

```
org.apache.ibatis.session.defaults.DefaultSqlSession@13deb50e
org.apache.ibatis.session.defaults.DefaultSqlSession@13deb50e
```

```
org.apache.ibatis.session.defaults.DefaultSqlSession@13deb50e
org.apache.ibatis.session.defaults.DefaultSqlSession@13deb50e
org.apache.ibatis.session.defaults.DefaultSqlSession@13deb50e
org.apache.ibatis.session.defaults.DefaultSqlSession@13deb50e
```

这说明获得的 SqlSession 对象是同一个。

1.12.6 添加记录及异常回滚的测试

添加记录，代码如下：

```
public class Insert1 {
    public static void main(String[] args) throws IOException {
        SqlSession session = null;
        try {
            Userinfo userinfo = new Userinfo();
            userinfo.setUsername("中国 111");
            userinfo.setPassword("中国人 111");
            userinfo.setAge(100L);
            userinfo.setInsertdate(new Date());

            session = GetSqlSession.getSqlSession();
            UserinfoMapper mapper = session.getMapper(UserinfoMapper.class);
            mapper.insert(userinfo);
        } catch (Exception e) {
            e.printStackTrace();
            GetSqlSession.rollback();
        } finally {
            GetSqlSession.commit();
        }
    }
}
```

程序运行后，成功地在数据表中增加了一条记录。

再来测试异常回滚的情况，示例代码如下：

```
public class Insert2 {
    public static void main(String[] args) throws IOException {
        SqlSession session = null;
        try {
            Userinfo userinfo1 = new Userinfo();
            userinfo1.setUsername("中国 111");
            userinfo1.setPassword("中国人 111");
            userinfo1.setAge(100L);
            userinfo1.setInsertdate(new Date());

            Userinfo userinfo2 = new Userinfo();
            userinfo2.setUsername("中国 111");
            userinfo2.setPassword(
                    "中国人 111 中国人 111 中国人 111 中国人 111 中国人 111 中国人 111 中国人 111 中国
人 111 中国人 111 中国人 111 中国人 111 中国人 111 中国人 111 中国人 111 中国人 111 中国人 111 中国人 111 中国人
111 中国人 111 中国人 111 中国人 111 中国人 111 中国人 111 中国人 111 中国人 111 中国人 111 中国人 111 中国人 111
中国人 111");
            userinfo2.setAge(100L);
            userinfo2.setInsertdate(new Date());

            session = GetSqlSession.getSqlSession();
```

```
            UserinfoMapper mapper = session.getMapper(UserinfoMapper.class);
            mapper.insert(userinfo1);
            mapper.insert(userinfo2);

        } catch (Exception e) {
            e.printStackTrace();
            GetSqlSession.rollback();
        } finally {
            GetSqlSession.commit();
        }
    }
}
```

程序运行后在控制台输出的异常信息如下：

```
Caused by: java.sql.SQLException: ORA-12899: 列 "Y2"."USERINFO"."PASSWORD" 的值太大 (实际值: 261, 最大值: 50)
```

程序出现异常，并且已经回滚，这是因为数据表 userinfo 中记录数量不变。其他业务方法不再重复进行测试。

ORM 框架 MyBatis 介绍到这里，读者应该能熟练地使用它进行数据库的 CURD 操作。

第 2 章　MyBatis 3 核心技术之实战技能

本章目标：

- 日志的处理
- 操作数据源 DataSource
- 动态 SQL 的使用
- 分页处理
- 级联操作
- 二级缓存的特性

2.1　结合 Log4j 实现输出日志

在 MyBatis 执行的过程中可以结合 Log4j 日志框架来实现输出日志，从而监测程序执行的细节。

在 MyBatis 中输出日志最主要的目的是监测 SQL 语句的执行情况，比如查看执行的 SQL 语句字符串、参数名以及参数值等信息。

Log4j 框架现在有两个版本，一个是 Log4j 1 版本，另一个就是比较新的 Log4j 2 版本。MyBatis 官方 zip 压缩包中包含 Log4j 1 和 Log4j 2 两个版本，但 Log4j 2 依赖 Log4j 1 的 JAR 包。

本节将要测试 Log4j 1 和 Log4j 2 版本的日志输出效果。

2.1.1　结合 Log4j 1 实现输出日志

创建测试用的项目 log4jTest，创建的过程中不要引用 MyBatisJAR 库，本实验要自己添加 JAR 包，防止 Log4j 1 和 Log4j 2 版本的 JAR 包混杂在一起。

添加 Log4j 1 版本的 JAR 包。

在 src 节点下创建 log4j.properties 文件，内容如下：

```
log4j.rootLogger=DEBUG,Console
log4j.appender.Console=org.apache.log4j.ConsoleAppender
log4j.appender.Console.layout=org.apache.log4j.PatternLayout
log4j.appender.Console.layout.ConversionPattern=%d [%t] %-5p [%c] - %m%n
log4j.logger.org.apache=INFO
```

在配置文件中添加 Log4j 1 版本的注册，代码如下：

```
<settings>
    <setting name="logImpl" value="LOG4J" />
</settings>
```

SQL 映射文件 userinfoMapping.xml 的代码如下：

```
<mapper namespace="mybatis.testcurd">
    <select id="getUserinfo" parameterType="entity.Userinfo"
        resultType="entity.Userinfo">
        select * from userinfo
        where
        username like #{username} or
        password like #{password}
    </select>
</mapper>
```

运行类代码如下：

```
public class GetUserinfo {
    public static void main(String[] args) {
        try {
            Userinfo queryUserinfo = new Userinfo();
            queryUserinfo.setUsername("%中国%");
            queryUserinfo.setPassword("%中国人%");

            SqlSession sqlSession = GetSqlSession.getSqlSession();
            List<Userinfo> listUserinfo = sqlSession.selectList("getUserinfo", queryUserinfo);
            for (int i = 0; i < listUserinfo.size(); i++) {
                Userinfo userinfo = listUserinfo.get(i);
                System.out.println(userinfo.getId() + " " + userinfo.getUsername() +
" " + userinfo.getPassword() + " "
                        + userinfo.getAge() + " " + userinfo.getInsertdate());
            }
            sqlSession.commit();
            sqlSession.close();
        } catch (Exception e) {
            e.printStackTrace();
        }
    }
}
```

程序运行的效果如下：

```
[main] DEBUG [mybatis.testcurd.getUserinfo] - ==>  Preparing: select * from userinfo w
here username like ? or password like ?
[main] DEBUG [mybatis.testcurd.getUserinfo] - ==> Parameters: %中国%(String), %中国人
%(String)
[main] DEBUG [mybatis.testcurd.getUserinfo] - <==      Total: 8
600419 中国 中国人 333 Tue Jun 05 18:21:32 CST 2018
600414 中国 中国人 333 Tue Jun 05 18:18:41 CST 2018
600417 中国 中国人 333 Tue Jun 05 18:21:31 CST 2018
600423 中国 中国人 333 Tue Jun 05 18:21:33 CST 2018
600425 中国 中国人 333 Tue Jun 05 18:21:34 CST 2018
600418 中国 中国人 333 Tue Jun 05 18:21:31 CST 2018
600420 中国 中国人 333 Tue Jun 05 18:21:32 CST 2018
600424 中国 中国人 333 Tue Jun 05 18:21:33 CST 2018
```

输出的日志中打印出了 SQL 语句，以及传给 SQL 语句的参数值。

2.1.2 结合 Log4j 2 实现输出日志

创建测试用的项目 log4j2Test，创建的过程中不要引用 MyBatisJAR 库，本实验要自己添加 JAR 包，防止 Log4j 1 和 Log4j 2 版本的 JAR 包混杂在一起。

添加 Log4j 2 版本的 jar 包，但 Log4j 2 版本需要依赖于 log4j-1.2.17.jar 文件，而且 Log4j 2 版本还依赖于 log4j2.xml 文件。

在配置文件中添加 Log4j 2 版本的注册，代码如下：

```
<settings>
    <setting name="logImpl" value="LOG4J2" />
</settings>
```

创建 log4j2.xml 配置文件，内容如下：

```
<?xml version="1.0" encoding="UTF-8"?>
<Configuration>
    <Appenders>
        <Console name="STDOUT" target="SYSTEM_OUT">
            <PatternLayout
                pattern="%d %-5p [%t] %C{2} (%F:%L) - %m%n" />
        </Console>
    </Appenders>
    <Loggers>
        <Logger name="org.apache.log4j.xml" level="info" />
        <Root level="debug">
            <AppenderRef ref="STDOUT" />
        </Root>
    </Loggers>
</Configuration>
```

运行类代码如下：

```
public class GetUserinfo {
    public static void main(String[] args) {
        try {
            Userinfo queryUserinfo = new Userinfo();
            queryUserinfo.setUsername("%中国%");
            queryUserinfo.setPassword("%中国人%");

            SqlSession sqlSession = GetSqlSession.getSqlSession();
            List<Userinfo> listUserinfo = sqlSession.selectList("getUserinfo", queryUserinfo);
            for (int i = 0; i < listUserinfo.size(); i++) {
                Userinfo userinfo = listUserinfo.get(i);
                System.out.println(userinfo.getId() + " " + userinfo.getUsername() +
" " + userinfo.getPassword() + " "
                        + userinfo.getAge() + " " + userinfo.getInsertdate());
            }
            sqlSession.commit();
            sqlSession.close();
        } catch (Exception e) {
            e.printStackTrace();
        }
    }
}
```

程序运行后在控制台输出日志信息如下：

```
DEBUG [main] logging.LogFactory (LogFactory.java:135) - Logging initialized using
'class org.apache.ibatis.logging.log4j2.Log4j2Impl' adapter.
DEBUG [main] logging.LogFactory (LogFactory.java:135) - Logging initialized using
'class org.apache.ibatis.logging.log4j.Log4jImpl' adapter.
DEBUG [main] pooled.PooledDataSource (PooledDataSource.java:335) - PooledDataSource
forcefully closed/removed all connections.
DEBUG [main] pooled.PooledDataSource (PooledDataSource.java:335) - PooledDataSource
forcefully closed/removed all connections.
DEBUG [main] pooled.PooledDataSource (PooledDataSource.java:335) - PooledDataSource
forcefully closed/removed all connections.
DEBUG [main] pooled.PooledDataSource (PooledDataSource.java:335) - PooledDataSource
forcefully closed/removed all connections.
DEBUG [main] jdbc.JdbcTransaction (JdbcTransaction.java:137) - Opening JDBC Connection
DEBUG [main] pooled.PooledDataSource (PooledDataSource.java:406) - Created connection
1836463382.
DEBUG [main] jdbc.JdbcTransaction (JdbcTransaction.java:101) - Setting autocommit to
false on JDBC Connection [oracle.jdbc.driver.T4CConnection@6d763516]
DEBUG [main] jdbc.BaseJdbcLogger (BaseJdbcLogger.java:159) - ==>  Preparing: select *
from userinfo where username like ? or password like ?
DEBUG [main] jdbc.BaseJdbcLogger (BaseJdbcLogger.java:159) - ==> Parameters: %中国
%(String), %中国人%(String)
DEBUG [main] jdbc.BaseJdbcLogger (BaseJdbcLogger.java:159) - <==      Total: 8
600419 中国 中国人 333 Tue Jun 05 18:21:32 CST 2018
600414 中国 中国人 333 Tue Jun 05 18:18:41 CST 2018
600417 中国 中国人 333 Tue Jun 05 18:21:31 CST 2018
600423 中国 中国人 333 Tue Jun 05 18:21:33 CST 2018
600425 中国 中国人 333 Tue Jun 05 18:21:34 CST 2018
600418 中国 中国人 333 Tue Jun 05 18:21:31 CST 2018
600420 中国 中国人 333 Tue Jun 05 18:21:32 CST 2018
600424 中国 中国人 333 Tue Jun 05 18:21:33 CST 2018
DEBUG [main] jdbc.JdbcTransaction (JdbcTransaction.java:123) - Resetting autocommit
to true on JDBC Connection [oracle.jdbc.driver.T4CConnection@6d763516]
DEBUG [main] jdbc.JdbcTransaction (JdbcTransaction.java:91) - Closing JDBC Connection
[oracle.jdbc.driver.T4CConnection@6d763516]
DEBUG [main] pooled.PooledDataSource (PooledDataSource.java:363) - Returned connection
1836463382 to pool.
```

日志中同样输出了 SQL 语句以及参数值。

2.2 SQL 语句中特殊符号的处理

如果 SQL 语句有一些特殊符号，那么必须使用如下的格式设计 SQL 语句：

```
<![CDATA[ 特殊符号 ]]>
```

创建测试用的项目 hasOtherChar。

SQL 映射文件 userinfoMapping.xml 的代码如下：

```
<mapper namespace="mybatis.testcurd">
    <select id="getUserinfo" parameterType="long"
        resultType="entity.Userinfo">
        select * from userinfo
        where
        id <![CDATA[<]]>
```

```
        #{idParam} order by id asc
    </select>
</mapper>
```

小于号“<”在 XML 文件中是特殊的符号，因此，要放在<![CDATA[特殊符号]]>中。

2.3 使用 typeAliases 配置别名

别名的作用就是使用短名称代替冗长的全名称，以简化配置。

别名分为系统预定义别名和自定义别名。

2.3.1 系统预定义别名

系统预定义别名其实早就接触过了，在配置文件中就使用到了别名，示例配置文件的代码如下：

```
<configuration>
    <environments default="development">
        <environment id="development">
            <transactionManager type="JDBC" />
            <dataSource type="POOLED">
                <property name="driver"
                    value="oracle.jdbc.driver.OracleDriver" />
                <property name="url"
                    value="jdbc:oracle:thin:@localhost:1521:orcl" />
                <property name="username" value="y2" />
                <property name="password" value="123" />
            </dataSource>
        </environment>
    </environments>
    <mappers>
        <mapper resource="userinfoMapping.xml" />
    </mappers>
</configuration>
```

配置文件中的值 type="JDBC"和 type="POOLED"都是具体类的别名，在 MyBatis 中已经对这些别名进行过注册，代码如下：

```
public Configuration() {
    typeAliasRegistry.registerAlias("JDBC", JdbcTransactionFactory.class);
    typeAliasRegistry.registerAlias("MANAGED", ManagedTransactionFactory.class);

    typeAliasRegistry.registerAlias("JNDI", JndiDataSourceFactory.class);
    typeAliasRegistry.registerAlias("POOLED", PooledDataSourceFactory.class);
    typeAliasRegistry.registerAlias("UNPOOLED", UnpooledDataSourceFactory.class);

    typeAliasRegistry.registerAlias("PERPETUAL", PerpetualCache.class);
    typeAliasRegistry.registerAlias("FIFO", FifoCache.class);
    typeAliasRegistry.registerAlias("LRU", LruCache.class);
    typeAliasRegistry.registerAlias("SOFT", SoftCache.class);
typeAliasRegistry.registerAlias("WEAK", WeakCache.class);

    typeAliasRegistry.registerAlias("DB_VENDOR", VendorDatabaseIdProvider.class);

    typeAliasRegistry.registerAlias("XML", XMLLanguageDriver.class);
```

```
    typeAliasRegistry.registerAlias("RAW", RawLanguageDriver.class);

    typeAliasRegistry.registerAlias("SLF4J", Slf4jImpl.class);
    typeAliasRegistry.registerAlias("COMMONS_LOGGING", JakartaCommonsLoggingImpl.class);
    typeAliasRegistry.registerAlias("LOG4J", Log4jImpl.class);
    typeAliasRegistry.registerAlias("LOG4J2", Log4j2Impl.class);
    typeAliasRegistry.registerAlias("JDK_LOGGING", Jdk14LoggingImpl.class);
    typeAliasRegistry.registerAlias("STDOUT_LOGGING", StdOutImpl.class);
    typeAliasRegistry.registerAlias("NO_LOGGING", NoLoggingImpl.class);

    typeAliasRegistry.registerAlias("CGLIB", CglibProxyFactory.class);
    typeAliasRegistry.registerAlias("JAVASSIST", JavassistProxyFactory.class);

    languageRegistry.setDefaultDriverClass(XMLLanguageDriver.class);
    languageRegistry.register(RawLanguageDriver.class);
}
```

而在 SQL 映射文件中同样也使用了别名，SQL 映射代码如下：

```
<select id="getUserinfo" parameterType="long"
    resultType="entity.Userinfo">
    select * from userinfo
    where
    id <![CDATA[<]]>
    #{idParam} order by id asc
</select>
```

SQL 映射中的 long 值就是全名称为 java.lang.Long 的别名，常见的数据类型都在 MyBatis 的源代码中进行了注册，代码如下：

```
public TypeAliasRegistry() {
  registerAlias("string", String.class);

  registerAlias("byte", Byte.class);
  registerAlias("long", Long.class);
  registerAlias("short", Short.class);
  registerAlias("int", Integer.class);
  registerAlias("integer", Integer.class);
  registerAlias("double", Double.class);
  registerAlias("float", Float.class);
  registerAlias("boolean", Boolean.class);

  registerAlias("byte[]", Byte[].class);
  registerAlias("long[]", Long[].class);
  registerAlias("short[]", Short[].class);
  registerAlias("int[]", Integer[].class);
  registerAlias("integer[]", Integer[].class);
  registerAlias("double[]", Double[].class);
  registerAlias("float[]", Float[].class);
  registerAlias("boolean[]", Boolean[].class);

  registerAlias("_byte", byte.class);
  registerAlias("_long", long.class);
  registerAlias("_short", short.class);
  registerAlias("_int", int.class);
  registerAlias("_integer", int.class);
  registerAlias("_double", double.class);
  registerAlias("_float", float.class);
```

```
    registerAlias("_boolean", boolean.class);

    registerAlias("_byte[]", byte[].class);
    registerAlias("_long[]", long[].class);
    registerAlias("_short[]", short[].class);
    registerAlias("_int[]", int[].class);
    registerAlias("_integer[]", int[].class);
    registerAlias("_double[]", double[].class);
    registerAlias("_float[]", float[].class);
    registerAlias("_boolean[]", boolean[].class);

    registerAlias("date", Date.class);
    registerAlias("decimal", BigDecimal.class);
    registerAlias("bigdecimal", BigDecimal.class);
    registerAlias("biginteger", BigInteger.class);
    registerAlias("object", Object.class);

    registerAlias("date[]", Date[].class);
    registerAlias("decimal[]", BigDecimal[].class);
    registerAlias("bigdecimal[]", BigDecimal[].class);
    registerAlias("biginteger[]", BigInteger[].class);
    registerAlias("object[]", Object[].class);

    registerAlias("map", Map.class);
    registerAlias("hashmap", HashMap.class);
    registerAlias("list", List.class);
    registerAlias("arraylist", ArrayList.class);
    registerAlias("collection", Collection.class);
    registerAlias("iterator", Iterator.class);

    registerAlias("ResultSet", ResultSet.class);
}
```

2.3.2 使用<typeAlias>单独配置自定义别名

在执行 select 查询或 insert 添加的 SQL 语句时，都要在 parameterType 或 resultType 属性写上完整的实体类路径，路径中需要包含完整的包名，代码如下：

```
<insert id="insertUserinfo" parameterType="sqlmapping.Userinfo">
    <selectKey resultType="java.lang.Long" keyProperty="id"
        order="BEFORE">
        select idauto.nextval from dual
    </selectKey>
    insert into
    userinfo(id,username,password,age)
    values(#{id},#{username},#{password},#{age})
</insert>
<select id="getUserinfoById" parameterType="long" resultType="sqlmapping.Userinfo">
    select * from
    userinfo where id=#{id}
</select>
```

如果包名嵌套层级较多，那么会出现大量冗余的配置代码，这时可以在 mybatis-config.xml 配置文件中使用<typeAliases>自定义别名标签来进行简化。

创建测试用的项目 typeAliasTest。

在 mybatis-config.xml 配置文件中添加<typeAlias>标签，核心代码如下：

```
<configuration>
    <typeAliases>
        <typeAlias type="entity.Userinfo" alias="userinfo" />
    </typeAliases>
    <environments default="development">
    </environments>
</configuration>
```

以下配置代码中的 type 属性值就是完整的实体类包路径，而属性 alias 就是实体类的别名：

```
<typeAlias alias="userinfo" type="entity.Userinfo" />
```

别名 userinfo 可以在 SQL 映射文件中使用以简化配置代码，SQL 映射代码如下：

```
<mapper namespace="mapping.UserinfoMapper">
    <insert id="insertUserinfo" parameterType="USERinfo">
        <selectKey order="BEFORE" resultType="java.lang.Long"
            keyProperty="id">
            select idauto.nextval from dual
        </selectKey>
        insert into userinfo(id,username,password,age,insertdate)
        values(#{id},#{username},#{password},#{age},#{insertdate})
    </insert>

    <select id="selectById" parameterType="long"
        resultType="USERINFo">
        select *
        from userinfo where id=#{userId}
    </select>
</mapper>
```

使用别名后，可以正常地操作数据库，配置代码也不再冗余了。

在引用别名时是不区分大小写的，比如如下代码也能正确运行：

```
parameterType="USERinfo"
resultType="USERINFo"
```

2.3.3 使用<package>批量配置自定义别名

虽然使用<typeAlias>可以实现配置别名，但是如果实体类的数量较多，则极易出现<typeAlias>配置“爆炸”。这种情况可以通过<package>标签来解决，它的原理就是扫描指定包下的类。这些类都被自动赋予了与类同名的别名，不区分大小写，别名中不包含包名。

创建名称为 typeAliasPackageTest 的项目，更改 mybatis-config.xml 配置文件，核心代码如下：

```
<configuration>
    <typeAliases>
        <package name="entity" />
    </typeAliases>
    <environments default="development">
    </environments>
</configuration>
```

在<typeAliases>标签中使用<package>子标签来扫描 entity 包中的类而自动创建出类的别名。

SQL 映射文件也需要进行更改，代码如下：

```
<mapper namespace="mapping.UserinfoMapper">
    <insert id="insertUserinfo" parameterType="USERinfo">
        <selectKey order="BEFORE" resultType="java.lang.Long"
            keyProperty="id">
            select idauto.nextval from dual
        </selectKey>
        insert into userinfo(id,username,password,age,insertdate)
        values(#{id},#{username},#{password},#{age},#{insertdate})
    </insert>

    <select id="selectById" parameterType="long"
        resultType="USERINFo">
        select *
        from userinfo where id=#{userId}
    </select>
</mapper>
```

在 SQL 映射文件中使用以下配置也能正确地操作数据库，说明使用<package>方式定义的别名也是不区分大小写的：

```
parameterType="usERInFO"
```

2.3.4 别名重复的解决办法

在使用<package>包扫描时：

```
<typeAliases>
    <package name="entity1" />
    <package name="entity2" />
</typeAliases>
```

如果在不同的包中出现相同实体类名，那么 MyBatis 解析 XML 配置文件时就会出现如下异常信息：

```
    Caused by: org.apache.ibatis.type.TypeException: The alias 'Userinfo' is already
mapped to the value 'entity1.Userinfo'.
```

要解决这个异常，可以在实体类上方使用@Alias 注解来对 entity2 包中的 Userinfo.java 类设置别名 userinfo2。

```
@Alias(value = "userinfo2")
public class Userinfo {
```

然后在 SQL 映射文件引用这个别名即可，映射代码如下：

```
<mapper namespace="mapping.UserinfoMapper">
    <select id="selectById" parameterType="long"
        resultType="userinfo2">
        select *
        from userinfo where id=#{userId}
    </select>
</mapper>
```

运行类代码如下：

```
public class SelectById {
    public static void main(String[] args) throws IOException {
        SqlSession session = GetSqlSession.getSqlSession();
        entity2.Userinfo userinfo = session.selectOne("selectById", 600419L);
        System.out.println(userinfo + " " + userinfo.getId() + " " + userinfo.
getUsername() + " "
                + userinfo.getPassword() + " " + userinfo.getAge() + " " + userinfo.
getInsertdate());
        session.commit();
        session.close();
    }
}
```

上面的源代码存放在项目 typeAliasPackageTest2 中。

2.4 使用 properties 保存数据库信息

在配置文件中可以使用<properties>标签或引用*.properties 属性文件来增强配置的灵活性。

2.4.1 使用<properties><property name="" value="" /></properties>配置内部属性值

可以在配置文件中使用<properties>保存 4 个变量，再结合${}进行引用。

创建测试用的项目 properties4value。

配置文件的代码如下：

```
<configuration>
    <properties>
        <property name="driverKey"
            value="oracle.jdbc.driver.OracleDriver" />
        <property name="urlKey"
            value="jdbc:oracle:thin:@localhost:1521:orcl" />
        <property name="usernameKey" value="y2" />
        <property name="passwordKey" value="123" />
    </properties>
    <environments default="development">
        <environment id="development">
            <transactionManager type="JDBC" />
            <dataSource type="POOLED">
                <property name="driver" value="${driverKey}" />
                <property name="url" value="${urlKey}" />
                <property name="username" value="${usernameKey}" />
                <property name="password" value="${passwordKey}" />
            </dataSource>
        </environment>
    </environments>
    <mappers>
        <mapper resource="userinfoMapping.xml" />
    </mappers>
</configuration>
```

2.4.2 使用<properties resource="">引用外部 properties 属性文件中的配置

配置文件 mybatis-config.xml 承载的配置信息过多，因此可以将连接数据库的信息转移到外部的 properties 属性文件中，这样便于代码的后期维护与管理。

创建名称为 propertiesFileSaveDBInfo 的项目。

创建 db.properties 文件，代码如下：

```
driver=oracle.jdbc.OracleDriver
url=jdbc:oracle:thin:@localhost:1521:orcl
username=y2
password=123
```

配置文件 mybatis-config.xml，代码如下：

```
<configuration>
    <properties resource="db.properties"></properties>
    <environments default="development">
        <environment id="development">
            <transactionManager type="JDBC" />
            <dataSource type="POOLED">
                <property name="driver" value="${driver}" />
                <property name="url" value="${url}" />
                <property name="username" value="${username}" />
                <property name="password" value="${password}" />
            </dataSource>
        </environment>
    </environments>
    <mappers>
        <mapper resource="mapping/UserinfoMapper.xml" />
    </mappers>
</configuration>
```

添加了如下配置代码：

```
<properties resource="db.properties"></properties>
```

使用 db.properties 文件中的连接数据库信息。

2.4.3 使用程序代码读取 properties 文件中的参数

前面章节示例中连接数据库的具体参数值是由 MyBatis 进行读取并使用${}赋值的，其实这些参数值还可以由程序员写的代码进行读取并赋值。

创建测试用的项目 propertiesMyCodeGetSet。

更改 mybatis-config.xml 配置文件，代码如下：

```
<configuration>
    <environments default="development">
        <environment id="development">
            <transactionManager type="JDBC" />
            <dataSource type="POOLED">
                <property name="driver" value="${driver}" />
```

```
                <property name="url" value="${url}" />
                <property name="username" value="${username}" />
                <property name="password" value="${password}" />
            </dataSource>
        </environment>
    </environments>
    <mappers>
        <mapper resource="mapping/UserinfoMapper.xml" />
    </mappers>
</configuration>
```

在 src 中创建 db.properties 属性文件，内容如下：

```
driver=oracle.jdbc.OracleDriver
url=jdbc:oracle:thin:@localhost:1521:orcl
username=y2
password=123
```

更改 GetSqlSession.java，代码如下：

```
public class GetSqlSession {
    public static SqlSession getSqlSession() throws IOException {
        String configFile = "mybatis-config.xml";
        String propFile = "db.properties";
        InputStream configStream = Resources.getResourceAsStream(configFile);
        InputStream propStream = Resources.getResourceAsStream(propFile);

        Properties prop = new Properties();
        prop.load(propStream);

        SqlSessionFactoryBuilder builder = new SqlSessionFactoryBuilder();
        SqlSessionFactory factory = builder.build(configStream, prop);
        return factory.openSession();
    }
}
```

在 SqlSessionFactoryBuilder 类的 build()方法中，将配置 inputStream 流与属性对象 prop 进行关联，就可以根据${username}表达式读取 properties 属性文件中的同名 key 对应的值了，从而成功读取连接数据库的 4 个必要参数。

2.4.4 数据库密码加/解密

前面由程序员写的代码对属性文件进行读取并将信息赋值给 MyBatis 框架，程序员对 properties 属性文件中的内容具有可控性，比如对 properties 属性文件中的内容进行加密与解密，这有助于增强连接数据库四大变量的安全性。

创建测试用的项目 properties_encode_decode。

属性文件 db.properties 中的代码如下：

```
driver=_oracle.jdbc.OracleDriver
url=_jdbc:oracle:thin:@localhost:1521:orcl
username=_y2
password=_123
```

属性文件中的值是模拟经过加密后的信息。

加密、解密工具类代码如下：

```
public class Tools {
    // 加密的算法
    public static String encodeString(String value) {
        return "_" + value;
    }

    // 解密的算法
    public static String decodeString(String value) {
        return value.substring(1);
    }
}
```

GetSqlSession 类的代码如下：

```
public class GetSqlSession {
    public static SqlSession getSqlSession() throws IOException {
        String configFile = "mybatis-config.xml";
        String dbProperties = "db.properties";
        InputStream configStream = Resources.getResourceAsStream(configFile);
        InputStream propStream = Resources.getResourceAsStream(dbProperties);
        Properties prop = new Properties();
        prop.load(propStream);

        String driver = prop.getProperty("driver");
        String url = prop.getProperty("url");
        String username = prop.getProperty("username");
        String password = prop.getProperty("password");

        System.out.println("解密之前 begin");
        System.out.println("driver=" + driver);
        System.out.println("url=" + url);
        System.out.println("username=" + username);
        System.out.println("password=" + password);
        System.out.println("解密之前   end");

        driver = Tools.decodeString(driver);
        url = Tools.decodeString(url);
        username = Tools.decodeString(username);
        password = Tools.decodeString(password);

        System.out.println();

        System.out.println("解密之后 begin");
        System.out.println("driver=" + driver);
        System.out.println("url=" + url);
        System.out.println("username=" + username);
        System.out.println("password=" + password);
        System.out.println("解密之后   end");

        prop.setProperty("driver", driver);
        prop.setProperty("url", url);
        prop.setProperty("username", username);
        prop.setProperty("password", password);

        SqlSessionFactoryBuilder builder = new SqlSessionFactoryBuilder();
        SqlSessionFactory factory = builder.build(configStream, prop);
```

```
            return factory.openSession();
        }
    }
```

在获得 SqlSessionFactory 对象前完成了解密的工作，并且顺利取得 SqlSession 对象来操作数据库。

2.5 配置多个连接数据库环境

可以在 mybatis-config.xml 配置文件中创建多个连接环境，方便切换欲操作的数据库。

2.5.1 实现多个连接数据库环境

创建名称为 moreEnvironment 的项目。

创建 db.properties 属性文件，内容如下：

```
oracle1.driver=oracle.jdbc.OracleDriver
oracle1.url=jdbc:oracle:thin:@localhost:1521:orcl
oracle1.username=y2
oracle1.password=123

oracle2.driver=oracle.jdbc.OracleDriver
oracle2.url=jdbc:oracle:thin:@localhost:1521:orcl
oracle2.username=y2
oracle2.password=##############################
```

从属性文件中提供的内容可以发现有两个 Oracle 数据库的连接信息，其中第 2 个 Oracle 的连接密码是错误的。

配置文件 mybatis-config.xml 的代码如下：

```
<configuration>
    <properties resource="db.properties"></properties>
    <environments default="oracle2">
        <environment id="oracle1">
            <transactionManager type="JDBC" />
            <dataSource type="POOLED">
                <property name="driver" value="${oracle1.driver}" />
                <property name="url" value="${oracle1.url}" />
                <property name="username" value="${oracle1.username}" />
                <property name="password" value="${oracle1.password}" />
            </dataSource>
        </environment>

        <environment id="oracle2">
            <transactionManager type="JDBC" />
            <dataSource type="POOLED">
                <property name="driver" value="${oracle2.driver}" />
                <property name="url" value="${oracle2.url}" />
                <property name="username" value="${oracle2.username}" />
                <property name="password" value="${oracle2.password}" />
            </dataSource>
        </environment>
```

```
    </environments>
    <mappers>
        <mapper resource="mapping/UserinfoMapper.xml" />
    </mappers>
</configuration>
```

配置<environments default="oracle2">代码中的 default 属性值是 oracle2，代表要连接的是<environment id="oracle2">中的配置信息，也就是在配置代码<environments default="oracle2">中来切换欲操作数据库的实例，从而达到配置多数据源的目的。

2.5.2 多个连接数据库环境与数据库加/解密

创建测试用的项目 mybatis20。

属性文件 db.properties 的代码如下：

```
a1=_com.mysql.jdbc.Driver
b1=_jdbc:mysql://localhost:3306/y2
c1=_root
d1=_123123

a2=_com.mysql.jdbc.Driver
b2=_jdbc:mysql://localhost:3306/y2
c2=_root
d2=_123123123123123123123123123123123123123123123123123123123123123123
```

属性文件中的值是模拟经过加密后的信息。

配置文件 mybatis-config.xml 的核心代码如下：

```
<environments default="mysql_2">
    <environment id="mysql_1">
        <transactionManager type="JDBC" />
        <dataSource type="POOLED">
            <property name="driver" value="${a1}" />
            <property name="url" value="${b1}" />
            <property name="username" value="${c1}" />
            <property name="password" value="${d1}" />
        </dataSource>
    </environment>

    <environment id="mysql_2">
        <transactionManager type="JDBC" />
        <dataSource type="POOLED">
            <property name="driver" value="${a2}" />
            <property name="url" value="${b2}" />
            <property name="username" value="${c2}" />
            <property name="password" value="${d2}" />
        </dataSource>
    </environment>
</environments>
```

GetSqlSessionFactory 类的代码如下：

```
package tools;

import java.io.IOException;
```

```java
import java.io.InputStream;
import java.util.Iterator;
import java.util.Properties;

import org.apache.ibatis.io.Resources;
import org.apache.ibatis.session.SqlSessionFactory;
import org.apache.ibatis.session.SqlSessionFactoryBuilder;

public class GetSqlSessionFactory {
    private static SqlSessionFactory factory;

    private GetSqlSessionFactory() {
    }

    synchronized public static SqlSessionFactory getSqlSessionFactory() throws IOException {
        if (factory == null) {
            String configFile = "mybatis-config.xml";

            String propFile = "db.properties";
            InputStream propStream = Resources.getResourceAsStream(propFile);
            Properties prop = new Properties();
            prop.load(propStream);

            Iterator iterator = prop.keySet().iterator();
            while (iterator.hasNext()) {
                String propName = "" + iterator.next();
                String propValue = prop.getProperty(propName);
                propValue = propValue.substring(1);
                prop.setProperty(propName, propValue);
            }

            InputStream inputStream = Resources.getResourceAsStream(configFile);
            SqlSessionFactoryBuilder builder = new SqlSessionFactoryBuilder();
            factory = builder.build(inputStream, prop);
        } else {
        }
        return factory;
    }
}
```

2.6 使用数据源

在 Web 开发中，如果使用 Tomcat 作为 Web 容器，那么 Tomcat 实现了获得连接（Connection）对象的数据源接口（DataSource），可以使用 DataSource 的 API 获得连接，并操作数据库。

2.6.1 DataSource 接口介绍

DataSource 接口的信息如图 2-1 所示。

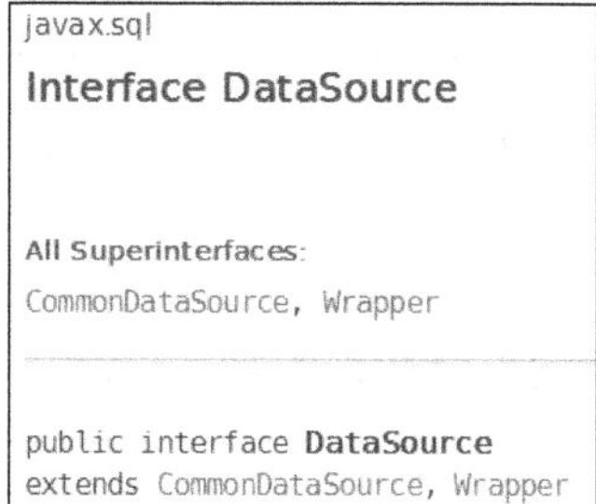

图 2-1 DataSource 接口的信息

DataSource 接口在 Java Doc 中的官方解释如下。

DataSource 是 DriverManager 工具的替代项，DataSource 对象是获取连接的首选方法。实现 DataSource 接口的对象通常在基于 Java Naming and Directory Interface（JNDI）API 的命名服务中注册。

从以上文字可以总结出 3 点。

（1）DataSource 是 DriverManager 工具的替代使用方式。

（2）DataSource 常常与 JNDI 规范联合使用。

（3）通过 DataSource 可以获得连接。

在以上 3 点中，陌生的知识点就是 JNDI，在后面的章节会有介绍。

使用 DataSource 获得 Connection 对象实现了 Java EE 规范的标准性，从 DataSource 获得 Connection 的示例代码在 OracleDataSourceGetConn 项目中，运行类代码如下：

```
public class Test {
    public static void main(String[] args) throws SQLException {
        OracleDataSource oracleDataSource = new OracleDataSource();
        oracleDataSource.setDriverType("oracle.jdbc.OracleDriver");
        oracleDataSource.setURL("jdbc:oracle:thin:@localhost:1521:orcl");
        oracleDataSource.setUser("y2");
        oracleDataSource.setPassword("123");

        DataSource dataSource = oracleDataSource;

        Connection connection = dataSource.getConnection();
        System.out.println(connection);
        connection.close();
    }
}
```

在项目中使用统一的接口获得 Connection 对象是值得推荐的。从 Web 容器中的 DataSource 获得 Connection 大多数的位置都是连接池（Connection Pool），使用数据库连接池可以提高获得 Connection 对象的速度，从而提升程序运行效率。

2.6.2 JNDI 介绍

在标准场景下，DataSource 常常会与 JNDI 规范联合使用。

什么是 JNDI？JNDI 全称是 Java Naming and Directory Interface，中文名称为 Java 命名目录接口。JNDI 规范的最终目的有两个。

（1）共享全局数据。

（2）访问全局数据 API 的统一性。

先来看一看第 1 个目的：共享全局数据。

为了更好地组织多个对象，可以把这些对象放到一个 Map 中，然后对这个 Map 中的数据进行 CURD。如果这样做只是在本项目中能访问这个 Map 对象中的数据，其他的项目并不能访问这个项目 Map 中的数据，那么也就意味着这个 Map 是私有的，只有本项目才能访问，这个 Map 并不是在多个项目中可以共享的，因此，单纯地使用 Map 对象组织数据就有些不适合数据在多个项目之间共享的需求了。Sun 公司针对这个情况发布了 JNDI 规范，JNDI 使用 JNDI 树来组织数据，JNDI 树中的数据可以在不同的项目中共享，这就避免了单纯地使用 Map 数据是私有性的缺点，这就是 JNDI 所具有的“共享全局数据”。JNDI 树结构如图 2-2 所示。

再来看一看第 2 个目的：访问全局数据 API 的统一性。

如果每个厂商都不遵守 JNDI 规范，而且每个厂商都使用自己的数据结构来组织数据，比如有的放入 List，有的放入 Map，有的创建自己的数据结构来保存对象，那么程序员需要掌握上面 3 种 API 来获取每个厂商提供的对象。

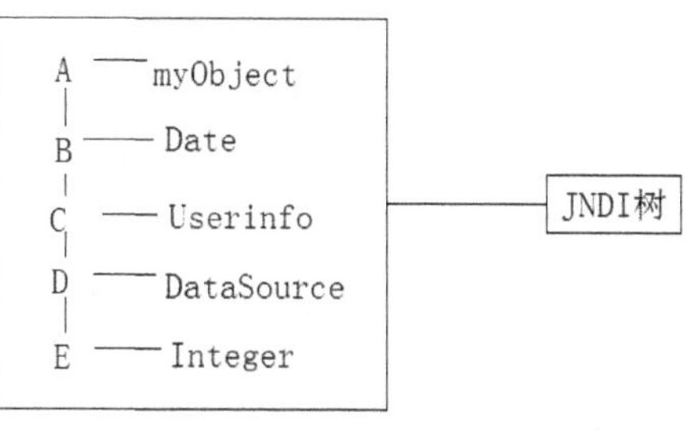

图 2-2　JNDI 树结构

（1）AMyList.getFromList（0）。

（2）BMyMap.getFromMapMapMap（key）。

（3）自定义的数据结构 C.getObject()。

从上面的示例代码中可以发现，每个厂商的类名不一样，方法名称也不一样，这样的写法大大增加程序员学习的成本，并且程序代码的移植性不高。为了达到获取对象的 API 命名的统一，Sun 公司发布了 JNDI 规范，让这些厂商实现 JNDI 规范。使用了 JNDI 规范后，无论使用什么样的方式保存对象，程序员只需要使用一种方式获得对象，就可以在不同厂商的不同的数据结构中获得指定的对象值。这个特点和使用一种 JDBC 接口就可以操作不同数据库的原理是一样的。

JNDI 统一了不同厂商的不同的 API，效果如图 2-3 所示。

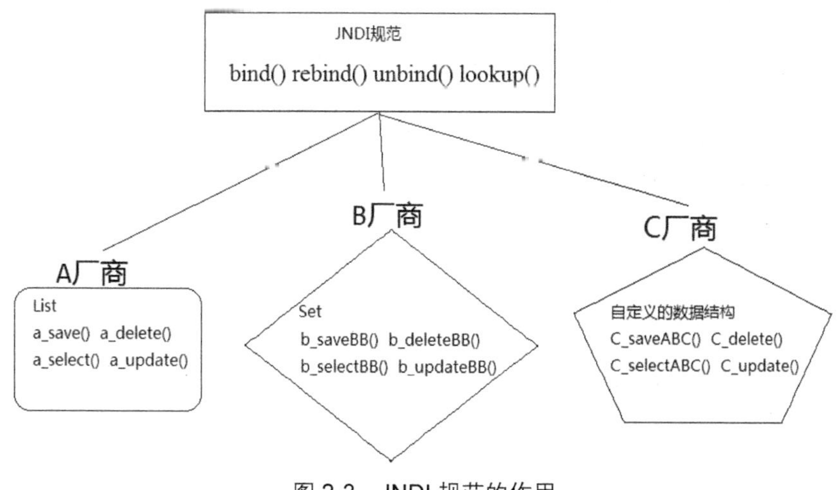

图 2-3　JNDI 规范的作用

JNDI 操作数据的规范主要由以下 4 个方法组成。

（1）bind()方法：增加数据。

（2）rebind()方法：修改数据。

（3）unbind()方法：删除数据。

（4）lookup()方法：查询数据。

2.6.3　DataSource 与 JNDI 的关系

知识点介绍到这里，DataSource 和 JNDI 有什么关系呢？可以在 JNDI 树中挂载 DataSource，效果如图 2-4 所示。

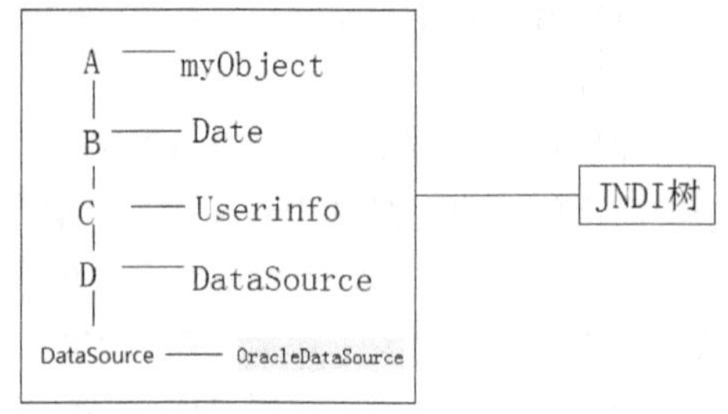

图 2-4 在 JNDI 树中挂载 DataSource

将 DataSource 挂载到 JNDI 树上的优点就是所有的项目都可以访问 JNDI 树上的资源，JNDI 树上的所有资源得到彻底的共享，包含 DataSource，这也是在一处集中提供数据共享的服务，在多处进行调用并使用的方式，通称服务中心化。

想要在 JNDI 树中获得 DataSource，就要使用 JNDI 规范中提供的 API 方法 lookup()。

接口 DataSource 主要作用是获得 Connection，它对在不同容器产品中获得 Connection 的 API 进行规范化，而接口 JNDI 主要作用是操作 JNDI 树中数据的 API，使其规范化。

简单总结如下。

（1）DataSource 规范是为了获得 Connection 规范化，因为获得 Connection 需要调用 DataSource 中的 getConnection()方法。

（2）JNDI 规范是为了获得 JNDI 树中数据的规范化，因为在操作 JNDI 树时使用的是 Context 接口，对树中的数据进行 CURD 时，调用的是 Context 接口中的 bind()、rebind()、unbind()和 lookup()方法。

2.6.4 使用 JNDI 接口操作 JNDI Tree 上的数据

下面来看一下在 Tomcat 容器中使用 JNDI 接口中的 API 来操作 JNDI 树节点与对应值的代码。

创建项目 jdbcJNDITest。

在 JNDI 树中绑定值的代码如下：

```
public class test1 extends HttpServlet {
    protected void doGet(HttpServletRequest request, HttpServletResponse response)
            throws ServletException, IOException {
        try {
            Context context = new InitialContext();
            context.bind("username", "中国人");
            context.close();
        } catch (NamingException e) {
            e.printStackTrace();
        }
    }
}
```

从 JNDI 树中取得值的代码如下：

```
public class test2 extends HttpServlet {
    protected void doGet(HttpServletRequest request, HttpServletResponse response)
            throws ServletException, IOException {
        try {
            Context context = new InitialContext();
            String value = (String) context.lookup("username");
            context.close();
            System.out.println(value);
```

```
        } catch (NamingException e) {
            e.printStackTrace();
        }
    }
}
```

节点值还可以是父子关系，代码如下：

```
public class test3 extends HttpServlet {
    protected void doGet(HttpServletRequest request, HttpServletResponse response)
            throws ServletException, IOException {
        try {
            Context context = new InitialContext();
            context = context.createSubcontext("y");
            context.bind("z", "我是中国人-父子节点测试");
            context.close();
        } catch (NamingException e) {
            e.printStackTrace();
        }
    }
}
```

取得节点是父子关系对应值的代码如下：

```
public class test4 extends HttpServlet {
    protected void doGet(HttpServletRequest request, HttpServletResponse response)
            throws ServletException, IOException {
        try {
            Context context = new InitialContext();
            context = (Context) context.lookup("y");
            System.out.println(context.lookup("z"));
            context.close();
        } catch (NamingException e) {
            e.printStackTrace();
        }
    }
}
```

删除节点的代码如下：

```
public class test5 extends HttpServlet {
    protected void doGet(HttpServletRequest request, HttpServletResponse response)
            throws ServletException, IOException {
        try {
            Context context = new InitialContext();
            context.unbind("y");
            context.close();
        } catch (NamingException e) {
            e.printStackTrace();
        }
    }
}
```

更新 JNDI 节点值的代码如下：

```
public class test6 extends HttpServlet {
    protected void doGet(HttpServletRequest request, HttpServletResponse response)
            throws ServletException, IOException {
```

```
            try {
                Context context = new InitialContext();
                context.rebind("username", "我是中国人-新值");
                context = (Context) context.lookup("y");
                context.rebind("z", "我是 z 的新值");
                context.close();
            } catch (NamingException e) {
                e.printStackTrace();
            }
        }
    }
```

上面的代码就是在 Tomcat 中操作 JNDI 树节点与对应的值，这些代码可以在不更改任何一行的情况下成功地在其他容器中移植运行，这就是 JNDI 规范的优点：API 统一，代码移植性好。

使用 JNDI 规范中的 API 就可以从 JNDI 树中找到对应的对象。大多数 Web 容器都支持 JNDI 规范，因为 JNDI 规范属于 Java EE 技术体系之一。使用 JNDI 规范后，只要 Web 容器支持 JNDI 规范，那么在操作 JDNI 树上的对象时使用的代码就是一样的。

2.6.5 在 JNDI 树中先获得 DataSource 再获得 Connection

大多数 Java Web 容器内部的 DataSource 提供的 Connection 对象都已经被池化。所谓的“池化”也就是预先在 Tomcat 中创建出一个空间，在这个空间中存储一些已经连接到数据库的 Connection，客户端在使用 Connection 时只需要使用 get 获取即可，省略了每次连接数据库时的 IP 寻址，以及 username 和 password 验证等重复的环节，大大提高了软件的运行效率，这个空间就是 Connection Pool。

那么 JNDI、DataSource 与 Connection Pool 之间到底是什么关系呢？如图 2-5 所示。

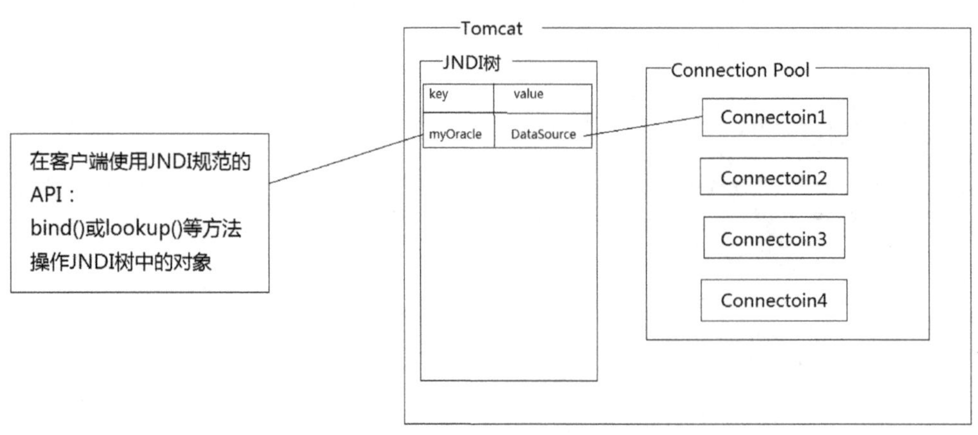

图 2-5　三者之间的关系

关系总结如下。

（1）JNDI 树：组织数据的一种数据结构。

（2）JNDI 规范：共两个作用，一是访问 JNDI 树上数据的 API 并使之达到统一，二是共享数据。

（3）DataSource：达到获得 Connection 对象 API 的统一性，统一使用 DataSource 接口的 getConnection()方法。

（4）Connection Pool：提高获取 Connection 连接对象的效率。

在 Tomcat 中获得 DataSource 之前，要在 Tomcat 的 conf 文件夹中的 context.xml 文件里配置 DataSource 信息，Tomcat 启动时会读取这些信息，然后将 DataSource 对象放到 Tomcat 中的 JNDI 树节点上，核心代码如下：

```
<Resource name="jdbc/myOracle" type="javax.sql.DataSource"
driverClassName="oracle.jdbc.OracleDriver" url="jdbc:oracle:thin:@localhost:1521:orcl"
username="y2" password="123" maxActive="200" maxIdle="10" maxWait="-1" />
```

<Resource>中的 name 值“jdbc/myOracle”就是 JNDI 树节点的名称，对应的值就是数据源 DataSource。

然后在 Tomcat 的 conf 文件夹中的 web.xml 文件里引用数据源，核心代码如下：

```
<resource-ref>
    <res-ref-name>jdbc/myOracle</res-ref-name>
    <res-type>javax.sql.DataSource</res-type>
</resource-ref>
```

先来看一看不使用 MyBatis 从数据源 DataSource 中获得 Connection，而是使用 Servlet 直接从 Tomcat 的 JNDI 树上先获得数据源 DataSource，再获得连接 Connection 对象。继续在 jdbcJNDITest 项目中进行测试，创建 Servlet 的代码如下：

```
public class test7 extends HttpServlet {
    protected void doGet(HttpServletRequest request, HttpServletResponse response)
            throws ServletException, IOException {
        try {
            Context context = new InitialContext();
            DataSource ds = (DataSource) context.lookup("java:/comp/env/jdbc/myOracle");
            Connection connection = ds.getConnection();
            ResultSet rs = connection.prepareStatement("select * from userinfo order
by id asc").executeQuery();
            while (rs.next()) {
                System.out.println(rs.getString("id") + " " + rs.getString("username"));
            }
            rs.close();
            connection.close();
        } catch (NamingException e) {
            e.printStackTrace();
        } catch (SQLException e) {
            e.printStackTrace();
        }
    }
}
```

执行此 Servlet 后在 Tomcat 的 JDNI 树上找到 DataSource 数据源对象，然后通过此对象获得 Connection 连接，进而操作数据库。

JNDI 是以统一的数据组织方式“JNDI 树”来组织数据的，以统一的 API 方式对数据进行存取，比如存储数据要使用 bind()方法，而取得数据要使用 lookup()方法。它是访问数据、共享数据的一种 Java EE 规范。

2.6.6 在 MyBatis 中从 JNDI 获得 DataSource

MyBatis 使用 DataSource 数据源的实验在项目 JNDIDataSourceTest 中进行。创建配置文件 mybatis-config.xml，代码如下：

```
<configuration>
    <properties resource="dbinfo.properties">
    </properties>
    <environments default="jndi1">
        <environment id="jndi1">
            <transactionManager type="JDBC" />
            <dataSource type="JNDI">
                <property name="data_source" value="${jndiName}" />
            </dataSource>
        </environment>
        <environment id="jndi2">
            <transactionManager type="JDBC" />
            <dataSource type="JNDI">
                <property name="data_source"
                    value="java:/comp/env/jdbc/myOracle" />
            </dataSource>
        </environment>
    </environments>
    <mappers>
        <mapper resource="mapping/UserinfoMapper.xml" />
    </mappers>
</configuration>
```

从 JNDI 树中根据 JNDI 节点名称 name 取得对应的 DataSource 数据源 value 值，再从 DataSource 中取得 Connection 对象，以上就是在 Web 环境中使用 JNDI 的过程。JNDI 的代码需要程序员自己输入，而 MyBatis 封装了这个过程。

2.6.7 在 MyBatis 中使用第三方的 HikariCP 连接池

在第 1 章中介绍 MyBatis 时使用了如下配置来创建 Connection Pool：

```
<environment id="oracle1">
    <transactionManager type="JDBC" />
    <dataSource type="POOLED">
        <property name="driver" value="value1" />
        <property name="url" value=" value2" />
        <property name="username" value=" value3" />
        <property name="password" value=" value4" />
    </dataSource>
</environment>
```

这个连接池是 MyBatis 自己创建的，在效率上得到了保障。而 MyBatis 还支持第三方连接池，其在获得 Connection 时会更加高效。本节就将现阶段主流的 HikariCP 连接池与 MyBatis 进行整合。

创建名称为 use_HikariDataSource 的项目，MyDataSourceFactory 类的代码如下：

```
public class MyDataSourceFactory implements DataSourceFactory {

    private Properties prop;
```

```
    @Override
    public DataSource getDataSource() {
        System.out.println("2 public DataSource getDataSource()");

        HikariDataSource ds = new HikariDataSource();

        ds.setDriverClassName(prop.getProperty("a"));
        ds.setUsername(prop.getProperty("b"));
        ds.setPassword(prop.getProperty("c"));
        ds.setJdbcUrl(prop.getProperty("d"));

        ds.setMaximumPoolSize(10);
        ds.setMinimumIdle(10);

        return ds;
    }

    @Override
    public void setProperties(Properties arg0) {
        System.out.println("1 public void setProperties(Properties arg0) " + arg0.size());
        prop = arg0;
    }
}
```

配置文件 mybatis-config.xml 的代码如下：

```
<configuration>
    <properties resource="dbinfo.properties">
    </properties>
    <environments default="jdbc1">
        <environment id="jdbc1">
            <transactionManager type="JDBC" />
            <dataSource type="datasourcefactory.MyDataSourceFactory">
                <property name="a" value="${driver}" />
                <property name="b" value="${username}" />
                <property name="c" value="${password}" />
                <property name="d" value="${url}" />
            </dataSource>
        </environment>
    </environments>
    <mappers>
        <mapper resource="mapping/UserinfoMapper.xml" />
    </mappers>
</configuration>
```

运行类代码如下：

```
public class test1 {
    public static void main(String[] args) throws IOException {
        try {
            String configFile = "mybatis-config.xml";
            InputStream configStream = Resources.getResourceAsStream(configFile);
            SqlSessionFactoryBuilder builder = new SqlSessionFactoryBuilder();
            SqlSessionFactory factory = builder.build(configStream);
            SqlSession session = factory.openSession();

            UserinfoMapper mapper = session.getMapper(UserinfoMapper.class);
```

```
            Userinfo userinfo1 = mapper.selectById(1L);
            Userinfo userinfo2 = mapper.selectById(2L);
            Userinfo userinfo3 = mapper.selectById(3L);
            Userinfo userinfo4 = mapper.selectById(4L);

            System.out.println(userinfo1.getId() + " " + userinfo1.getUsername());
            System.out.println(userinfo2.getId() + " " + userinfo1.getUsername());
            System.out.println(userinfo3.getId() + " " + userinfo1.getUsername());
            System.out.println(userinfo4.getId() + " " + userinfo1.getUsername());

            session.commit();
            session.close();

            DataSource ds = factory.getConfiguration().getEnvironment().getDataSource();
            for (int i = 0; i < 15; i++) {
                Connection connection = ds.getConnection();
                System.out.println(connection);
                connection.close();
            }
        } catch (SQLException e) {
            e.printStackTrace();
        }
    }
}
```

程序运行后控制台输出的信息如下：

```
1 public void setProperties(Properties arg0) 4
2 public DataSource getDataSource()
1 中国
2 中国
3 中国
4 中国
HikariProxyConnection@343856911 wrapping oracle.jdbc.driver.T4CConnection@8d7d54d
HikariProxyConnection@102617125 wrapping oracle.jdbc.driver.T4CConnection@8d7d54d
HikariProxyConnection@306980751 wrapping oracle.jdbc.driver.T4CConnection@8d7d54d
HikariProxyConnection@363988129 wrapping oracle.jdbc.driver.T4CConnection@8d7d54d
HikariProxyConnection@1997963191 wrapping oracle.jdbc.driver.T4CConnection@8d7d54d
HikariProxyConnection@534906248 wrapping oracle.jdbc.driver.T4CConnection@8d7d54d
HikariProxyConnection@1826699684 wrapping oracle.jdbc.driver.T4CConnection@8d7d54d
HikariProxyConnection@1769193365 wrapping oracle.jdbc.driver.T4CConnection@8d7d54d
HikariProxyConnection@769429195 wrapping oracle.jdbc.driver.T4CConnection@8d7d54d
HikariProxyConnection@580718781 wrapping oracle.jdbc.driver.T4CConnection@8d7d54d
HikariProxyConnection@1196695891 wrapping oracle.jdbc.driver.T4CConnection@8d7d54d
HikariProxyConnection@867148091 wrapping oracle.jdbc.driver.T4CConnection@8d7d54d
HikariProxyConnection@2051853139 wrapping oracle.jdbc.driver.T4CConnection@8d7d54d
HikariProxyConnection@815674463 wrapping oracle.jdbc.driver.T4CConnection@8d7d54d
HikariProxyConnection@1453774246 wrapping oracle.jdbc.driver.T4CConnection@8d7d54d
```

从控制台输出的信息来看，使用的是同一个 Connection 连接，证明连接池中的 Connection 被复用。

2.7 不同数据库执行不同 SQL 语句的支持

在设计软件系统时，经常需要考虑多数据库的支持，也就是在 Java 代码与 SQL 映射代码都不变的情况下来执行不同的 SQL 语句。新版 MyBatis 支持这个功能。

2.7.1 使用<databaseIdProvider type="DB_VENDOR">实现执行不同的 SQL 语句

创建名称为 databaseIdTest 的项目。

SQL 映射文件 UserinfoMapper.xml 的代码如下：

```
<mapper namespace="mapping.UserinfoMapper">
    <select id="selectAll" resultType="entity.Userinfo"
        databaseId="xxxOracle">
        select * from userinfo order
        by id asc
    </select>
    <select id="selectAll" resultType="entity.Userinfo"
        databaseId="yyyMySQL">
        select * from userinfo order
        by id desc
    </select>
</mapper>
```

配置文件 mybatis-config.xml 的代码如下：

```
<configuration>
    <environments default="oracle">
        <environment id="oracle">
            <transactionManager type="JDBC" />
            <dataSource type="POOLED">
                <property name="driver" value="oracle.jdbc.OracleDriver" />
                <property name="url"
                    value="jdbc:oracle:thin:@localhost:1521:orcl" />
                <property name="username" value="y2" />
                <property name="password" value="123" />
            </dataSource>
        </environment>
        <environment id="mysql">
            <transactionManager type="JDBC" />
            <dataSource type="POOLED">
                <property name="driver" value="com.mysql.jdbc.Driver" />
                <property name="url" value="jdbc:mysql://localhost:3306/y2" />
                <property name="username" value="root" />
                <property name="password" value="123123" />
            </dataSource>
        </environment>
    </environments>
    <databaseIdProvider type="DB_VENDOR">
        <property name="Oracle" value="xxxOracle" />
        <property name="MySQL" value="yyyMySQL" />
    </databaseIdProvider>
    <mappers>
        <mapper resource="mapping/UserinfoMapper.xml" />
    </mappers>
</configuration>
```

配置代码<property name="Oracle" value="xxxOracle" />中的 name 属性值是通过如下代码获得的：

```
    System.out.println("getDatabaseProductName()=" + factory.getConfiguration().
getEnvironment().getDataSource()
            .getConnection().getMetaData().getDatabaseProductName());
```

属性 name 的值不能随意添写，并且还要注意大小写，（区分大小写）。属性 name 代表数据库的产品名称，而 value 属性代表这个数据库产品名称的别名，value 的属性可以随意命名，但尽量命名得有意义。然后，在 SQL 映射文件使用如下代码：

```
<select id="selectAll" resultType="entity.Userinfo"
    databaseId="xxxOracle">
    select * from userinfo order
    by id asc
</select>
<select id="selectAll" resultType="entity.Userinfo"
    databaseId="yyyMySQL">
    select * from userinfo order
    by id desc
</select>
```

在不同的 SQL 映射上引用不同的数据库别名 xxxOracle 和 yyyMySQL 即可，这样就可以达到 SQL 映射的 id 值一样，但在不同的数据库中可以执行不同 SQL 语句的目的了。

使用不同的数据源可以执行不同的 SQL 语句，Oracle 是正序，而 MySQL 是倒序。

2.7.2 在 SQL 映射的 id 值相同的情况下有无 databaseId 的优先级判断

下面测试一下如果 SQL 映射的 id 值相同，一个 SQL 映射有 databaseId，另一个 SQL 映射无 databaseId，那么二者有无优先级呢？

创建测试用的项目 databaseIdTest2。

SQL 映射文件 UserinfoMapper.xml 的代码如下：

```
<mapper namespace="mapping.UserinfoMapper">
    <select id="selectAll" resultType="entity.Userinfo">
        select *
        from userinfo order
        by
        id desc
    </select>

    <select id="selectAll" resultType="entity.Userinfo"
        databaseId="xxxOracle">
        select * from userinfo order
        by id asc
    </select>

</mapper>
```

配置文件 mybatis-config.xml 的代码如下：

```
<configuration>
    <environments default="oracle">
        <environment id="oracle">
            <transactionManager type="JDBC" />
```

```
            <dataSource type="POOLED">
                <property name="driver" value="oracle.jdbc.OracleDriver" />
                <property name="url"
                    value="jdbc:oracle:thin:@localhost:1521:orcl" />
                <property name="username" value="y2" />
                <property name="password" value="123" />
            </dataSource>
        </environment>
    </environments>
    <databaseIdProvider type="DB_VENDOR">
        <property name="Oracle" value="xxxOracle" />
    </databaseIdProvider>
    <mappers>
        <mapper resource="mapping/UserinfoMapper.xml" />
    </mappers>
</configuration>
```

运行类代码如下：

```
public class test1 {
    public static void main(String[] args) throws IOException, SQLException {
        SqlSession session = GetSqlSession.getSqlSession();
        UserinfoMapper mapper = session.getMapper(UserinfoMapper.class);
        List<Userinfo> list = mapper.selectAll();
        for (int i = 0; i < list.size(); i++) {
            Userinfo userinfo = list.get(i);
            System.out.println(userinfo.getId() + " " + userinfo.getUsername());
        }
        session.commit();
        session.close();
    }
}
```

程序运行的结果是正序效果，输出内容如下：

```
getDatabaseProductName()=Oracle
1 中国
2 中国
3 中国
4 中国
5 中国
6 中国
7 中国
8 中国
9 a
```

结果表明具有 databaseId="xxxOracle"属性的 SQL 映射的优先级高。

2.8 获取 Mapper 的多种方式

创建名称为 moreMapperTest 的项目。

在 mybatis-config.xml 配置文件中可以使用 4 种方式来获取 Mapper 映射，代码如下：

```
<!-- 第 1 种写法： 直接获取 SQL 映射 XML 文件-->
<mapper resource="sqlmapping/UserinfoMapper.xml" />

<!-- 第 2 种写法： 扫描包，注意接口的名称和 XML 文件的主文件名必须一致-->
```

```
<package name="sqlmapping" />

<!-- 第 3 种写法： 指定接口名称，注意接口的名称和 XML 文件的主文件名必须一致-->
<mapper class="sqlmapping.UserinfoMapper" />

<!-- 第 4 种写法： 使用绝对路径-->
<mapper
    url="file:\\\C:\Users\gaohongyan\workspace\moreMapperTest\src\sqlmapping\Userinfo
Mapper.xml" />
```

建议不要混用这 4 种方式。

2.9 <transactionManager type="" />中 type 为 JDBC 和 MANAGED 时的区别

配置代码<transactionManager type="JDBC" />的主要作用就是程序员需要使用显式代码进行事务的提交，也就是程序员必须要调用 SqlSession 对象的 commit()方法才会更改数据库。

还有另外一个写法：

```
<transactionManager type="MANAGED" />
```

此写法的执行效果是每执行 1 条 SQL 语句，事务进行自动提交，而值 MANAGED 的本意却是由容器来管理事务。

如果读者正在使用 Spring 和 MyBatis，那么没有必要配置事务管理器，因为 Spring 模块会使用自带的管理器来覆盖前面的配置。

创建测试用的项目 transactionManagerTypeValueDiff。

创建 Servlet 类 Test1，代码如下：

```
public class test1 extends HttpServlet {
    protected void doGet(HttpServletRequest request, HttpServletResponse response)
            throws ServletException, IOException {

        Userinfo userinfo1 = new Userinfo();
        userinfo1.setUsername("中国 1");
        userinfo1.setPassword("中国人 1");
        userinfo1.setAge(100L);
        userinfo1.setInsertdate(new Date());

        Userinfo userinfo2 = new Userinfo();
        userinfo2.setUsername("中国 2");
        userinfo2.setPassword("中国人 2");
        userinfo2.setAge(100L);
        userinfo2.setInsertdate(new Date());

        SqlSession session = GetSqlSession.getSqlSession();
        UserinfoMapper mapper = session.getMapper(UserinfoMapper.class);
        mapper.insertUserinfo(userinfo1);
        mapper.insertUserinfo(userinfo2);
        // 此案例没有提交
        // session.commit();
        session.close();
    }
}
```

程序运行后在数据表 userinfo 中并没有添加任何一条记录，因为事务使用的是<transactionManager type="JDBC" />显式处理。

更改配置文件 mybatis-config.xml，代码如下：

```
<environments default="jndi1">
    <environment id="jndi1">
        <transactionManager type="MANAGED" />
        <dataSource type="JNDI">
            <property name="data_source" value="${jndiName}" />
        </dataSource>
    </environment>
```

再次执行 Test1.java 类，程序执行完毕后在数据表 userinfo 中添加了两条记录，每执行 1 条 SQL 语句后事务进行了自动提交。

以上测试的结果说明，在程序代码没有错误的情况下，以下两段配置代码的主要区别就是事务是否自动提交：

```
<transactionManager type="MANAGED" />
<transactionManager type="JDBC" />
```

下面再来看一下在程序代码出现错误的情况下，这两者是如何处理事务的。

首先要将 userinfo 数据表中的全部记录删除。

更改配置文件 mybatis-config.xml，代码如下：

```
<environments default="jndi1">
    <environment id="jndi1">
        <transactionManager type="JDBC" />
        <dataSource type="JNDI">
            <property name="data_source" value="${jndiName}" />
        </dataSource>
    </environment>
```

创建 Servlet 类 Test2，代码如下：

```
public class test2 extends HttpServlet {
    protected void doGet(HttpServletRequest request, HttpServletResponse response)
            throws ServletException, IOException {
        Userinfo userinfo1 = new Userinfo();
        userinfo1.setUsername("中国 1");
        userinfo1.setPassword("中国人 1");
        userinfo1.setAge(100L);
        userinfo1.setInsertdate(new Date());

        Userinfo userinfo2 = new Userinfo();
        userinfo2.setUsername("中国 2");
        userinfo2.setPassword(
                "中国人 2 中国人 2 中国人 2 中国人 2 中国人 2 中国人 2 中国人 2 中国人 2 中国人 2 中国人 2
中国人 2 中国人 2 中国人 2 中国人 2 中国人 2 中国人 2 中国人 2 中国人 2 中国人 2 中国人 2 中国人 2 中国人 2 中
国人 2 中国人 2 中国人 2 中国人 2 中国人 2 中国人 2 中国人 2");
        userinfo2.setAge(100L);
        userinfo2.setInsertdate(new Date());

        SqlSession session = GetSqlSession.getSqlSession();
        UserinfoMapper mapper = session.getMapper(UserinfoMapper.class);
```

```
        mapper.insertUserinfo(userinfo1);
        mapper.insertUserinfo(userinfo2);
        // 此案例没有提交
        // session.commit();
        session.close();
    }
}
```

程序运行后数据表 userinfo 中并没有添加任何一条记录，程序代码出现异常导致事务整体回滚，因为以下配置代码会使用 autocommit 值为 false 的配置：

```
<transactionManager type="JDBC" />
```

而下面的配置代码会使用 autocommit 值为 true 的配置：

```
<transactionManager type="MANAGED" />
```

继续更改配置文件 mybatis-config.xml，代码如下：

```
<environments default="jndi1">
    <environment id="jndi1">
        <transactionManager type="MANAGED" />
        <dataSource type="JNDI">
            <property name="data_source" value="${jndiName}" />
        </dataSource>
    </environment>
```

再次执行名称为 Test2.java 的 Servlet，程序执行完毕后，数据表 userinfo 中只出现 1 条记录，因为第 2 个 insert 操作失败。因为<transactionManager type="MANAGED" />的配置是每执行一个 SQL 就自动提交事务，所以第 2 条 insert 语句由于出现异常并没有成功添加到 userinfo 数据表中。

2.10 动态 SQL

因为 MyBatis 框架是基于 SQL 映射的，所以 SQL 映射文件在此框架中的位置非常重要，而动态 SQL 是 MyBatis 提供的根据指定的条件来执行指定的 SQL 语句，使 SQL 映射文件中的 SQL 语句在执行时具有动态性，但好在 SQL 映射文件与动态 SQL 被设计得非常简单。本节将介绍在 SQL 映射文件中的常用实例的使用。

2.10.1 <resultMap>标签的基本使用

如果数据表中字段的名称和 Java 实体类中属性的名称不一致，就要使用<resultMap>标签来做一个映射。

创建名称为 resultMapTest 的 Java 项目，SQL 映射配置文件 UserinfoMapper.xml 的代码如下：

```
<mapper namespace="mapping.UserinfoMapper">

    <resultMap type="entity.UserinfoABC" id="userinfo">
        <result column="id" property="idABC" />
        <result column="username" property="usernameABC" />
        <result column="password" property="passwordABC" />
```

```
        <result column="age" property="ageABC" />
        <result column="insertdate" property="insertdateABC" />
    </resultMap>

    <select id="selectAll" resultMap="userinfo">
        select * from userinfo order
        by
        id asc
    </select>
</mapper>
```

在<select>标签中使用 resultMap 属性来引用<resultMap>的 id 属性值，形成映射关系。

实体类 UserinfoABC 的代码如下：

```
public class UserinfoABC {

    private long idABC;
    private String usernameABC;
    private String passwordABC;
    private Long ageABC;
    private Date insertdateABC;

    public UserinfoABC() {
        System.out.println("public UserinfoABC()");
    }

     //省略 get 和 set
}
```

创建运行类，代码如下：

```
public class SelectAll {
    public static void main(String[] args) throws IOException {
        SqlSession session = GetSqlSession.getSqlSession();
        UserinfoMapper mapper = session.getMapper(UserinfoMapper.class);
        List<UserinfoABC> listUserinfo = mapper.selectAll();
        for (int i = 0; i < listUserinfo.size(); i++) {
            UserinfoABC userinfo = listUserinfo.get(i);
            System.out.println(userinfo.getIdABC() + " " + userinfo.getUsernameABC() +
" " + userinfo.getPasswordABC()
                        + " " + userinfo.getAgeABC() + " " + userinfo.getInsertdateABC());
        }
        session.commit();
        session.close();
    }
}
```

程序运行后的效果如下：

```
public UserinfoABC()
public UserinfoABC()
public UserinfoABC()
public UserinfoABC()
public UserinfoABC()
public UserinfoABC()
public UserinfoABC()
public UserinfoABC()
public UserinfoABC()
```

```
1 中国 中国人 333 Tue Jun 05 18:21:32 CST 2018
2 中国 中国人 333 Tue Jun 05 18:18:41 CST 2018
3 中国 中国人 333 Tue Jun 05 18:21:31 CST 2018
4 中国 中国人 333 Tue Jun 05 18:21:33 CST 2018
5 中国 中国人 333 Tue Jun 05 18:21:34 CST 2018
6 中国 中国人 333 Tue Jun 05 18:21:31 CST 2018
7 中国 中国人 333 Tue Jun 05 18:21:32 CST 2018
8 中国 中国人 333 Tue Jun 05 18:21:33 CST 2018
9 a aa 999999 Tue Jun 05 18:23:14 CST 2018
```

实例化实体类 UserinfoABC.java 时执行的是无参构造方法。

如果想执行有参构造方法呢？可以查看下面的内容。

2.10.2 <resultMap>标签与有参构造方法

创建名称为 resultMapConstructParam 的 Java 项目。

SQL 映射配置文件 UserinfoMapper.xml 的代码如下：

```
<resultMap type="entity.UserinfoABC" id="userinfo">
    <constructor>
        <arg column="id" javaType="long" />
        <arg column="username" javaType="String" />
        <arg column="password" javaType="String" />
        <arg column="age" javaType="long" />
        <arg column="insertdate" javaType="java.util.Date" />
    </constructor>
</resultMap>

<select id="selectAll" resultMap="userinfo">
    select * from userinfo order
    by
    id asc
</select>
```

在<select>标签中使用 resultMap 属性来引用<resultMap>的 id 属性值，形成映射关系。

实体类 UserinfoABC 的代码如下：

```
public class UserinfoABC {

    private Long idABC;
    private String usernameABC;
    private String passwordABC;
    private Long ageABC;
    private Date insertdateABC;

    public UserinfoABC() {
        System.out.println("public UserinfoABC()");
    }

    public UserinfoABC(Long idABC, String usernameABC, String passwordABC, Long ageABC,
Date insertdateABC) {
        this.idABC = idABC;
        this.usernameABC = usernameABC;
        this.passwordABC = passwordABC;
        this.ageABC = ageABC;
```

```
        this.insertdateABC = insertdateABC;
        System.out.println("执行了 public UserinfoABC 有参构造方法");
    }

    //省略 get 和 set 方法
}
```

创建运行类，代码如下：

```
public class SelectAll {
    public static void main(String[] args) throws IOException {
        SqlSession session = GetSqlSession.getSqlSession();
        UserinfoMapper mapper = session.getMapper(UserinfoMapper.class);
        List<UserinfoABC> listUserinfo = mapper.selectAll();
        for (int i = 0; i < listUserinfo.size(); i++) {
            UserinfoABC userinfo = listUserinfo.get(i);
            System.out.println(userinfo.getIdABC() + " " + userinfo.getUsernameABC()
+ " " + userinfo.getPasswordABC()
                    + " " + userinfo.getAgeABC() + " " + userinfo.getInsertdateABC());
        }
        session.commit();
        session.close();
    }
}
```

程序运行后的效果如下：

```
执行了 public UserinfoABC 有参构造方法
执行了 public UserinfoABC 有参构造方法
执行了 public UserinfoABC 有参构造方法
执行了 public UserinfoABC 有参构造方法
执行了 public UserinfoABC 有参构造方法
执行了 public UserinfoABC 有参构造方法
执行了 public UserinfoABC 有参构造方法
执行了 public UserinfoABC 有参构造方法
执行了 public UserinfoABC 有参构造方法
1 中国 中国人 333 Tue Jun 05 18:21:32 CST 2018
2 中国 中国人 333 Tue Jun 05 18:18:41 CST 2018
3 中国 中国人 333 Tue Jun 05 18:21:31 CST 2018
4 中国 中国人 333 Tue Jun 05 18:21:33 CST 2018
5 中国 中国人 333 Tue Jun 05 18:21:34 CST 2018
6 中国 中国人 333 Tue Jun 05 18:21:31 CST 2018
7 中国 中国人 333 Tue Jun 05 18:21:32 CST 2018
8 中国 中国人 333 Tue Jun 05 18:21:33 CST 2018
9 a aa 999999 Tue Jun 05 18:23:14 CST 2018
```

实例化 UserinfoABC.java 实体类时执行的是有参构造方法。

2.10.3 使用 ${} 拼接 SQL 语句

#{} 代表向 Statement 传入参数值，而 ${} 是拼接字符串。

新建名称为 sqlStringVar 的 Java 项目，SQL 映射文件 UserinfoMapper.xml 的代码如下：

```
<mapper namespace="mapping.UserinfoMapper">
    <select id="selectAll" parameterType="map"
        resultType="entity.Userinfo">
```

```
        select * from userinfo where id>#{userId} order by
        ${orderSQL}
    </select>
</mapper>
```

运行类代码如下：

```
public class SelectAll {
    public static void main(String[] args) throws IOException {
        Map map = new HashMap();
        map.put("userId", 4);
        map.put("orderSQL", " id desc");

        SqlSession session = GetSqlSession.getSqlSession();
        UserinfoMapper mapper = session.getMapper(UserinfoMapper.class);
        List<Userinfo> listUserinfo = mapper.selectAll(map);
        for (int i = 0; i < listUserinfo.size(); i++) {
            Userinfo userinfo = listUserinfo.get(i);
            System.out.println(userinfo.getId() + " " + userinfo.getUsername() + " "
+ userinfo.getPassword() + " "
                    + userinfo.getAge() + " " + userinfo.getInsertdate());
        }
        session.commit();
        session.close();
    }
}
```

注意： 使用 ${} 拼接 SQL 语句容易发生 SQL 注入，对系统安全有影响。

2.10.4 <sql>标签

重复的 SQL 语句永远不可避免，可以复用那些重复的 SQL 语句，<sql>标签就是用来解决这个问题的。

1. 静态传值

静态传值就是向<sql>标签传入常量值。

创建名称为 sqlTest 的 Java 项目，映射配置文件 UserinfoMapper.xml 的代码如下：

```
<mapper namespace="mapping.UserinfoMapper">

    <sql id="userinfo5Column1">id "id",username "username",password,age,insertdate</sql>
    <select id="selectAll1" resultType="entity.Userinfo">
        select
        <include refid="userinfo5Column1"></include>
        from userinfo order
        by id asc
    </select>

    <sql id="userinfo5Column2">${col1} ,${col2} </sql>
    <select id="selectAll2" resultType="entity.Userinfo">
        select
        <include refid="userinfo5Column2">
            <property name="col1" value="id" />
```

```
            <property name="col2" value="password" />
        </include>
        from userinfo order
        by id asc
    </select>

</mapper>
```

上面代码中的 id、username、password、age 和 insertdate 这 5 个字段在 SQL 映射文件中多次出现，因此，可以将这 5 个字段封装进<sql id="userinfo5Column1">标签中，以减少配置的代码量。

以下配置代码的作用是向<sql id="userinfo5Column2">传递参数：

```
<include refid="userinfo5Column2">
    <property name="col1" value="id" />
    <property name="col2" value="password" />
</include>
```

如果<select>映射的 resultType 属性值是“map”，那么代表是从 map 里使用 map.get（字段名称）方法的形式取得字段对应的值，但字段名称在 Oracle 中是大写的形式，因此要使用 map.get（大写字段名称）的写法获得列值。为了支持小写，可以在 SQL 映射文件中定义的 SQL 语句中为字段另起一个别名：

```
select id "id",username "username",password "password",age "age" from userinfo
```

还可以使用<sql>标签来进行声明，示例代码如下：

```
<sql id="userinfo5Column1">id "id",username "username",password,age,insertdate</sql>
```

这样从 map 中就可以以小写的形式取得字段值了。

2. 动态传值

动态传值就是向<sql>标签传入变量值，变量值来自于 parameterType 属性。

前面使用如下配置代码以静态的方式向<sql>标签的参数传入常量参数值，在 MyBatis 中还支持动态地传入参数值：

```
<include refid="userinfo5Column2">
    <property name="col1" value="id" />
    <property name="col2" value="password" />
</include>
```

创建示例项目 mybatis39。

创建实体类 Userinfo，代码如下：

```
package entity;

import java.util.Date;

public class Userinfo {
    private Long id;
    private String username;
    private String password;
    private Long age;
```

```
    private Date insertdate;

    public Userinfo() {
        super();
        // TODO Auto-generated constructor stub
    }

    public Userinfo(Long id, String username, String password, Long age, Date insertdate) {
        super();
        this.id = id;
        this.username = username;
        this.password = password;
        this.age = age;
        this.insertdate = insertdate;
    }

    //省略 set 和 get 方法

}
```

创建实体类 UserinfoABC，代码如下：

```
package entity;

import java.util.Date;

public class UserinfoABC {
    private Long idABC;
    private String usernameABC;
    private String passwordABC;
    private Long ageABC;
    private Date insertdateABC;

    public UserinfoABC() {
        System.out.println("无参构造");
    }

    public UserinfoABC(Long idABC, String usernameABC) {
        super();
        this.idABC = idABC;
        this.usernameABC = usernameABC;
    }

    public UserinfoABC(Long idABC, String usernameABC, String passwordABC, Long ageABC,
Date insertdateABC) {
        super();
        System.out.println("有参构造 " + usernameABC);
        this.idABC = idABC;
        this.usernameABC = usernameABC;
        this.passwordABC = passwordABC;
        this.ageABC = ageABC;
        this.insertdateABC = insertdateABC;
    }

    //省略 set 和 get 方法

}
```

创建 SQL 映射接口代码如下：

```
package sqlmapping;

import java.util.List;
import java.util.Map;

import entity.Userinfo;
import entity.UserinfoABC;

public interface IUserinfoMapping {
    public List<UserinfoABC> selectTest(Map map);

    public List<Userinfo> selectTest2(Map map);
}
```

创建 SQL 映射文件，代码如下：

```
<mapper namespace="sqlmapping.IUserinfoMapping">
    <resultMap type="entity.UserinfoABC" id="userinfo">
        <constructor>
            <arg column="id" name="arg0" jdbcType="INTEGER" javaType="long" />
            <arg column="username" name="arg1" jdbcType="VARCHAR"
                javaType="String" />
        </constructor>
    </resultMap>

    <sql id="sqlTemplate">${col1},${col2}</sql>

    <select id="selectTest" parameterType="map" resultMap="userinfo">
        select
        <include refid="sqlTemplate">
            <property name="col1" value="${idCol}" />
            <property name="col2" value="${usernameCol}" />
        </include>
        from userinfo order
        by id asc
    </select>

    <select id="selectTest2" parameterType="map"
        resultType="entity.Userinfo">
        select
        <include refid="sqlTemplate">
            <property name="col1" value="${idCol}" />
            <property name="col2" value="${usernameCol}" />
        </include>
        from userinfo order
        by id desc
    </select>
</mapper>
```

创建运行类 Test1，代码如下：

```
public class Test1 {
    public static void main(String[] args) throws IOException {
        Map map = new HashMap();
        map.put("idCol", "id");
        map.put("usernameCol", "username");
```

```
        SqlSession session = GetSqlSession.getSqlSession();
        IUserinfoMapping mapping = session.getMapper(IUserinfoMapping.class);
        List<UserinfoABC> listUserinfo = mapping.selectTest(map);
        for (int i = 0; i < listUserinfo.size(); i++) {
            UserinfoABC userinfo = listUserinfo.get(i);
            System.out.println(userinfo.getIdABC() + " " + userinfo.getUsernameABC()
+ " " + userinfo.getPasswordABC()
                    + " " + userinfo.getAgeABC() + " " + userinfo.getInsertdateABC());
        }
        session.commit();
        session.close();
    }
}
```

创建运行类 Test2，代码如下：

```
public class Test2 {
    public static void main(String[] args) throws IOException {
        Map map = new HashMap();
        map.put("idCol", "id");
        map.put("usernameCol", "username");

        SqlSession session = GetSqlSession.getSqlSession();
        IUserinfoMapping mapping = session.getMapper(IUserinfoMapping.class);
        List<Userinfo> listUserinfo = mapping.selectTest2(map);
        for (int i = 0; i < listUserinfo.size(); i++) {
            Userinfo userinfo = listUserinfo.get(i);
            System.out.println(userinfo.getId() + " " + userinfo.getUsername() + " "
+ userinfo.getPassword() + " "
                    + userinfo.getAge() + " " + userinfo.getInsertdate());
        }
        session.commit();
        session.close();
    }
}
```

2.10.5 插入 null 值的第 1 种方法——JdbcType

创建名称为 dynSqlTest 的 Java 项目。

创建 SQL 映射，代码如下：

```
<insert id="insertUserinfo" parameterType="entity.Userinfo">
    <selectKey order="BEFORE" resultType="java.lang.Long"
        keyProperty="id">
        select idauto.nextval from dual
    </selectKey>
    insert into userinfo(id,username,password,age,insertdate)
    values(#{id},#{username},#{password},#{age},#{insertdate})
</insert>
```

执行添加功能的代码如下：

```
public class Insert1 {
    public static void main(String[] args) throws IOException {
        Userinfo userinfo = new Userinfo();
        userinfo.setUsername("中国");
```

```
        userinfo.setPassword(null);
        userinfo.setAge(100L);
        userinfo.setInsertdate(new Date());

        SqlSession session = GetSqlSession.getSqlSession();
        session.insert("insertUserinfo", userinfo);
        session.commit();
        session.close();
    }
}
```

程序运行后出现的异常信息如下：

```
Caused by: org.apache.ibatis.type.TypeException: Error setting null for parameter #3
with JdbcType OTHER . Try setting a different JdbcType for this parameter or a different
jdbcTypeForNull configuration property. Cause: java.sql.SQLException: 无效的列类型: 1111
```

从出错信息中可以看到，是将 null 值对 password 字段进行了设置，造成 MyBatis 无法识别出应该插入默认的数据类型值。这种情况可以通过设置映射的默认数据类型来解决，更改映射配置代码如下：

```
<insert id="insertUserinfoNew" parameterType="entity.Userinfo">
    <selectKey order="BEFORE" resultType="java.lang.Long"
        keyProperty="id">
        select idauto.nextval from dual
    </selectKey>
    insert into userinfo(id,username,password,age,insertdate)
    values(#{id},#{username,jdbcType=VARCHAR},#{password,jdbcType=VARCHAR},#{age,jdbc
Type=INTEGER},#{insertdate,jdbcType=TIMESTAMP})
</insert>
```

也就是在 #{} 格式中加入了默认数据类型的声明，这样可以明确地告诉 MyBatis 框架遇到 null 值该如何处理。

再次运行程序，成功插入数据表。

2.10.6 插入 null 值的第 2 种方法——<if>

通过在 #{} 格式中加入 JdbcType 即可避免插入 null 值时的异常，其实使用动态 SQL 标签也可以达到同样的效果。

继续在 dynSqlTest 项目中添加代码。

在 SQL 映射文件 userinfoMapping.xml 中添加如下代码：

```
<insert id="insertUserinfoIf" parameterType="entity.Userinfo">
    <selectKey order="BEFORE" resultType="java.lang.Long"
        keyProperty="id">
        select idauto.nextval from dual
    </selectKey>
    <if test="password!=null">
        insert into userinfo(id,username,password,age,insertdate)
        values(#{id},#{username},#{password},#{age},#{insertdate})
    </if>
    <if test="password==null">
        insert into userinfo(id,username,age,insertdate)
        values(#{id},#{username},#{age},#{insertdate})
```

```
    </if>
</insert>
```

创建运行类，代码如下：

```
public class Insert3 {
    public static void main(String[] args) throws IOException {

        Userinfo userinfo1 = new Userinfo();
        userinfo1.setUsername("中国");
        userinfo1.setPassword("中国人");
        userinfo1.setAge(100L);
        userinfo1.setInsertdate(new Date());

        Userinfo userinfo2 = new Userinfo();
        userinfo2.setUsername("中国");
        userinfo2.setAge(100L);
        userinfo2.setInsertdate(new Date());

        SqlSession session = GetSqlSession.getSqlSession();
        session.insert("insertUserinfoIf", userinfo1);
        session.insert("insertUserinfoIf", userinfo2);
        session.commit();
        session.close();

    }
}
```

程序运行后在数据表中成功插入两条记录。

标签<if>在判断字符串为空时还可以更加具体，示例代码如下：

```
<if test="password!=null and password!=''">
```

也就是 null 值与“ ”（空字符串）一同进行判断。

2.10.7 <where>标签

标签<where>的主要作用是生成 where 语句，可以使用在 delete、update 及 select 语句中。

创建测试用的项目 whereTAG，创建 SQL 映射文件 UserinfoMapper.xml，代码如下：

```
<mapper namespace="mapping.UserinfoMapper">
    <select id="selectAll" parameterType="map"
        resultType="entity.Userinfo">
        select * from userinfo
        <where>
            <if test="username!=null">and username like #{username}</if>
            <if test="password!=null">and password like #{password}</if>
        </where>
    </select>
</mapper>
```

运行类代码如下：

```
public class SelectAll {
    public static void main(String[] args) throws IOException {
        Map map = new HashMap();
        map.put("username", "%中国%");
```

```
        map.put("password", "%中国人%");

        SqlSession session = GetSqlSession.getSqlSession();
        List<Userinfo> listUserinfo = session.selectList("selectAll", map);
        for (int i = 0; i < listUserinfo.size(); i++) {
            Userinfo userinfo = listUserinfo.get(i);
            System.out.println(userinfo.getId() + " " + userinfo.getUsername() + " "
+ userinfo.getPassword() + " "
                    + userinfo.getAge() + " " + userinfo.getInsertdate());
        }
        session.commit();
        session.close();
    }
}
```

如果在代码中不对 username 传递值，<where>标签也能自动去掉语句“and password like #{password}”中的 and 关键字而成功执行查询的 SQL 语句。

2.10.8 <choose>标签的使用

<choose>标签的作用是在众多的条件中选择出一条，有些类似于 Java 语言中的 switch+break 语句的作用。

创建测试用的项目 chooseTAG，创建 SQL 映射文件 UserinfoMapper.xml，代码如下：

```
<mapper namespace="mapping.UserinfoMapper">
    <select id="selectAll2" parameterType="map"
        resultType="entity.Userinfo">
        select * from userinfo where 1=1
        <choose>
            <when test="username!=null"> and username like #{username}</when>
            <when test="password!=null"> and password like #{password}</when>
            <otherwise>and age=100</otherwise>
        </choose>
    </select>
</mapper>
```

运行类代码如下：

```
public class SelectAll {
    public static void main(String[] args) throws IOException {
        Map map1 = new HashMap();
        map1.put("username", "%中国%");

        Map map2 = new HashMap();
        map2.put("password", "%中国人%");

        Map map3 = new HashMap();

        SqlSession session = GetSqlSession.getSqlSession();
        List<Userinfo> listUserinfo = session.selectList("selectAll2", map1);
        for (int i = 0; i < listUserinfo.size(); i++) {
            Userinfo userinfo = listUserinfo.get(i);
            System.out.println(userinfo.getId() + " " + userinfo.getUsername() + " "
+ userinfo.getPassword() + " "
                    + userinfo.getAge() + " " + userinfo.getInsertdate());
        }
```

```
            System.out.println();
            listUserinfo = session.selectList("selectAll2", map2);
            for (int i = 0; i < listUserinfo.size(); i++) {
                Userinfo userinfo = listUserinfo.get(i);
                System.out.println(userinfo.getId() + " " + userinfo.getUsername() + " "
+ userinfo.getPassword() + " "
                        + userinfo.getAge() + " " + userinfo.getInsertdate());
            }
            System.out.println();
            listUserinfo = session.selectList("selectAll2", map3);
            for (int i = 0; i < listUserinfo.size(); i++) {
                Userinfo userinfo = listUserinfo.get(i);
                System.out.println(userinfo.getId() + " " + userinfo.getUsername() + " "
+ userinfo.getPassword() + " "
                        + userinfo.getAge() + " " + userinfo.getInsertdate());
            }
            session.commit();
            session.close();
        }
    }
```

标签<choose>的作用同 switch+break 语句一样。

2.10.9　<set>标签的使用

<set>标签可以在 update 语句中使用，作用是动态指定要更新的列。

创建测试用的项目 setTAG，创建 SQL 映射文件 UserinfoMapper.xml 的代码如下：

```
<mapper namespace="mapping.UserinfoMapper">
    <update id="updateUserinfoById" parameterType="map">
        update userinfo
        <set>
            <if test="username!=null">username=#{username},</if>
            <if test="password!=null">password=#{password},</if>
            <if test="age!=null">age=#{age},</if>
            <if test="insertdate!=null">insertdate=#{insertdate},</if>
        </set>
        where id=#{id}
    </update>
</mapper>
```

最后一个 if 条件中的逗号可以保留，<set>标签可以自动删除。

创建运行类，代码如下：

```
public class UpdateUserinfoById {
    public static void main(String[] args) throws IOException {
        Map map1 = new HashMap();
        map1.put("id", 2182498L);
        map1.put("username", "xxxxx");
        map1.put("password", null);
        map1.put("age", 999);
        map1.put("insertdate", new Date());

        SqlSession session = GetSqlSession.getSqlSession();
        session.update("updateUserinfoById", map1);
        session.commit();
```

```
        session.close();
    }
}
```

运行效果就是 password 列未更新，其他列都更新了。

2.10.10 <foreach>标签的使用

<foreach>标签有循环的功能，可以用来生成有规律的 SQL 语句。

<foreach>标签主要的属性有 item、index、collection、open、separator 和 close。

item 表示集合中每一个元素进行迭代时的别名；index 指定一个名字，用于表示在迭代过程中每次迭代到的位置；open 表示该语句以什么开始；separator 表示在每次进行迭代之间以什么符号作为分隔符；close 表示该语句以什么结束。

创建测试用的项目 foreachTest。

在 SQL 映射文件 UserinfoMapper.xml 中添加配置，代码如下：

```
<mapper namespace="mapping.UserinfoMapper">
    <select id="selectAll1" parameterType="list"
        resultType="entity.Userinfo">
        select *
        from userinfo where id in
        <foreach item="eachId" collection="list" open="("
            separator="," close=")">
            #{eachId}
        </foreach>
    </select>

    <select id="selectAll2" parameterType="queryentity.QueryEntity"
        resultType="entity.Userinfo">
        select *
        from userinfo where id in
        <foreach item="eachId" collection="xxxxxxxxx" open="("
            separator="," close=")">
            #{eachId}
        </foreach>
    </select>

    <select id="selectAll3" parameterType="map"
        resultType="entity.Userinfo">
        select *
        from userinfo where id in
        <foreach item="eachId" collection="yyyyyyyyyyyy" open="("
            separator="," close=")">
            #{eachId}
        </foreach>
    </select>

    <select id="selectAll4" parameterType="map"
        resultType="entity.Userinfo">
        select *
        from userinfo where username like
        '%'||#{username}||'%'
    </select>
</mapper>
```

在 Oracle 数据库中，字符串的拼接要使用 || 符号。

运行类代码如下：

```
public class SelectAll {
    public static void main(String[] args) throws IOException {
        SqlSession session = GetSqlSession.getSqlSession();
        UserinfoMapper mapper = session.getMapper(UserinfoMapper.class);

        List idList = new ArrayList();
        idList.add(1);
        idList.add(2);

        List<Userinfo> listUserinfo = mapper.selectAll1(idList);
        for (int i = 0; i < listUserinfo.size(); i++) {
            Userinfo userinfo = listUserinfo.get(i);
            System.out.println(userinfo.getId() + " " + userinfo.getUsername() + " "
+ userinfo.getPassword() + " "
                    + userinfo.getAge() + " " + userinfo.getInsertdate());
        }
        System.out.println();
        QueryEntity entity = new QueryEntity();
        idList.add(3);
        entity.setXxxxxxxxx(idList);
        listUserinfo = mapper.selectAll2(entity);
        for (int i = 0; i < listUserinfo.size(); i++) {
            Userinfo userinfo = listUserinfo.get(i);
            System.out.println(userinfo.getId() + " " + userinfo.getUsername() + " "
+ userinfo.getPassword() + " "
                    + userinfo.getAge() + " " + userinfo.getInsertdate());
        }
        System.out.println();
        HashMap map = new HashMap();
        idList.add(4);
        map.put("yyyyyyyyyyyy", idList);
        listUserinfo = mapper.selectAll3(map);
        for (int i = 0; i < listUserinfo.size(); i++) {
            Userinfo userinfo = listUserinfo.get(i);
            System.out.println(userinfo.getId() + " " + userinfo.getUsername() + " "
+ userinfo.getPassword() + " "
                    + userinfo.getAge() + " " + userinfo.getInsertdate());
        }
        System.out.println();
        HashMap map4 = new HashMap();
        map4.put("username", "中国");
        listUserinfo = mapper.selectAll4(map4);
        for (int i = 0; i < listUserinfo.size(); i++) {
            Userinfo userinfo = listUserinfo.get(i);
            System.out.println(userinfo.getId() + " " + userinfo.getUsername() + " "
+ userinfo.getPassword() + " "
                    + userinfo.getAge() + " " + userinfo.getInsertdate());
        }
        session.commit();
        session.close();
    }
}
```

2.10.11 使用<foreach>执行批量插入

如何实现批量 insert 插入操作呢？由于批量插入的 SQL 语句在每种数据库中都不一样，比如 Oracle 是使用如下格式的 SQL 语句：

```
INSERT INTO userinfo (id,
                   username,
                   password,
                   age,
                   insertdate)
   SELECT idauto.NEXTVAL,
         username,
         password,
         age,
         insertdate
     FROM (SELECT 'a' username,
                'aa' password,
                1 age,
                TO_DATE ('2000-1-1', 'yyyy-MM-dd') insertdate
            FROM DUAL
          UNION ALL
          SELECT 'b' username,
                'bb' password,
                1 age,
                TO_DATE ('2000-1-1', 'yyyy-MM-dd') insertdate
            FROM DUAL)
```

而 MySQL 是使用如下格式的 SQL 语句来实现批量 insert 插入：

```
insert into userinfo(username,password,age,insertdate)
values('a','aa',1,'2000-1-1'),
('b','aa',1,'2000-1-1'),
('c','aa',1,'2000-1-1'),
('d','aa',1,'2000-1-1'),
('e','aa',1,'2000-1-1')
```

由于批量 insert 插入的 SQL 语句在每种数据库中都不一样，因此在 MyBatis 中 SQL 映射文件的写法也不一样。

下面来看一看针对 Oracle 数据库的 SQL 映射文件的代码：

```
<insert id="insertOracle" parameterType="list">
    INSERT INTO userinfo (id,
    username,
    password,
    age,
    insertdate)
    select
    idauto.nextval,username,password,age,insertdate from (
    <foreach collection="list" item="eachUserinfo" separator="union all">
        select
        #{eachUserinfo.username} username,#{eachUserinfo.password}
        password,#{eachUserinfo.age} age,#{eachUserinfo.insertdate}
        insertdate
        from dual
    </foreach>
```

```
    )
</insert>
```

Java 代码如下：

```
public class Insert1 {
    public static void main(String[] args) throws IOException {

        List list = new ArrayList();

        for (int i = 0; i < 5; i++) {
            Userinfo userinfo = new Userinfo();
            userinfo.setUsername("中国" + (i + 1));
            userinfo.setPassword("中国人" + (i + 1));
            userinfo.setAge(100L);
            userinfo.setInsertdate(new Date());
            list.add(userinfo);
        }

        SqlSession session = GetSqlSession.getSqlSession();
        session.insert("insertOracle", list);
        session.commit();
        session.close();

    }
}
```

下面来看一看针对 MySQL 数据库的 SQL 映射文件的代码：

```
    <insert id="insertMySQL" parameterType="list">
    INSERT INTO userinfo (
    username,
    password,
    age,
    insertdate)
    values
    <foreach collection="list" item="eachUserinfo" separator=",">
        (
        #{eachUserinfo.username},#{eachUserinfo.password},
        #{eachUserinfo.age},#{eachUserinfo.insertdate}
        )
    </foreach>
</insert>
```

Java 代码如下：

```
public class Insert1 {
    public static void main(String[] args) throws IOException {

        List list = new ArrayList();

        for (int i = 0; i < 5; i++) {
            Userinfo userinfo = new Userinfo();
            userinfo.setUsername("中国 MYSQL" + (i + 1));
            userinfo.setPassword("中国人 MYSQL" + (i + 1));
            userinfo.setAge(100L);
            userinfo.setInsertdate(new Date());
            list.add(userinfo);
        }
```

```
        SqlSession session = GetSqlSession.getSqlSession();
        session.insert("insertMySQL", list);
        session.commit();
        session.close();

    }
}
```

2.10.12 使用<bind>标签对 like 语句进行适配

在 SQL 语句中使用 like 查询时，MySQL 在拼接字符串时使用 concat()方法，而 Oracle 却使用 || 运算符，两者的 SQL 语句并不相同，这就需要创建两个 SQL 映射语句。下面创建实验用的项目 mysql_oracle_2_sql，SQL 映射文件示例代码如下：

```
<mapper namespace="mapping.UserinfoMapper">
    <select id="selectAllOracle" parameterType="map"
        resultType="entity.Userinfo">
        select * from userinfo
        where username like
        '%'||#{username}||'%'
    </select>
    <select id="selectAllMySQL" parameterType="map"
        resultType="entity.Userinfo">
        select * from userinfo
        where username like
        concat('%',#{username},'%')
    </select>
</mapper>
```

通过在运行代码中更改 selectList() 方法的第 1 个参数值来切换执行不同的 SQL 语句，示例代码如下：

```
List<Userinfo> listUserinfo = session.selectList("selectAllOracle", map);
List<Userinfo> listUserinfo = session.selectList("selectAllMySQL", map);
```

这样会更改软件的源代码，不利于软件的稳定性。这时可以使用<bind>标签来在多个数据库之间进行适配。

创建名称为 bindTAGNew 的项目，SQL 映射文件的代码如下：

```
<mapper namespace="mapping.UserinfoMapper">
    <select id="selectAll1" parameterType="string"
        resultType="entity.Userinfo">
        <bind name="querySQL" value="'%'+_parameter+'%'" />
        select * from userinfo where username like #{querySQL}
    </select>
    <select id="selectAll2" parameterType="entity.Userinfo"
        resultType="entity.Userinfo">
        <bind name="querySQL" value="'%'+_parameter.getUsername()+'%'" />
        select * from userinfo where username like #{querySQL}
    </select>
    <select id="selectAll3" parameterType="entity.Userinfo"
        resultType="entity.Userinfo">
        <bind name="querySQL" value="'%'+username+'%'" />
        select * from userinfo where username like #{querySQL}
```

```
    </select>
    <select id="selectAll4" parameterType="entity.Userinfo"
        resultType="entity.Userinfo">
        <bind name="querySQL" value="'%'+_parameter.username+'%'" />
        select * from userinfo where username like #{querySQL}
    </select>
    <select id="selectAll5" parameterType="entity.Userinfo"
        resultType="entity.Userinfo">
        <bind name="querySQL" value="'%'+#root._parameter.username+'%'" />
        select * from userinfo where username like #{querySQL}
    </select>
    <select id="selectAll6" parameterType="map"
        resultType="entity.Userinfo">
        <bind name="querySQL" value="'%'+username+'%'" />
        select * from userinfo where username like #{querySQL}
    </select>
</mapper>
```

id 为 selectAll6 的<select>标签的子标签<bind>中的代码 value="'%'+username+'%'"的含义是从 map 中根据 key 名称为 username 找到对应的值，然后再拼接成 %value% 的形式。

因为<bind>标签的 value 属性值中的运算是使用 OGNL 表达式，OGNL 表达式的底层是使用 Java 语言，因此，在 Java 语言中可以直接使用+号进行字符串的拼接，并不像在 Oracle 或 MySQL 中要使用 || 或 concat()。

运行类代码如下：

```
public class SelectAll {
    public static void main(String[] args) throws IOException {
        SqlSession session = GetSqlSession.getSqlSession();
        UserinfoMapper mapper = session.getMapper(UserinfoMapper.class);
        List<Userinfo> listUserinfo = mapper.selectAll1("中国");
        for (int i = 0; i < listUserinfo.size(); i++) {
            Userinfo userinfo = listUserinfo.get(i);
            System.out.println(userinfo.getId() + " " + userinfo.getUsername() + " "
+ userinfo.getPassword() + " "
                    + userinfo.getAge() + " " + userinfo.getInsertdate());
        }
        System.out.println();
        System.out.println();
        Userinfo userinfo = new Userinfo();
        userinfo.setUsername("中国");
        listUserinfo = mapper.selectAll2(userinfo);
        for (int i = 0; i < listUserinfo.size(); i++) {
            Userinfo eachUserinfo = listUserinfo.get(i);
            System.out.println(eachUserinfo.getId() + " " + eachUserinfo.getUsername() + " "
                    + eachUserinfo.getPassword() + " " + eachUserinfo.getAge() + " "
+ eachUserinfo.getInsertdate());
        }
        System.out.println();
        System.out.println();
        listUserinfo = mapper.selectAll3(userinfo);
        for (int i = 0; i < listUserinfo.size(); i++) {
            Userinfo eachUserinfo = listUserinfo.get(i);
            System.out.println(eachUserinfo.getId() + " " + eachUserinfo.getUsername() + " "
                    + eachUserinfo.getPassword() + " " + eachUserinfo.getAge() + " "
+ eachUserinfo.getInsertdate());
```

```
            }
            System.out.println();
            System.out.println();
            listUserinfo = mapper.selectAll4(userinfo);
            for (int i = 0; i < listUserinfo.size(); i++) {
                Userinfo eachUserinfo = listUserinfo.get(i);
                System.out.println(eachUserinfo.getId() + " " + eachUserinfo.getUsername() + " "
                        + eachUserinfo.getPassword() + " " + eachUserinfo.getAge() + " "
+ eachUserinfo.getInsertdate());
            }
            System.out.println();
            System.out.println();
            listUserinfo = mapper.selectAll5(userinfo);
            for (int i = 0; i < listUserinfo.size(); i++) {
                Userinfo eachUserinfo = listUserinfo.get(i);
                System.out.println(eachUserinfo.getId() + " " + eachUserinfo.getUsername() + " "
                        + eachUserinfo.getPassword() + " " + eachUserinfo.getAge() + " "
+ eachUserinfo.getInsertdate());
            }
            System.out.println();
            System.out.println();
            Map map = new HashMap();
            map.put("username", "中国");
            listUserinfo = mapper.selectAll6(map);
            for (int i = 0; i < listUserinfo.size(); i++) {
                Userinfo eachUserinfo = listUserinfo.get(i);
                System.out.println(eachUserinfo.getId() + " " + eachUserinfo.getUsername() + " "
                        + eachUserinfo.getPassword() + " " + eachUserinfo.getAge() + " "
+ eachUserinfo.getInsertdate());
            }
            session.commit();
            session.close();
        }
    }
```

2.10.13 使用<trim>标签规范 SQL 语句

创建实验用的项目 trim_test。

SQL 映射代码如下：

```
<select id="selectAll1" parameterType="map"
    resultType="entity.Userinfo">
    select * from userinfo where 1=1
    <if test="username!=null">and username like #{username}</if>
</select>
```

创建运行类，代码如下：

```
public class SelectAll1 {
    public static void main(String[] args) throws IOException {
        Map map = new HashMap();
        map.put("username", "%法国%");

        SqlSession session = GetSqlSession.getSqlSession();
        List<Userinfo> listUserinfo = session.selectList("selectAll1", map);
        for (int i = 0; i < listUserinfo.size(); i++) {
            Userinfo userinfo = listUserinfo.get(i);
```

```
                System.out.println(userinfo.getId() + " " + userinfo.getUsername() + " "
+ userinfo.getPassword() + " "
                        + userinfo.getAge() + " " + userinfo.getInsertdate());
            }
            System.out.println();
            System.out.println();
            listUserinfo = session.selectList("selectAll1", new HashMap());
            for (int i = 0; i < listUserinfo.size(); i++) {
                Userinfo userinfo = listUserinfo.get(i);
                System.out.println(userinfo.getId() + " " + userinfo.getUsername() + " "
+ userinfo.getPassword() + " "
                        + userinfo.getAge() + " " + userinfo.getInsertdate());
            }
            session.commit();
            session.close();
        }
    }
```

程序运行后可以得出正确的结果，但在 SQL 语句中使用“where 1=1”总有些不规范的感觉，因此，创建新的 SQL 映射，代码如下：

```
<select id="selectAll2" parameterType="map"
    resultType="entity.Userinfo">
    select * from userinfo where
    <if test="username!=null">and username like #{username}</if>
</select>
```

运行类代码如下：

```
public class SelectAll2 {
    public static void main(String[] args) throws IOException {
        Map map = new HashMap();
        map.put("username", "%法国%");

        SqlSession session = GetSqlSession.getSqlSession();
        List<Userinfo> listUserinfo = session.selectList("selectAll2", map);
        for (int i = 0; i < listUserinfo.size(); i++) {
            Userinfo userinfo = listUserinfo.get(i);
            System.out.println(userinfo.getId() + " " + userinfo.getUsername() + " "
+ userinfo.getPassword() + " "
                    + userinfo.getAge() + " " + userinfo.getInsertdate());
        }
        session.commit();
        session.close();
    }
}
```

程序运行后出现的异常信息如下：

```
### SQL: select * from userinfo where    and username like ?
### Cause: java.sql.SQLSyntaxErrorException: ORA-00936: 缺失表达式
```

异常信息提示 SQL 语句不正确，原因是 where 语句的后面有一个 and，去掉第 1 个 and 可以使用<trim>标签，新创建的 SQL 映射代码如下：

```
<select id="selectAll3" parameterType="map"
    resultType="entity.Userinfo">
    select * from userinfo
```

```
        <where>
            <trim prefixOverrides="and">
                <if test="username!=null">and username like #{username}</if>
            </trim>
        </where>
    </select>
```

运行类代码如下：

```
public class SelectAll3 {
    public static void main(String[] args) throws IOException {
        Map map = new HashMap();
        map.put("username", "%法国%");

        SqlSession session = GetSqlSession.getSqlSession();
        List<Userinfo> listUserinfo = session.selectList("selectAll3", map);
        for (int i = 0; i < listUserinfo.size(); i++) {
            Userinfo userinfo = listUserinfo.get(i);
            System.out.println(userinfo.getId() + " " + userinfo.getUsername() + " "
+ userinfo.getPassword() + " "
                        + userinfo.getAge() + " " + userinfo.getInsertdate());
        }
        System.out.println();
        System.out.println();
        listUserinfo = session.selectList("selectAll3", new HashMap());
        for (int i = 0; i < listUserinfo.size(); i++) {
            Userinfo userinfo = listUserinfo.get(i);
            System.out.println(userinfo.getId() + " " + userinfo.getUsername() + " "
+ userinfo.getPassword() + " "
                        + userinfo.getAge() + " " + userinfo.getInsertdate());
        }
        session.commit();
        session.close();
    }
}
```

程序运行后不再出现异常。

如果向 parameterType 传入 String 类型，那么 SQL 映射的代码如下：

```
<select id="queryParamString" parameterType="string"
    resultType="entity.Userinfo">
    select * from userinfo
    <where>
        <trim prefixOverrides="and">
            <if test="_parameter!=null">and username like #{username}</if>
        </trim>
    </where>
</select>
```

2.11 读写 CLOB 类型的数据

MyBatis 框架对 Oracle 的 CLOB 支持也非常好，不需要特别的环境配置即可完成 CLOB 字段的读写操作。

创建名称为 bigCLOB 的项目，SQL 映射文件 BigTextMapper.xml 的代码如下：

```
<mapper namespace="mapping.BigTextMapper">
    <insert id="insertUserinfo" parameterType="entity.BigText">
        <selectKey order="BEFORE" resultType="java.lang.Long"
            keyProperty="id">
            select idauto.nextval from dual
        </selectKey>
        insert into bigtext(id,bigtext) values(#{id},#{bigtext})
    </insert>
    <select id="selectById1" resultType="entity.BigText">
        select *
        from bigtext where
        id=123
    </select>
    <select id="selectById2" resultType="map">
        select *
        from bigtext where
        id=123
    </select>
</mapper>
```

创建实体类，代码如下：

```
public class BigText {

    private long id;
    private String bigtext;

    public BigText() {
    }

    public BigText(long id, String bigtext) {
        super();
        this.id = id;
        this.bigtext = bigtext;
    }

    //省略 get 和 set 方法

}
```

数据表 bigtext 中 id 为 123 的 bigtext 列中存储了大文本数据。

进行查询操作，将数据封装进实体类的代码如下：

```
public class SelectById1 {
    public static void main(String[] args) throws IOException {
        SqlSession session = GetSqlSession.getSqlSession();
        BigTextMapper mapper = session.getMapper(BigTextMapper.class);
        BigText bigtext = mapper.selectById1();
        System.out.println(bigtext.getBigtext());
        System.out.println();
        session.close();
    }
}
```

程序运行后在控制台打印出了全部信息。

进行查询操作，将数据封装进 Map 的代码如下：

```java
public class SelectById2 {
    public static void main(String[] args) throws IOException, SQLException {
        SqlSession session = GetSqlSession.getSqlSession();
        BigTextMapper mapper = session.getMapper(BigTextMapper.class);
        Map map = mapper.selectById2();
        oracle.sql.CLOB clobRef = (oracle.sql.CLOB) map.get("BIGTEXT");
        Reader reader = clobRef.getCharacterStream();
        char[] charArray = new char[10000];
        int readLength = reader.read(charArray);
        while (readLength != -1) {
            String newString = new String(charArray, 0, readLength);
            System.out.println(newString);
            readLength = reader.read(charArray);
        }
        reader.close();
        session.close();
    }
}
```

程序运行后在控制台打印出了全部信息。

插入新的 CLOB 记录的代码如下：

```java
public class Insert {
    public static void main(String[] args) throws IOException {
        SqlSession session = GetSqlSession.getSqlSession();
        BigTextMapper mapper = session.getMapper(BigTextMapper.class);
        BigText bigtext = mapper.selectById1();
        mapper.insertUserinfo(bigtext);
        session.commit();
        session.close();

    }
}
```

在数据表中插入了另一条大文本记录。

2.12　处理分页

想要实现分页功能，就要先算出起始位置，起始位置的算法如下：

起始位置=（目标到达的页数-1）×一页显示记录的条数

示例代码如下：

```java
public class SelectAll1 {
    public static void main(String[] args) throws IOException {
        String gotoPage = "4";
        int gotoPageInt = 1;
        try {
            gotoPageInt = Integer.parseInt(gotoPage);
            if (gotoPageInt <= 0) {
                gotoPageInt = 1;
            }
        } catch (NumberFormatException e) {
            gotoPageInt = 1;
        }
```

```
            System.out.println(gotoPageInt);
        }
    }
```

2.12.1 使用 SqlSession 对象对查询的数据进行分页

创建名称为 pageTest 的项目。

SQL 映射文件的配置代码如下：

```
<mapper namespace="mapping.UserinfoMapper">
    <select id="selectAll" resultType="entity.Userinfo">
        select * from userinfo order
        by id asc
    </select>
</mapper>
```

使用 MyBatis 实现分页功能的代码如下：

```
public class SelectAll2 {
    public static void main(String[] args) throws IOException {

        String gotoPage = "1";
        int gotoPageInt = 1;
        try {
            gotoPageInt = Integer.parseInt(gotoPage);
            if (gotoPageInt <= 0) {
                gotoPageInt = 1;
            }
        } catch (NumberFormatException e) {
            gotoPageInt = 1;
        }
        int pageSize = 2;
        int beginPosition = (gotoPageInt - 1) * pageSize;

        SqlSession session = GetSqlSession.getSqlSession();
        List<Userinfo> listUserinfo = session.selectList("selectAll", null, new
RowBounds(beginPosition, pageSize));
        for (int i = 0; i < listUserinfo.size(); i++) {
            Userinfo userinfo = listUserinfo.get(i);
            System.out.println(userinfo.getId() + " " + userinfo.getUsername() + " "
+ userinfo.getPassword() + " "
                    + userinfo.getAge() + " " + userinfo.getInsertdate());
        }
        session.commit();
        session.close();
    }
}
```

2.12.2 使用 Mapper 接口对查询的数据进行分页

创建名称为 InterfaceMapping_RowBoundsTest 的项目。

SQL 映射文件的配置代码如下：

```
<mapper namespace="mapping.IUserinfoMapping">
    <select id="getUserinfo" resultType="orm.Userinfo">
        select * from userinfo
```

```
        where
        username = #{username} order by id asc
    </select>
</mapper>
```

Mapper 接口代码如下：

```
public interface IUserinfoMapping {
    public List<Userinfo> getUserinfo(String username, RowBounds rowBounds);
}
```

使用 MyBatis 实现分页功能的代码如下：

```
public class GetUserinfo {
    public static void main(String[] args) {
        try {
            SqlSession sqlSession = GetSqlSession.getSqlSession();
            IUserinfoMapping mapping = sqlSession.getMapper(IUserinfoMapping.class);

            List<Userinfo> listUserinfo = mapping.getUserinfo("中国", new RowBounds(1, 2));
            for (int i = 0; i < listUserinfo.size(); i++) {
                Userinfo userinfo = listUserinfo.get(i);
                System.out.println(userinfo.getId() + " " + userinfo.getUsername() +
" " + userinfo.getPassword());
            }

            sqlSession.commit();
            sqlSession.close();
        } catch (Exception e) {
            e.printStackTrace();
        }
    }
}
```

但 MyBatis 提供的分页功能在执行效率上是比较低的。它的算法是先将数据表中符合条件的全部记录放入内存，然后在内存中进行分页，造成内存占用率较高。推荐使用第三方的 MyBatis 分页插件 PageHelper 来优化分页执行的效率。

2.13 实现批处理

创建名称为 BatchTest 的项目。

在获得 SqlSession 时传入参数来决定是否具有批处理特性，代码如下：

```
public class GetSqlSession {
    public static SqlSession getSqlSession() throws IOException {
        String configFile = "mybatis-config.xml";
        InputStream configStream = Resources.getResourceAsStream(configFile);
        SqlSessionFactoryBuilder builder = new SqlSessionFactoryBuilder();
        SqlSessionFactory factory = builder.build(configStream);
        return factory.openSession(ExecutorType.BATCH);
    }
}
```

使用批处理与未使用批处理技术向数据库插入 100000 条记录。从运行时间上来看，还是使用批处理技术在执行速度上更快一点。

如果在 Oracle 数据库中对序列进行池化，那么两者的执行速度还会大幅提升。如果使用固态硬盘，那么运行速度差距较小；如果使用机械硬盘，那么时间差距较大。

2.14　实现一对一级联

什么是级联？级联就是 A 实体类中有 B 实体类的属性，这种情况是一对一级联，代码结构如图 2-6 所示。

在取得 A 对象时就可以取得与 A 对象关联的 B 对象。

当 A 实体类中有 B 实体类的集合属性时，这是一对多级联，代码结构如图 2-7 所示。

A.java

```java
package test1;

public class A {
    private B b = new B();
}
```

B.java

```java
package test1;

public class B {

}
```

图 2-6　一对一级联

A.java

```java
package test2;

import java.util.ArrayList;
import java.util.List;

public class A {
    private List<B> listB = new ArrayList<>();
}
```

B.java

```java
package test2;

public class B {

}
```

图 2-7　一对多级联

MyBatis 支持一对一级联和一对多级联操作。

2.14.1　数据表结构和内容以及关系

先来测试一下一对一级联，创建测试用的项目 one_one。

项目中使用的数据表、数据表内容，以及关联关系如图 2-8 所示。

用户信息表

USERINFO:
Primary Key: ID

ID	USERNAME	PASSWORD	AGE	INSERTDATE	CARDIDCOLUMN
1	中国	中国人	333	2000/1/1	101
2	中国	中国人	333	2000/1/1	102
3	中国	中国人	333	2000/1/1	103
4	中国	中国人	333	2000/1/1	104

身份证表

IDCARD:
Primary Key: ID

ID	CARDNO
101	AA
102	BB
103	CC
104	DD

图 2-8　项目中使用的数据表

每个人有一个身份证号，人和身份证号之间属于一对一级联。

2.14.2 创建实体类

创建用户信息实体类，代码如下：

```
public class Userinfo {
    private long id;
    private String username;
    private String password;
    private long age;
    private Date insertdate;
    private IDCard idCard;

    //省略 set 和 get 方法
}
```

创建身份证号实体类，代码如下：

```
public class IDCard {
    private long id;
    private String cardNo;

    //省略 set 和 get 方法
}
```

2.14.3 创建 SQL 映射文件

创建身份证号对应的 SQL 映射文件，代码如下：

```
<mapper namespace="mybatis.testcurd">
    <select id="selectIDCardById" parameterType="long"
        resultType="entity.IDCard">
        select * from idcard where id=#{id}
    </select>
</mapper>
```

创建用户信息对应的 SQL 映射文件，代码如下：

```
<mapper namespace="mybatis.testcurd">
    <resultMap type="entity.Userinfo" id="userinfoMap">
        <result column="username" property="username" />
        <result column="password" property="password" />
        <result column="age" property="age" />
        <result column="insertdate" property="insertdate" />
        <association property="idCard" column="CARDIDCOLUMN"
            select="mybatis.testcurd.selectIDCardById"></association>
    </resultMap>

    <select id="getUserinfoById" parameterType="long"
        resultMap="userinfoMap">
        select * from
        userinfo where id=#{id}
    </select>
    <select id="getAllUserinfo" resultMap="userinfoMap">
        select * from userinfo
```

```
        order by id asc
    </select>
</mapper>
```

2.14.4　级联解析

实现一对一级联的映射关系如图 2-9 所示。

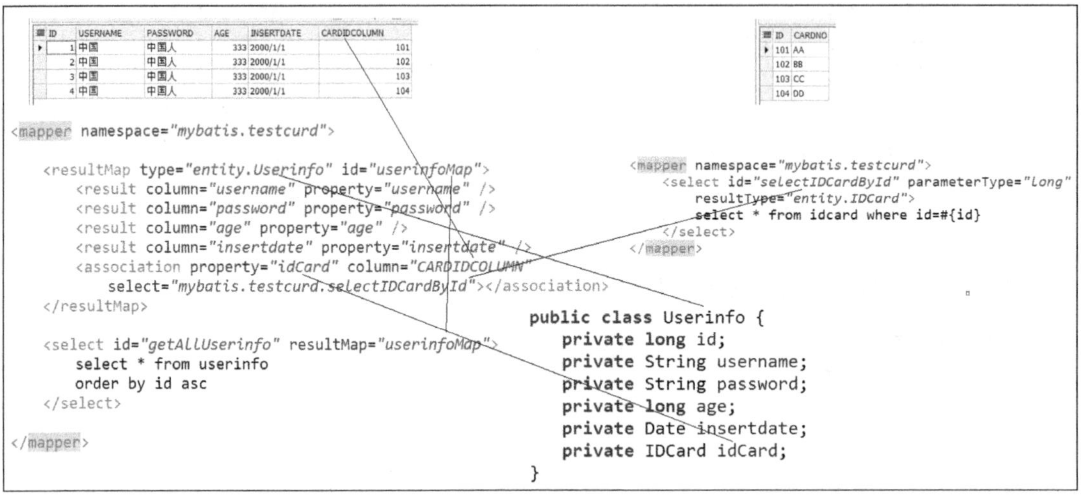

ID	USERNAME	PASSWORD	AGE	INSERTDATE	CARDIDCOLUMN
1	中国	中国人	333	2000/1/1	101
2	中国	中国人	333	2000/1/1	102
3	中国	中国人	333	2000/1/1	103
4	中国	中国人	333	2000/1/1	104

ID	CARDNO
101	AA
102	BB
103	CC
104	DD

```
<mapper namespace="mybatis.testcurd">

    <resultMap type="entity.Userinfo" id="userinfoMap">
        <result column="username" property="username" />
        <result column="password" property="password" />
        <result column="age" property="age" />
        <result column="insertdate" property="insertdate" />
        <association property="idCard" column="CARDIDCOLUMN"
            select="mybatis.testcurd.selectIDCardById"></association>
    </resultMap>

    <select id="getAllUserinfo" resultMap="userinfoMap">
        select * from userinfo
        order by id asc
    </select>

</mapper>
```

```
<mapper namespace="mybatis.testcurd">
    <select id="selectIDCardById" parameterType="Long"
        resultType="entity.IDCard">
        select * from idcard where id=#{id}
    </select>
</mapper>
```

```
public class Userinfo {
    private long id;
    private String username;
    private String password;
    private long age;
    private Date insertdate;
    private IDCard idCard;
}
```

图 2-9　级联原理

2.14.5　根据 ID 查询记录

创建运行类，代码如下：

```
public class SelectById {
    public static void main(String[] args) throws IOException {
        SqlSession session = DBTools.getSqlSession();
        Userinfo userinfo = session.selectOne("getUserinfoById", 3L);
        System.out.println(userinfo.getId() + " " + userinfo.getUsername() + " " +
userinfo.getPassword() + " "
                + userinfo.getAge() + " " + userinfo.getInsertdate() + " " + userinfo
.getIdCard().getId() + " "
                + userinfo.getIdCard().getCardNo());
        session.commit();
        session.close();
    }
}
```

程序运行后控制台输出的结果如下：

```
[main] DEBUG [mybatis.testcurd.getUserinfoById] - ==>  Preparing: select * from
userinfo where id=?
[main] DEBUG [mybatis.testcurd.getUserinfoById] - ==> Parameters: 3(Long)
[main] DEBUG [mybatis.testcurd.selectIDCardById] - ====>  Preparing: select * from
idcard where id=?
[main] DEBUG [mybatis.testcurd.selectIDCardById] - ====> Parameters: 103(Long)
[main] DEBUG [mybatis.testcurd.selectIDCardById] - <====      Total: 1
[main] DEBUG [mybatis.testcurd.getUserinfoById] - <==      Total: 1
3 中国 中国人 333 Sat Jan 01 00:00:00 CST 2000 103 CC
```

从控制台输出的日志信息来看，使用的是两个 SQL 语句，那么查询更多的记录会不会出现执行的 SQL 语句次数增长呢？继续测试。

2.14.6 查询所有记录

创建运行类，代码如下：

```
public class SelectAll {
    public static void main(String[] args) throws IOException {
        SqlSession session = DBTools.getSqlSession();
        List<Userinfo> listUserinfo = session.selectList("getAllUserinfo");
        for (int i = 0; i < listUserinfo.size(); i++) {
            Userinfo userinfo = listUserinfo.get(i);
            System.out.println(userinfo.getId() + " " + userinfo.getUsername() + " "
+ userinfo.getPassword() + " "
                    + userinfo.getAge() + " " + userinfo.getInsertdate() + " " +
userinfo.getIdCard().getId() + " "
                    + userinfo.getIdCard().getCardNo());
        }
        session.commit();
        session.close();
    }
}
```

程序运行后控制台输出的结果如下：

```
[main] DEBUG [mybatis.testcurd.getAllUserinfo] - ==>  Preparing: select * from
userinfo order by id asc
[main] DEBUG [mybatis.testcurd.getAllUserinfo] - ==> Parameters:
[main] DEBUG [mybatis.testcurd.selectIDCardById] - ====>  Preparing: select * from
idcard where id=?
[main] DEBUG [mybatis.testcurd.selectIDCardById] - ====> Parameters: 101(Long)
[main] DEBUG [mybatis.testcurd.selectIDCardById] - <====      Total: 1
[main] DEBUG [mybatis.testcurd.selectIDCardById] - ====>  Preparing: select * from
idcard where id=?
[main] DEBUG [mybatis.testcurd.selectIDCardById] - ====> Parameters: 102(Long)
[main] DEBUG [mybatis.testcurd.selectIDCardById] - <====      Total: 1
[main] DEBUG [mybatis.testcurd.selectIDCardById] - ====>  Preparing: select * from
idcard where id=?
[main] DEBUG [mybatis.testcurd.selectIDCardById] - ====> Parameters: 103(Long)
[main] DEBUG [mybatis.testcurd.selectIDCardById] - <====      Total: 1
[main] DEBUG [mybatis.testcurd.selectIDCardById] - ====>  Preparing: select * from
idcard where id=?
[main] DEBUG [mybatis.testcurd.selectIDCardById] - ====> Parameters: 104(Long)
[main] DEBUG [mybatis.testcurd.selectIDCardById] - <====      Total: 1
[main] DEBUG [mybatis.testcurd.getAllUserinfo] - <==      Total: 4
1 中国 中国人 333 Sat Jan 01 00:00:00 CST 2000 101 AA
2 中国 中国人 333 Sat Jan 01 00:00:00 CST 2000 102 BB
3 中国 中国人 333 Sat Jan 01 00:00:00 CST 2000 103 CC
4 中国 中国人 333 Sat Jan 01 00:00:00 CST 2000 104 DD
```

果然，从控制台输出的日志信息来看，使用更多的 SQL 语句来进行查询，会造成客户端与数据库的通信次数过多，降低运行效率，来看一下优化的代码。

2.14.7 对 SQL 语句执行次数进行优化

创建 one_one2 项目。

创建用户信息对应的 SQL 映射文件，代码如下：

```
<mapper namespace="mybatis.testcurd">

    <resultMap type="entity.Userinfo" id="userinfoMap">
        <result column="id" property="id" />
        <result column="username" property="username" />
        <result column="password" property="password" />
        <result column="age" property="age" />
        <result column="insertdate" property="insertdate" />
        <association property="idCard" javaType="entity.IDCard">
            <result column="cardid" property="id" />
            <result column="cardNo" property="cardNo" />
        </association>
    </resultMap>

    <select id="getAllUserinfo" resultMap="userinfoMap">
        select u.*,card.id
        cardid,card.cardNo
        from userinfo
        u,idcard card where
        u.cardidcolumn=card.id
        order by u.id
        asc
    </select>

    <select id="getUserinfoById" parameterType="long"
        resultMap="userinfoMap">
        select u.*,card.id cardid,card.cardNo
        from userinfo
        u,idcard
        card where
        u.cardidcolumn=card.id and u.id=#{id}
        order by u.id asc
    </select>
</mapper>
```

创建运行类，代码如下：

```
public class SelectById {
    public static void main(String[] args) throws IOException {
        SqlSession session = DBTools.getSqlSession();
        Userinfo userinfo = session.selectOne("getUserinfoById", 3L);
        System.out.println(userinfo.getId() + " " + userinfo.getUsername() + " " +
userinfo.getPassword() + " "
                + userinfo.getAge() + " " + userinfo.getInsertdate() + " " + userinfo
.getIdCard().getId() + " "
                + userinfo.getIdCard().getCardNo());
        session.commit();
        session.close();
    }
}
```

程序运行后控制台输出的结果如下：

```
[main] DEBUG [mybatis.testcurd.getUserinfoById] - ==>  Preparing: select u.*,card.id cardid,card.cardNo from userinfo u,idcard card where u.cardidcolumn=card.id and u.id=? order by u.id asc
[main] DEBUG [mybatis.testcurd.getUserinfoById] - ==> Parameters: 3(Long)
[main] DEBUG [mybatis.testcurd.getUserinfoById] - <==      Total: 1
3 中国 中国人 333 Sat Jan 01 00:00:00 CST 2000 103 CC
```

使用 1 个 SQL 语句进行查询，提高了运行效率。

如果查询全部信息，也会使用 1 个 SQL 语句，运行类代码如下：

```java
public class SelectAll {
    public static void main(String[] args) throws IOException {
        SqlSession session = DBTools.getSqlSession();
        List<Userinfo> listUserinfo = session.selectList("getAllUserinfo");
        for (int i = 0; i < listUserinfo.size(); i++) {
            Userinfo userinfo = listUserinfo.get(i);
            System.out.println(userinfo.getId() + " " + userinfo.getUsername() + " " + userinfo.getPassword() + " "
                    + userinfo.getAge() + " " + userinfo.getInsertdate() + " " + userinfo.getIdCard().getId() + " "
                    + userinfo.getIdCard().getCardNo());
        }
        session.commit();
        session.close();
    }
}
```

程序运行后控制台输出的结果如下：

```
[main] DEBUG [mybatis.testcurd.getAllUserinfo] - ==>  Preparing: select u.*,card.id cardid, card.cardNo from userinfo u,idcard card where u.cardidcolumn=card.id order by u.id asc
[main] DEBUG [mybatis.testcurd.getAllUserinfo] - ==> Parameters: 
[main] DEBUG [mybatis.testcurd.getAllUserinfo] - <==      Total: 4
1 中国 中国人 333 Sat Jan 01 00:00:00 CST 2000 101 AA
2 中国 中国人 333 Sat Jan 01 00:00:00 CST 2000 102 BB
3 中国 中国人 333 Sat Jan 01 00:00:00 CST 2000 103 CC
4 中国 中国人 333 Sat Jan 01 00:00:00 CST 2000 104 DD
```

也是使用 1 个 SQL 语句进行查询，运行效率得到了提高。

2.15 实现一对多级联

MyBatis 支持一对多级联。接下来创建测试用的项目 one_more。

2.15.1 数据表的结构、内容以及关系

项目中使用的数据表结构、内容，以及关联关系如图 2-10 所示。

每一个省有多个市，省和市之间属于一对多级联。

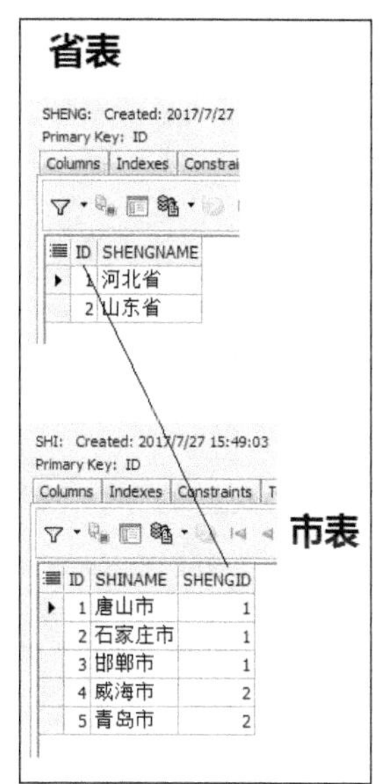

图 2-10 项目中使用的数据表

2.15.2 创建实体类

创建“Sheng”实体类，代码如下：

```
public class Sheng {
    private long id;
    private String shengname;
    private List<Shi> shiList;

    //省略 get 和 set 方法
}
```

创建“Shi”实体类，代码如下：

```
public class Shi {
    private long id;
    private String shiname;
    private long shengid;

    //省略 get 和 set 方法
}
```

2.15.3 创建 SQL 映射文件

创建省对应的 SQL 映射文件，代码如下：

```
<mapper namespace="mybatis.testcurd">
    <resultMap type="entity.Sheng" id="shengMap">
        <result property="id" column="id" />
        <result property="shengname" column="shengname" />
        <collection property="shiList"
            select="mybatis.testcurd.getShiByShengId" column="id"></collection>
    </resultMap>

    <select id="getAllSheng" resultMap="shengMap">
        select * from sheng order by
        id asc
    </select>

    <select id="getShengById" parameterType="long"
        resultMap="shengMap">
        select * from
        userinfo where id=#{id}
    </select>
</mapper>
```

创建市对应的 SQL 映射文件，代码如下：

```
<mapper namespace="mybatis.testcurd">
    <select id="getShiByShengId" parameterType="long"
        resultType="entity.Shi">
        select * from
        shi where shengid=#{id}
    </select>
</mapper>
```

2.15.4　级联解析

实现一对多级联的映射关系，如图 2-11 所示。

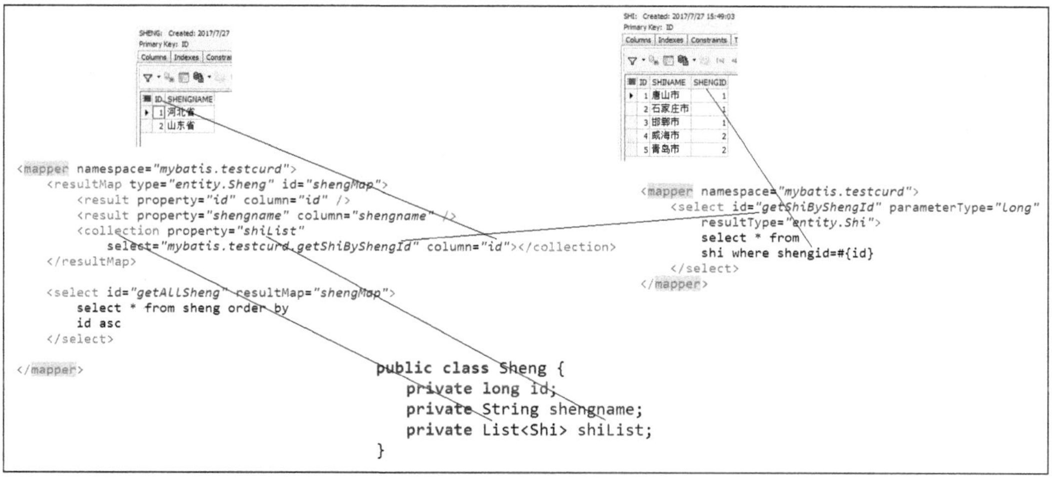

ID	SHENGNAME
1	河北省
2	山东省

ID	SHINAME	SHENGID
1	唐山市	1
2	石家庄市	1
3	邯郸市	1
4	威海市	2
5	青岛市	2

```
<mapper namespace="mybatis.testcurd">
    <resultMap type="entity.Sheng" id="shengMap">
        <result property="id" column="id" />
        <result property="shengname" column="shengname" />
        <collection property="shiList"
            select="mybatis.testcurd.getShiByShengId" column="id"></collection>
    </resultMap>

    <select id="getAllSheng" resultMap="shengMap">
        select * from sheng order by
        id asc
    </select>

</mapper>
```

```
<mapper namespace="mybatis.testcurd">
    <select id="getShiByShengId" parameterType="Long"
        resultType="entity.Shi">
        select * from
        shi where shengid=#{id}
    </select>
</mapper>
```

```
public class Sheng {
    private long id;
    private String shengname;
    private List<Shi> shiList;
}
```

图 2-11　级联原理

2.15.5　根据 ID 查询记录

创建运行类，代码如下：

```
public class SelectById {
    public static void main(String[] args) throws IOException {
        SqlSession session = DBTools.getSqlSession();
        Sheng sheng = session.selectOne("getShengById", 1L);
        System.out.println(sheng.getId() + " " + sheng.getShengname());
        List<Shi> listShi = sheng.getShiList();
        for (int j = 0; j < listShi.size(); j++) {
            Shi shi = listShi.get(j);
            System.out.println(" " + shi.getId() + " " + shi.getShiname());
        }
        session.commit();
        session.close();
    }
}
```

程序运行后控制台输出的结果如下：

```
[main] DEBUG [mybatis.testcurd.getShengById] - ==>  Preparing: select * from
userinfo where id=?
[main] DEBUG [mybatis.testcurd.getShengById] - ==> Parameters: 1(Long)
[main] DEBUG [mybatis.testcurd.getShiByShengId] - ====>  Preparing: select * from
shi where shengid=?
[main] DEBUG [mybatis.testcurd.getShiByShengId] - ====> Parameters: 1(Long)
[main] DEBUG [mybatis.testcurd.getShiByShengId] - <====      Total: 3
[main] DEBUG [mybatis.testcurd.getShengById] - <==      Total: 1
1 null
 1 唐山市
```

```
2 石家庄市
3 邯郸市
```

这里使用 2 个 SQL 语句进行查询。

2.15.6 查询所有记录

创建运行类，代码如下：

```
public class SelectAll {
    public static void main(String[] args) throws IOException {
        SqlSession session = DBTools.getSqlSession();
        List<Sheng> listSheng = session.selectList("getAllSheng");
        for (int i = 0; i < listSheng.size(); i++) {
            Sheng sheng = listSheng.get(i);
            System.out.println(sheng.getId() + " " + sheng.getShengname());
            List<Shi> listShi = sheng.getShiList();
            for (int j = 0; j < listShi.size(); j++) {
                Shi shi = listShi.get(j);
                System.out.println(" " + shi.getId() + " " + shi.getShiname());
            }
        }

        session.commit();
        session.close();
    }
}
```

程序运行后控制台输出的结果如下：

```
    [main] DEBUG [mybatis.testcurd.getAllSheng] - ==>  Preparing: select * from sheng
order by id asc
    [main] DEBUG [mybatis.testcurd.getAllSheng] - ==> Parameters:
    [main] DEBUG [mybatis.testcurd.getShiByShengId] - ====>  Preparing: select * from
shi where shengid=?
    [main] DEBUG [mybatis.testcurd.getShiByShengId] - ====> Parameters: 1(Long)
    [main] DEBUG [mybatis.testcurd.getShiByShengId] - <====      Total: 3
    [main] DEBUG [mybatis.testcurd.getShiByShengId] - ====>  Preparing: select * from
shi where shengid=?
    [main] DEBUG [mybatis.testcurd.getShiByShengId] - ====> Parameters: 2(Long)
    [main] DEBUG [mybatis.testcurd.getShiByShengId] - <====      Total: 2
    [main] DEBUG [mybatis.testcurd.getAllSheng] - <==      Total: 2
    1 河北省
     1 唐山市
     2 石家庄市
     3 邯郸市
    2 山东省
     4 威海市
     5 青岛市
```

这里使用 3 个 SQL 进行查询。

前面两个实验分别使用 2 个和 3 个 SQL 语句进行查询，其实还可以进一步优化，使 SQL 语句执行的次数变得更少。

2.15.7 对 SQL 语句的执行次数进行优化

创建 one_more2 项目。

创建省对应的 SQL 映射文件，代码如下：

```
<mapper namespace="mybatis.testcurd">
    <resultMap type="entity.Sheng" id="shengMap">
        <result property="id" column="id" />
        <result property="shengname" column="shengname" />
        <collection property="shiList" ofType="entity.Shi">
            <result property="id" column="shiId" />
            <result property="shiname" column="shiname" />
        </collection>
    </resultMap>

    <select id="getAllSheng" resultMap="shengMap">
        select sheng.*,shi.id
        shiId,shi.shiname from sheng,shi where sheng.id=shi.shengid
        order by
        sheng.id
        asc
    </select>

    <select id="getShengById" parameterType="long"
        resultMap="shengMap">
        select sheng.*,shi.id
        shiId,shi.shiname from sheng,shi where
        sheng.id=shi.shengid and sheng.id=#{id}
    </select>
</mapper>
```

以下配置代码中的 ofType 属性代表 List 中存储的是 Shi 对象：

```
<collection property="shiList" ofType="entity.Shi">
```

创建运行类，代码如下：

```
public class SelectById {
    public static void main(String[] args) throws IOException {
        SqlSession session = DBTools.getSqlSession();
        Sheng sheng = session.selectOne("getShengById", 1L);
        System.out.println(sheng.getId() + " " + sheng.getShengname());
        List<Shi> listShi = sheng.getShiList();
        for (int j = 0; j < listShi.size(); j++) {
            Shi shi = listShi.get(j);
            System.out.println(" " + shi.getId() + " " + shi.getShiname());
        }
        session.commit();
        session.close();
    }
}
```

程序运行后控制台输出的结果如下：

```
[main] DEBUG [mybatis.testcurd.getShengById] - ==>  Preparing: select sheng.*,shi.id
shiId,shi.shiname from sheng,shi where sheng.id=shi.shengid and sheng.id=?
```

```
[main] DEBUG [mybatis.testcurd.getShengById] - ==> Parameters: 1(Long)
[main] DEBUG [mybatis.testcurd.getShengById] - <==      Total: 3
1 河北省
 1 唐山市
 2 石家庄市
 3 邯郸市
```

使用 1 个 SQL 语句进行查询，提高了运行效率。

如果查询全部信息，也可以使用 1 个 SQL 语句，运行类代码如下：

```
public class SelectAll {
    public static void main(String[] args) throws IOException {
        SqlSession session = DBTools.getSqlSession();
        List<Sheng> listSheng = session.selectList("getAllSheng");
        for (int i = 0; i < listSheng.size(); i++) {
            Sheng sheng = listSheng.get(i);
            System.out.println(sheng.getId() + " " + sheng.getShengname());
            List<Shi> listShi = sheng.getShiList();
            for (int j = 0; j < listShi.size(); j++) {
                Shi shi = listShi.get(j);
                System.out.println(" " + shi.getId() + " " + shi.getShiname());
            }
        }

        session.commit();
        session.close();
    }
}
```

程序运行后控制台输出的结果如下：

```
[main] DEBUG [mybatis.testcurd.getAllSheng] - ==>  Preparing: select sheng.*,shi.id
shiId,shi.shiname from sheng,shi where sheng.id=shi.shengid order by sheng.id asc
[main] DEBUG [mybatis.testcurd.getAllSheng] - ==> Parameters:
[main] DEBUG [mybatis.testcurd.getAllSheng] - <==      Total: 5
1 河北省
 1 唐山市
 2 石家庄市
 3 邯郸市
2 山东省
 4 威海市
 5 青岛市
```

这里也是使用 1 个 SQL 语句进行查询，但运行效率得到了提高。

2.16 延迟加载

什么是延迟加载？延迟加载的反义词就是立即加载，立即加载是指在取得“省”对象时，省中所有的“市”对象都已经提取到内存中，因此其非常耗费 CPU 和内存资源。有时只需要省信息，而并不需要获得市信息，但立即加载还是把市信息也一同提取出来，在这种情况下就得使用延迟加载了。延迟加载就是在使用到市对象时，再把其加载到内存中，如果不使用市对象，就不加载到内存中，这就是延迟加载特性。

2.16.1 默认立即加载策略

MyBatis 默认的行为是立即加载。

将 one_one 项目和 one_more 项目中的资源复制到新建的 lazyTest 项目中，以进行延迟加载特性的测试。

创建测试类，代码如下：

```
public class SelectAllUserinfo_OneOne {
    public static void main(String[] args) throws IOException {
        SqlSession session = DBTools.getSqlSession();
        List<Userinfo> listUserinfo = session.selectList("getAllUserinfo");
        for (int i = 0; i < listUserinfo.size(); i++) {
            Userinfo userinfo = listUserinfo.get(i);
            System.out.println(userinfo.getId() + " " + userinfo.getUsername() + " " +
userinfo.getPassword() + " "
                    + userinfo.getAge() + " " + userinfo.getInsertdate() + " " +
userinfo.getIdCard().getId() + " "
                    + userinfo.getIdCard().getCardNo());
        }
        session.commit();
        session.close();
    }
}
```

程序运行后在 for 语句处设置断点，调试效果如图 2-12 所示。

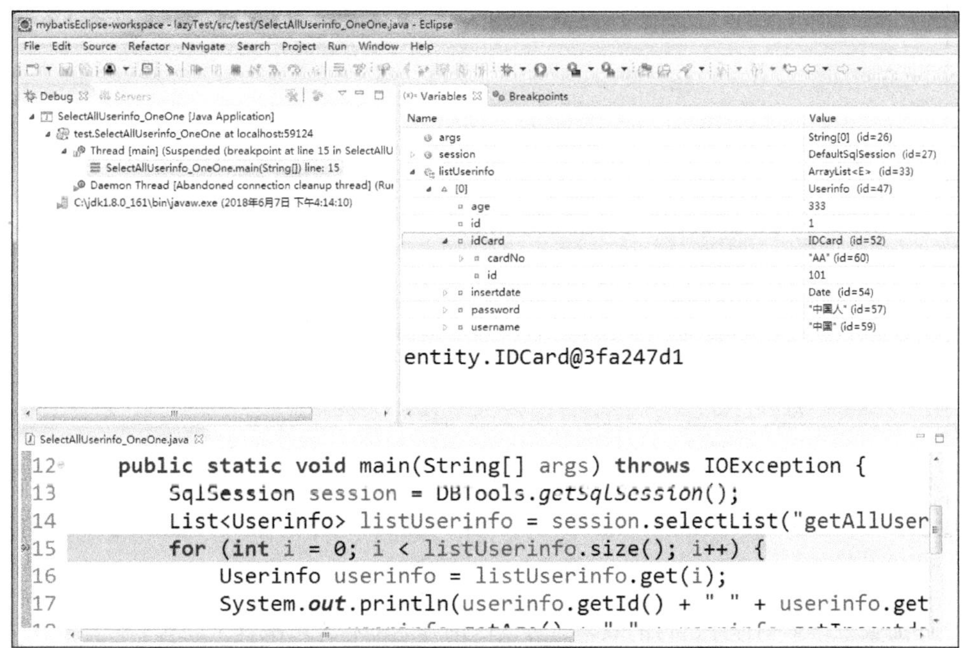

图 2-12 开始调试

执行 selectList()方法后，List 中 Userinfo 对象的 idCard 属性值并不是 Null，说明在取得 Userinfo 对象时，采用立即加载策略，立即将 Userinfo 对象对应的 IDCard 对象加载到内存中，在一对多级联时，默认的情况下也是立即加载，创建测试，代码如下：

```
public class SelectAllSheng_OneMore {
    public static void main(String[] args) throws IOException {
        SqlSession session = DBTools.getSqlSession();
        List<Sheng> listSheng = session.selectList("getAllSheng");
        for (int i = 0; i < listSheng.size(); i++) {
            Sheng sheng = listSheng.get(i);
            System.out.println(sheng.getId() + " " + sheng.getShengname());
            List<Shi> listShi = sheng.getShiList();
            for (int j = 0; j < listShi.size(); j++) {
                Shi shi = listShi.get(j);
                System.out.println(" " + shi.getId() + " " + shi.getShiname());
            }
        }
        session.commit();
        session.close();
    }
}
```

还是在 for 语句处设置断点，效果如图 2-13 所示。从调试信息中可以发现，List 中 Sheng 对象中的 shiList 集合中已经存在 Shi 对象，说明在获得 Sheng 对象时，采用立即加载策略，将 Sheng 对象对应的 Shi 集合放入内存。

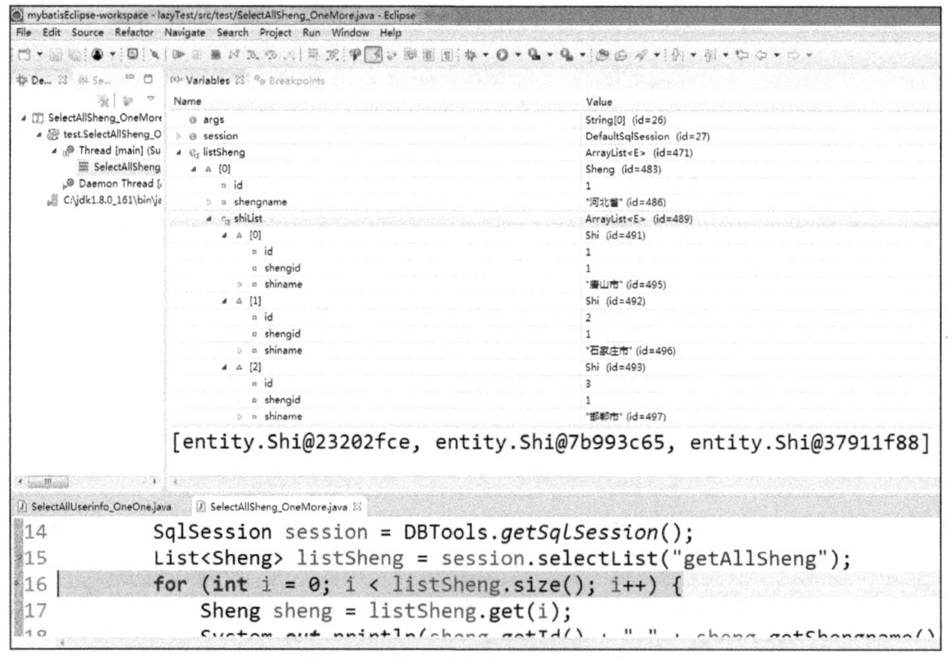

图 2-13 开始调试

立即加载策略影响程序运行的效率，可以使用延迟加载来解决这个问题。

2.16.2 使用全局延迟加载策略与两种加载方式

在配置文件中使用如下配置代码开始延迟加载：

```
<settings>
    <setting name="logImpl" value="LOG4J" />
    <setting name="lazyLoadingEnabled" value="true" />
</settings>
```

但延迟加载具体的实现方式分为两种。

（1）获得子表时再进行加载。

（2）打印主表信息时加载子表。

这两种延迟加载的方式可以使用以下配置来进行切换，默认情况下是第 1 种：

```
<settings>
    <setting name="aggressiveLazyLoading" value="" />
</settings>
```

更改配置文件，代码如下：

```
<settings>
    <setting name="logImpl" value="LOG4J" />
    <setting name="lazyLoadingEnabled" value="true" />
</settings>
```

对 SelectAllSheng_OneMore.java 类进行调试，代码如下：

```
public class SelectAllSheng_OneMore {
    public static void main(String[] args) throws IOException {
        SqlSession session = DBTools.getSqlSession();
        List<Sheng> listSheng = session.selectList("getAllSheng");
        for (int i = 0; i < listSheng.size(); i++) {
            Sheng sheng = listSheng.get(i);
            System.out.println(sheng.getId() + " " + sheng.getShengname());
            List<Shi> listShi = sheng.getShiList();
            for (int j = 0; j < listShi.size(); j++) {
                Shi shi = listShi.get(j);
                System.out.println(" " + shi.getId() + " " + shi.getShiname());
            }
        }
        session.commit();
        session.close();
    }
}
```

注意：一定要使用 Variables 面板查看对象属性值，不要用鼠标悬停的方式来查看变量值。另外，要在执行 SQL 语句次数多的项目中进行实验，才可以出现延迟加载的效果。

从调试中可以发现，直到执行以下代码 Shi 集合才被加载到内存中：

```
List<Shi> listShi = sheng.getShiList();
```

这个特性和以下配置代码效果是一样的：

```
<settings>
    <setting name="logImpl" value="LOG4J" />
    <setting name="lazyLoadingEnabled" value="true" />
    <setting name="aggressiveLazyLoading" value="false" />
</settings>
```

也就是说，配置 aggressiveLazyLoading（侵犯延迟加载）的默认值就是 false，意思是不侵犯延迟加载，保持执行 getSub 取得子表时再加载子表数据的特性。

如果使用以下配置：

```
<settings>
    <setting name="logImpl" value="LOG4J" />
    <setting name="lazyLoadingEnabled" value="true" />
    <setting name="aggressiveLazyLoading" value="true" />
</settings>
```

那么在调试的过程中，执行以下代码就可将 Shi 集合加载到内存中：

```
System.out.println(sheng.getId() + " " + sheng.getShengname());
```

2.16.3 使用 fetchType 属性设置局部加载策略

使用配置选项 lazyLoadingEnabled 和 aggressiveLazyLoading 设置的是全局加载行为。如果想针对某一个映射来决定是延迟加载还是立即加载，那么可以在<collection>配置中使用 fetchType 属性来决定，属性值 eager 代表立即加载，属性值 lazy 代表延迟加载。

以下配置文件代码开启延迟加载：

```
<settings>
    <setting name="logImpl" value="LOG4J" />
    <setting name="lazyLoadingEnabled" value="true" />
</settings>
```

映射文件 sheng.xml 配置代码取得 Shi 对象时开启立即加载：

```
<resultMap type="entity.Sheng" id="shengMap">
    <result property="id" column="id" />
    <result property="shengname" column="shengname" />
    <collection property="shiList"
        select="mybatis.testcurd.getShiByShengId" column="id"
        fetchType="eager"></collection>
</resultMap>
```

对 SelectAllSheng_OneMore.java 进行调试，执行以下代码后，List 中就存在 Shi 对象了，说明可以自定义某一个映射立即加载。

```
List<Sheng> listSheng = session.selectList("getAllSheng");
```

2.17 缓存的使用

用户可以将查询到的实体类放到缓存中，使后面的 select 查询得以复用，这会提高程序运行效率。

MyBatis 缓存分为一级缓存和二级缓存。

（1）一级缓存由 SqlSession 对象管理，每个 SqlSession 有自己所属的一级缓存。

（2）二级缓存由 SqlSessionFactory 对象管理，是 Application 级别。

2.17.1 一级缓存

一级缓存是 MyBatis 默认提供并开启的，每个 SqlSession 都有自己所属的一级缓存。

创建测试用的项目 cacheTest1。

SQL 映射文件的代码如下：

```
<mapper namespace="mybatis.testcurd">
    <select id="getUserinfoById" parameterType="long"
        resultType="entity.Userinfo">
        select * from
        userinfo where id=#{id}
    </select>
</mapper>
```

运行类代码如下：

```
public class SelectUserinfoById1 {
    public static void main(String[] args) throws IOException {
        String configFile = "mybatis-config.xml";
        InputStream configStream = Resources.getResourceAsStream(configFile);
        SqlSessionFactoryBuilder builder = new SqlSessionFactoryBuilder();
        SqlSessionFactory factory = builder.build(configStream);
        SqlSession session = factory.openSession();

        Userinfo userinfo1 = session.selectOne("getUserinfoById", 4L);
        System.out.println(userinfo1.getId() + " " + userinfo1.getUsername() + " " +
userinfo1.getPassword() + " "
                + userinfo1.getAge() + " " + userinfo1.getInsertdate());

        Userinfo userinfo2 = session.selectOne("getUserinfoById", 4L);
        System.out.println(userinfo2.getId() + " " + userinfo2.getUsername() + " " +
userinfo2.getPassword() + " "
                + userinfo2.getAge() + " " + userinfo2.getInsertdate());

        session.commit();
        session.close();
    }
}
```

程序运行的结果如下：

```
==>  Preparing: select * from userinfo where id=?
==> Parameters: 4(Long)
<==      Total: 1
4 中国 中国人 333 Sat Jan 01 00:00:00 CST 2000
4 中国 中国人 333 Sat Jan 01 00:00:00 CST 2000
```

说明在同一个 SqlSession 中复用了缓存中的 Userinfo 对象。

但不同 SqlSession 对象缓存中的数据不可以共享，创建如下示例代码：

```
public class SelectUserinfoById2 {
    public static void main(String[] args) throws IOException {
        String configFile = "mybatis-config.xml";
```

```
            InputStream configStream = Resources.getResourceAsStream(configFile);
            SqlSessionFactoryBuilder builder = new SqlSessionFactoryBuilder();
            SqlSessionFactory factory = builder.build(configStream);
            SqlSession session1 = factory.openSession();
            SqlSession session2 = factory.openSession();

            Userinfo userinfo1 = session1.selectOne("getUserinfoById", 4L);
            System.out.println(userinfo1.getId() + " " + userinfo1.getUsername() + " " +
userinfo1.getPassword() + " "
                    + userinfo1.getAge() + " " + userinfo1.getInsertdate());

            Userinfo userinfo2 = session2.selectOne("getUserinfoById", 4L);
            System.out.println(userinfo2.getId() + " " + userinfo2.getUsername() + " " +
userinfo2.getPassword() + " "
                    + userinfo2.getAge() + " " + userinfo2.getInsertdate());

            session1.commit();
            session2.commit();
            session1.close();
            session2.close();
        }
    }
```

程序运行的结果如下：

```
==>  Preparing: select * from userinfo where id=?
==> Parameters: 4(Long)
<==      Total: 1
4 中国 中国人 333 Sat Jan 01 00:00:00 CST 2000
==>  Preparing: select * from userinfo where id=?
==> Parameters: 4(Long)
<==      Total: 1
4 中国 中国人 333 Sat Jan 01 00:00:00 CST 2000
```

如果想实现不同 SqlSession 缓存中的数据可以共享，那么需要使用二级缓存。

2.17.2 二级缓存

创建二级缓存只需要在 SQL 映射文件中添加配置代码：

```
<cache></cache>
```

在 SQL 映射文件中使用了<cache></cache>配置后具有如下两个特性。

（1）当为 select 语句时：flushCache 默认为 false，表示任何时候语句被调用，都不会清空本地缓存和二级缓存，因为查询操作并没有将数据进行更改，不需要刷新缓存中的数据。useCache 属性默认为 true，表示会将当前 select 语句查询的结果放入二级缓存中。

（2）当为 insert、update、delete 语句时：flushCache 默认为 true，表示任何时候语句被调用，都会导致本地缓存和二级缓存被清空，因为数据被更改了。

先来看看二级缓存为 Application 级的特性，创建测试用的项目 cacheTest2。

SQL 映射文件的代码如下：

```
<mapper namespace="mybatis.testcurd">
    <cache></cache>
```

```
    <select id="getUserinfoById" resultType="entity.Userinfo">
        select * from
        userinfo
        where id=1
    </select>

    <update id="updateUserinfoById">
        update userinfo set
        username='a' where id=1
    </update>
</mapper>
```

在 SQL 映射文件中使用<cache></cache>配置开启二级缓存。

实体类 Userinfo 要实现 Serializable 序列化接口，部分代码如下：

```
public class Userinfo implements Serializable {
    private long id;
    private String username;
    private String password;
    private long age;
    private Date insertdate;
}
```

运行类代码如下：

```
public class SelectUserinfoById1 {
    public static void main(String[] args) throws IOException {
        // Userinfo.java 实体类必须实现 Serializable 接口
        String configFile = "mybatis-config.xml";
        InputStream configStream = Resources.getResourceAsStream(configFile);
        SqlSessionFactoryBuilder builder = new SqlSessionFactoryBuilder();
        SqlSessionFactory factory = builder.build(configStream);
        SqlSession session1 = factory.openSession();
        SqlSession session2 = factory.openSession();

        Userinfo userinfo1 = session1.selectOne("getUserinfoById");
        System.out.println(userinfo1.getId() + " " + userinfo1.getUsername() + " " +
userinfo1.getPassword() + " "
                + userinfo1.getAge() + " " + userinfo1.getInsertdate());
        session1.commit();
        session1.close();
        //注意：在查询完 userinfo1 之后要将 sqlsession1 进行提交和关闭
        //目的是将 userinfo1 对象放入二级缓存中，以备让新的 sqlsession2 对象从二级缓存中复用对象

        Userinfo userinfo2 = session2.selectOne("getUserinfoById");
        System.out.println(userinfo2.getId() + " " + userinfo2.getUsername() + " " +
userinfo2.getPassword() + " "
                + userinfo2.getAge() + " " + userinfo2.getInsertdate());
        session2.commit();
        session2.close();
    }
}
```

程序运行的结果如下：

```
Cache Hit Ratio [mybatis.testcurd]: 0.0
==>  Preparing: select * from userinfo where id=1
==> Parameters:
<==      Total: 1
1 中国 中国人 333 Sat Jan 01 00:00:00 CST 2000
Cache Hit Ratio [mybatis.testcurd]: 0.5
1 中国 中国人 333 Sat Jan 01 00:00:00 CST 2000
```

运行结果体现出来的特性就是不同的 SqlSession 可以共享二级缓存中的实体类。

2.17.3 验证 update 语句具有清除二级缓存的特性

使用<cache></cache>配置后，insert、delete 或 update 语句都具有清除二级缓存的特性。下面来看一下执行 update 语句清除二级缓存的方法。

数据表 userinfo 内容如图 2-14 所示。

ID	USERNAME	PASSWORD	AGE	INSERTDATE	CARDIDCOLUMN
1	中国	中国人	333	2000/1/1	101
2	中国	中国人	333	2000/1/1	102
3	中国	中国人	333	2000/1/1	103
4	中国	中国人	333	2000/1/1	104

图 2-14 数据表内容

运行类代码如下：

```
public class SelectUserinfoById2 {
    public static void main(String[] args) throws IOException {
        String configFile = "mybatis-config.xml";
        InputStream configStream = Resources.getResourceAsStream(configFile);
        SqlSessionFactoryBuilder builder = new SqlSessionFactoryBuilder();
        SqlSessionFactory factory = builder.build(configStream);
        SqlSession session = factory.openSession();
        Userinfo userinfo1 = session.selectOne("getUserinfoById");
        System.out.println(userinfo1.getId() + " " + userinfo1.getUsername() + " " +
userinfo1.getPassword() + " "
                + userinfo1.getAge() + " " + userinfo1.getInsertdate());

        session.update("updateUserinfoById");

        Userinfo userinfo2 = session.selectOne("getUserinfoById");
        System.out.println(userinfo2.getId() + " " + userinfo2.getUsername() + " " +
userinfo2.getPassword() + " "
                + userinfo2.getAge() + " " + userinfo2.getInsertdate());
        session.commit();
        session.close();
    }
}
```

程序运行的结果如下：

```
Cache Hit Ratio [mybatis.testcurd]: 0.0
==>  Preparing: select * from userinfo where id=1
==> Parameters:
```

```
<==      Total: 1
1 中国 中国人 333 Sat Jan 01 00:00:00 CST 2000
==>  Preparing: update userinfo set username='a' where id=1
==> Parameters:
<==    Updates: 1
Cache Hit Ratio [mybatis.testcurd]: 0.0
==>  Preparing: select * from userinfo where id=1
==> Parameters:
<==      Total: 1
1 a 中国人 333 Sat Jan 01 00:00:00 CST 2000
```

执行 update 语句后，二级缓存中 id 为 1 的 Userinfo 对象被清除，再执行第 2 个 select 语句时重新发起了 SQL 语句到数据库。

第 3 章　Spring 5 核心技术之 IoC

本章目标：

- 什么是 IoC
- 什么是 IoC 容器
- 什么是 DI
- 反射与 DI 的关系
- 实现装配 JavaBean

3.1　Spring 框架简介

Spring 框架简化了 Java EE 开发的流程，它是为了解决中国级应用开发复杂性问题而创建的，其强大之处在于对 Java EE 开发进行了全方位的简化，对大部分常用的功能进行了封装，比如管理 JavaBean，提供基于 Web 的 MVC 分层框架，支持数据库操作、安全验证等。但这些功能的实现要依赖于两种技术原理：IoC 和 AOP。本书的目的就是学习并掌握 Spring 中的这两种核心技术，并能在实际的软件开发中对其进行运用。

Spring 是一个开放源代码的 Java EE 框架，主要是为了解决中国级应用程序在开发及维护时的复杂性问题而创建的，使用 Spring 简化了 Java EE 的开发，提高了 Java EE 软件项目的开发效率。提高开发效率的办法就是使用模块架构，每个模块处理一个功能或者业务，模块架构允许程序员选择使用哪一个模块参与开发，同时为 Java EE 应用程序开发提供集成的容器。在 Spring 框架中提供了 1 个 JavaBean 容器（可以暂时将容器理解成为 1 个 List）。在该容器中存储不同数据类型的 JavaBean 对象，容器中可以将具有多种不同功能的 JavaBean 进行整合、集成，来达到多个技术综合应用的目的。

本节主要介绍 Spring 框架核心的原理。

（1）控制反转（Inversion of Control，IoC）。

（2）面向切面编程（Aspect Oriented Programming，AOP）中的控制反转。

3.2　Spring 框架的模块组成

Spring 框架发展多年，现在已经是一个初具规模的 Java EE 开发平台，在 Spring 5 版本中主要的模块如下。

（1）Core（核心）模块：dependency injection（依赖注入）、events（事件处理）、resources（资源访问）、i18n（国际化）、validation（验证）、data binding（数据绑定）、type conversion（数据类型转换）、SpEL（表达式语言）和 AOP（面向切面编程）。

（2）Testing（测试）模块：支持 mock objects、TestContext Framework、Spring MVC Test、WebTestClient 测试框架。

（3）Data Access（数据访问）模块：transactions（事务处理）、DAO support（数据访问对象）、JDBC（数据库连接）、ORM（映射）和 Marshalling XML（处理 XML）。

（4）Spring MVC 和 Spring WebFlux Web frameworks 模块。

（5）Integration（集成）模块：remoting（远程访问）、JMS（消息服务）、JCA（加解密）、JMX（管理扩展）、email（邮件处理）、tasks（任务）、scheduling（执行计划）、cache（缓存）。

（6）Languages（支持语言）：支持使用 Kotlin、Groovy 和 dynamic languages 等语言进行开发。

Spring 框架的功能可以用在任何 Java EE 服务器中，其核心要点是保证相同代码在不同 Java EE 容器的可移植性。

3.3　IoC 和 DI

在没有 Spring 框架的时候，如果在 A.java 类中使用 B.java 类，那么必须在 A.java 类中 new 实例化出 B.java 类的对象。这就造成了 A 类和 B 类的紧耦合，A 类完全依赖于 B 类的功能实现，这样的情况就是典型的“侵入式开发”。随着软件业务复杂度的提升，当原有的 B 类不能满足 A 类的功能实现时，就需要创建更为高级的 BExt.java 类，结果是把所有实例化 B.java 类的代码替换成 new BExt()代码，这就造成了源代码的改动，不利于软件运行的稳定性，并不符合商业软件的开发与维护流程。IoC 技术就可以解决这种问题，其解决的办法是使用“反射”技术，动态地对一个类中的属性进行反射赋值，这使得 Spring 形成了一个模块，模块的功能非常强大，并且 Spring 把这种机制命名为控制反转，即 IoC。

什么是 IoC？IoC 想要达到的目的就是将调用者与被调用者分离，让类与类之间的关系解耦，解耦的原理如图 3-1 所示。

图 3-1　使用 IoC 实现解耦

Spring 框架中的 IoC 技术可以实现 A 类和 B 类的解耦，在 A 类中不再出现 new B() 的情况，实例化 new B() 类的任务由 Spring 框架来进行处理，Spring 框架再使用反射的机制将 B 类的对象赋值给 A 类中的 b 对象。原来 A 类是主动实例化 B 类的对象，控制方是 A 类，而现在以被动的方式由 Spring 框架来实现，控制方现在变成了 Spring 框架，实现了反转，因此，此种技术被称为“控制反转”。

IoC 是一个理念，是一种设计思想，A 类中的 b 对象的值是需要被赋值的，实现的方式是 DI（Dependency Injection，依赖注入）。DI 是 IoC 思想的实现，其侧重于实现，A 类依赖于 B 类，B 类的对象由容器进行创建，容器再对 A 类中的 B 属性进行对象值的注入。DI 在 Java 中的底层技术原理就是“反射”，使用反射技术对某一个类中的属性值进行动态赋值，来达到 A 类和 B 类之间的解耦。

在 Spring 中管理 JavaBean 的容器官方定义为 IoC 容器（IoC container），而在 Spring 中对 IoC 的主要实现方式就是 DI。

3.4 IoC 容器

什么是 IoC 容器？前面介绍过 Spring 的 DI 其实就是对 JavaBean 的属性使用反射技术进行赋值。当有很多的 JavaBean 需要这样的操作时，这些 JavaBean 的管理就成了问题，因为某些 JavaBean 之间需要关联，而某些 JavaBean 之间并不需要关联，而且所有这些 JavaBean 的创建、销毁都要统一地调度，由 Spring 框架处理它们的生命周期。为了方便管理，Spring 框架提供了 IoC 容器，对 JavaBean 进行统一组织，便于后期代码的维护。曾经在任意的位置进行 new 实例化任何类对象的情况一去不复返了，所有 new 实例化的任务都要交给 IoC 容器实现，这时，对 JavaBean 的管理就需要更加规范了。

另外，Spring 的 IoC 容器完全脱离了平台，可以在任何支持 Java 语言的环境中运行，具有极好的移植性，其用极为简单的接口与实现分离的原理，对组件的调配提供很好的支持。IoC 容器用于管理 JavaBean，创建 JavaBean 的一个内存区，在这个内存区中可以将操作 JavaBean 的代码以面向接口的方式进行开发与管理，这样使接口的多态性与 DI 技术相结合，程序结构的分层更加灵活化，维护和扩展也很方便。

IoC/DI 从编程技术上来讲就是将接口和实现相分离，然后使用反射技术为类中的属性动态赋值，而 IoC 容器管理 JavaBean 的生命周期以及多个 JavaBean 之间的注入关系。

由以上内容可大概了解 IoC、IoC 容器以及 DI 的作用与使用场景了，程序员随意地在任何位置创建任何类的对象在 Spring 框架中是不规范的，Spring 框架对 JavaBean 的管理更加具有规划性，比如创建、销毁，还可以动态地为一个属性注入值。通过使用 Spring 的 IoC 容器，软件项目对 JavaBean 的管理更加统一了。

3.5 面向切面编程

在没有 AOP（Aspect Oriented Programming，面向切面编程）技术时，如果想为软件项目记录日志，那么必须要在关键的业务点写上记录日志的程序代码。日志的信息包含“开始时间”“结束时间”“执行人”以及“角色”等信息。随着软件项目越来越稳定，曾经的日志代码需要删除，因为输出日志会影响程序运行的效率，这时就要在 Java 源代码中删除记录日志的程序代码。另外，在未来有可能需要记录日志时，还要重新加入日志程序代码，这使源代码反复地被更改，不利于软件运行的稳定。这时使用 Spring 的 AOP 技术就可以解决这些具有“通用性”的问题。Spring 的 AOP 功能模块具有可插拔性，因此，几乎不需要大幅地更改代码即可完成前面想要实现的功能。

DI 使用的技术原理是“反射”，AOP 使用的技术原理则是“动态代理”。动态代理是 23 个标准设计模式中的一个，动态代理解决的问题是在不改变原有代码的基础上，对原有的模块进行功能上的加强和扩展，使扩展的功能与被扩展的模块充分解耦，利于软件项目模块化设计。应用 AOP 的场景，比如可以在不改变 Servlet 代码的基础上加入日志的功能，在不改变 Spring MVC 框架的控制层代码的基础上加入数据库事务的功能等。

第 4 章会介绍 AOP 的详细内容。

3.6 初步体会 IoC 的优势

前面一直在讨论使用 IoC 可以实现解耦，DI 是 IoC 思想的实现，那么如何在 Spring 框架中实现解耦呢？在继续实现解耦话题之前，先看一下使用传统的方法实现数据保存功能的弊端。

创建测试用的项目 firstSaveTest，创建业务类，代码如下：

```
public class SaveDBService {
    public void saveMethod() {
        System.out.println("将数据保存到 DB 数据库");
    }
}
```

运行类代码如下：

```
public class Test {
    private SaveDBService service = new SaveDBService();

    public SaveDBService getService() {
        return service;
    }

    public void setService(SaveDBService service) {
        this.service = service;
    }

    public static void main(String[] args) {
        Test test = new Test();
        test.getService().saveMethod();
    }
}
```

程序运行的结果如下：

```
将数据保存到 DB 数据库
```

控制台打印出正确的结果，这样的代码结构在大多数的软件项目中都在使用，而且有些已经应用到商业项目中。虽然打印出的结果是正确的，但是从项目的整合设计结构上来看很明显是不合理的，比如以下几点。

（1）源代码被反复修改：在本项目中会将数据保存到数据库，如果换成保存到 XML 文件中，那么就不得不更改程序，将 SaveDBService.java 中的程序改成保存到 XML 的文件代码，或者创建新的 SaveXMLService.java 类，然后更改代码变成实例化“private SaveXMLService service = new SaveXMLService();”对象，这就造成了程序的改动，不利于项目运行的稳定性。

（2）出现紧耦合：Test.java 和 SaveDBService.java 类产生了紧耦合，类 Test.java 负责创建 SaveDBService.java 类的对象，不利于软件功能的扩展、测试与复用。

（3）无法保证单例性：SaveDBService.java 无法在单实例的情况下被重用，因为它的声明在 Test.java 类中，也就是在 Test.java 类中可以随意地创建出很多 SaveDBService.java 类的实例，无法保证该类实例的单例性。

（4）无法保证资源被正确地释放：如果从 SaveDBService.java 类中获取一些资源，比如数据库的连接（Connection）、数据库的 JNDI、输入/输出流（Stream）等，那么 Test.java 不得不维护这些资源的开启（open）和关闭（close）。如果忘记 close，那么资源不能被有效地释放，将会造成资源占用，影响项目运行的稳定性。

典型的 4 个缺点就造成了设计的失败，那么根据面向对象三大特性中的“多态”技术，可以把 SaveDBService.java 类改成接口与实现类的模式吗？也就是在 Test.java 类中声明接口，然后再实例化这个接口的实现类（测试代码在 firstSaveTest2 项目中）。这样的设计的确比上面的示例要灵活一些，也符合面向接口编程的方式，但仅仅是将接口和实现进行分离，也没有完全解决业务变更后源代码还要被修改的问题。那么，如果在项目中经常有这样耦合的结构，该如何解决呢？Spring 是如何解决的呢？

在 IoC 容器的帮助下可以实现松耦合，并且模块之间是分离的，彼此可以互相访问。使用 DI 技术结合 IoC 容器之后，创建对象的操作是由 IoC 容器来控制的，并且完全基于接口（interface）和实现类（imple）的分离开发。这样将 SaveDBService.java 类的控制权从原来在 Test.java 类中转变成在 IoC 容器中，接口的实现类是依赖 IoC 容器进行注入赋值的。下面就使用 Spring 框架来解决这两个类之间的紧耦合问题，把这个问题画上一个句号。

创建测试用的项目 springTest11，在 classpath 中添加 Spring 的 JAR 包。

创建一个保存数据的接口，代码如下：

```
public interface ISaveService {
    public void saveMethod();
}
```

创建一个将数据保存到 DB 的实现类，代码如下：

```
public class SaveDBService implements ISaveService {
    public SaveDBService() {
        System.out.println("Spring 通过反射机制来实例化 SaveDBService 类的对象 " + this);
    }

    @Override
    public void saveMethod() {
        System.out.println("将数据保存到 DB 数据库");
```

```
    }
}
```

创建一个将数据保存到 XML 的实现类，代码如下：

```
public class SaveXMLService implements ISaveService {
    public SaveXMLService() {
        System.out.println("Spring 通过反射机制来实例化 SaveXMLService 类的对象 " + this);
    }

    @Override
    public void saveMethod() {
        System.out.println("将数据保存到 XML 文件");
    }
}
```

运行类代码如下：

```
public class Test {
    private ISaveService service;
    private String username;

    public ISaveService getService() {
        return service;
    }

    public void setService(ISaveService service) {
        this.service = service;
        System.out.println("setService(ISaveService service) service=" + service);
    }

    public String getUsername() {
        return username;
    }

    public void setUsername(String username) {
        this.username = username;
    }

    public static void main(String[] args) {
        ApplicationContext context = new ClassPathXmlApplicationContext
("applicationContext.xml");
        Test test = context.getBean(Test.class);
        test.getService().saveMethod();
        System.out.println(test.getUsername());
    }
}
```

接口 ApplicationContext 相当于 IoC 容器，创建容器可以通过使用 ClassPathXmlApplicationContext 类加载 applicationContext.xml 配置文件来实现，然后再使用 getBean() 方法从容器中获得 Java 对象。

文件 applicationContext.xml 在本项目中存放在 src 节点下。

配置文件 applicationContext.xml，代码如下：

```
<bean id="saveDBService" class="service.SaveDBService"></bean>
<bean id="saveXMLService" class="service.SaveXMLService"></bean>
<bean id="test" class="test.Test">
```

```
    <property name="service" ref="saveDBService"></property>
    <property name="username" value="我是 username 的值"></property>
</bean>
```

配置文件 applicationContext.xml 中的代码可以从帮助文档中获取再进行修改。

以下配置代码的作用是通过反射机制在 IoC 容器中创建 service.SaveDBService 类的对象，对象的别名就是 id 属性值 saveDBService。

```
<bean id="saveDBService" class="service.SaveDBService"></bean>
```

配置代码如下：

```
<bean id="test" class="test.Test">
    <property name="service" ref="saveDBService"></property>
    <property name="username" value="我是 username 的值"></property>
</bean>
```

其作用是创建 test 包中 Test.java 类的对象，并且对 Test 类对象中的 service 属性进行反射赋值，术语就是 DI（依赖注入），赋值的方式是调用 setService（ISaveService service）。赋予的值来自于 ref="saveDBService"配置代码。ref 代表反射赋予的值是复杂对象，它是引用类型，ref 属性值“saveDBService”是配置代码的 id 值：

```
<bean id="saveDBService" class="service.SaveDBService"></bean>
```

配置代码如下：

```
<property name="username" value="我是 username 的值"></property>
```

其作用是为 username 属性注入值。在 IoC 容器中，如果注入的数据类型是简单数据类型和 String 数据类型，则要使用 value 属性。ref 属性注入的是引用数据类型，这是 value 和 ref 属性的区别。

注入关系如图 3-2 所示。

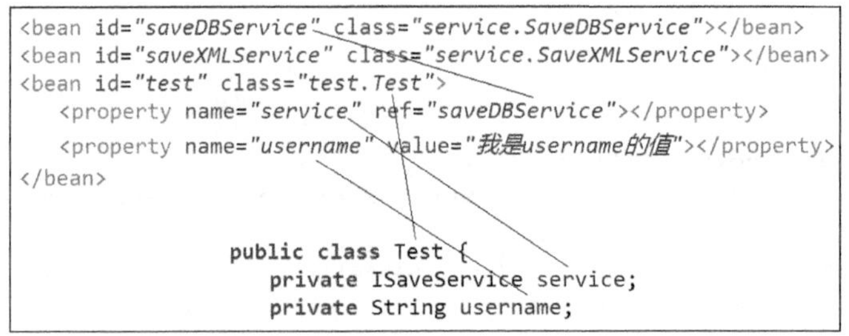

图 3-2 注入关系

图 3-2 要说明的是对 test 包的 Test 类中的 service 属性注入别名为 saveDBService 的对象，对 username 属性注入字符串“我是 username 的值”。

程序执行的结果如下：

```
Spring 通过反射机制来实例化 SaveDBService 类的对象 service.SaveDBService@548b7f67
Spring 通过反射机制来实例化 SaveXMLService 类的对象 service.SaveXMLService@59fa1d9b
setService(ISaveService service) service=service.SaveDBService@548b7f67
将数据保存到 DB 数据库
我是 username 的值
```

接口 ISaveService.java 的实现类不再由 Test.java 类创建，而是由 Spring 的 IoC 容器创建与管理，接口 ISaveService.java 的实现类与 Test.java 类完全解耦。

使用 Spring 框架，如果业务变更，则应该用 XML 文件来保存数据，只需要创建一个将数据保存到 XML 文件的实现类，并且对 service 属性进行注入。这样就可以非常灵活地将数据保存到 DB 或 XML 文件中，符合“开闭原则”的设计思想。

另外，通过 IoC 容器来对创建的 JavaBean 对象进行管理，可以独立地对各个模块进行依赖注入。这样的使用所带来的结果就是软件的 JavaBean 被完全重用了。因此，使用 Spring 的 IoC 技术，在面对复杂的业务逻辑、灵活变化的模块关系、烦恼的软件后期维护及扩展等方面得到很大程度上的解决。这也是 Spring 的哲学思想，用极为简单的技术原理“反射”来解决复杂的问题。

在前面总结了使用传递方式保存数据的弊端有 4 种，但使用 Spring 框架后，这 4 种弊端都一一被解决了，具体如下。

（1）源代码被反复修改：依赖关系不需要修改*.java 文件，只需要修改 ac.xml 配置文件。

（2）出现紧耦合：A 类和 B 类彻底解耦，B 类由 Spring 框架创建，再向 A 类进行注入。

（3）无法保证单例性：IoC 容器可以保证 JavaBean 的单例性，这一点在后面的章节会有验证。

（4）无法保证资源被正确地释放：结合 Spring 框架中的 AOP 技术，会自动执行 close()方法，资源被自动释放。

Spring 框架中的 IoC 技术使用了大量的反射，可以查看一下堆栈信息，证明使用反射技术创建对象，效果如图 3-3 所示。

图 3-3　使用反射技术创建对象

继续使用反射技术实现依赖注入，效果如图 3-4 所示。

下面就对依赖注入的细节进行进一步挖掘，深入理解 IoC 的理念与实现方式。

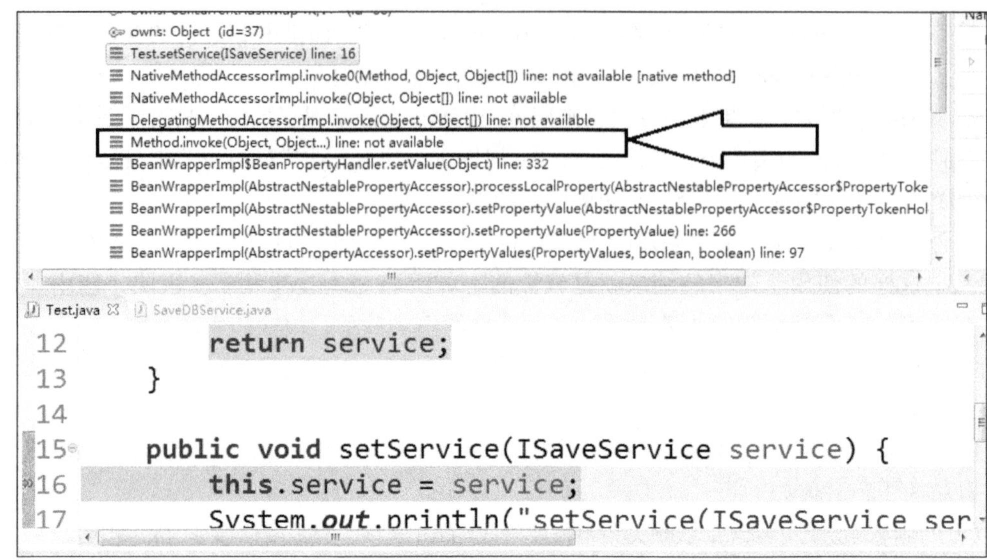

图 3-4 使用反射技术实现依赖注入

3.7 在 Spring 中创建 JavaBean

在 Spring 中并不使用传统的 new Object() 方式来创建 JavaBean 的实例，而是使用其他多种途径来创建出类的对象。

如果使用 new Object() 方式创建对象，虽然目的是达成了，但对象的管理却非常不方便。因此，在 Spring 框架中将创建出来的 JavaBean 对象放入 IoC 容器中，在容器中对其进行统一管理。

下面演示如何在 Spring 框架中使用两种方式创建对象，分别是 XML 声明法以及 Annotation 注解法。

3.7.1 使用 XML 声明法创建对象

从最初的 Spring 发布以来，经历了以下 3 种声明方式。

（1）XML 配置文件法。

（2）注解法。

（3）JavaConfig 类。

这 3 种声明方式都会在本节进行介绍。使用 XML 声明法创建对象是原始且有效的，它通过 DOM4J 和反射技术来生成对象，在学习通过 Spring 创建 JavaBean 以及注入时，从 XML 声明法入手是掌握此技术的捷径。

1. 创建对象

创建 Java 项目 useXMLCreateObject，添加 Spring 框架需要的 JAR 包，在 src 节点下创建名称为 applicationContext.xml 配置文件，代码如下：

```
<bean id="userinfo1" class="entity.Userinfo"></bean>
```

在 Spring 框架中，因为对象存放在 IoC 容器中，所以要在 applicationContext.xml 文件中进行声明。声明的作用就是告诉 Spring 创建对象，使用<bean>声明对象的配置代码：

```
<bean id="userinfo1" class="entity.Userinfo"></bean>
```

以上代码的作用就是创建出 entity 包中 Userinfo 类的对象，该对象在容器中的 id 是 userinfo1。

创建 Userinfo.java 类，代码如下：

```
public class Userinfo {
    public Userinfo() {
        System.out.println("类 Userinfo 被实例化=" + this);
    }
}
```

创建运行类，代码如下：

```
public class Test1 {
    public static void main(String[] args) {
        ApplicationContext context = new ClassPathXmlApplicationContext
("applicationContext.xml");
    }
}
```

程序运行后在控制台输出的信息如下：

```
类 Userinfo 被实例化=entity.Userinfo@3c756e4d
```

大多数人采用 ClassPathXmlApplicationContext 或 FileSystemXmlApplicationContext 创建 ApplicationContext 的实例。

类 ClassPathXmlApplicationContext 负责在类路径上找到 XML 配置资源并加载。

类 FileSystemXmlApplicationContext 负责在文件系统上找到 XML 配置资源并加载，运行类示例代码如下：

```
public class Test2 {
    public static void main(String[] args) {
        ApplicationContext context = new FileSystemXmlApplicationContext(
            "C:\\mybatisEclipse-workspace\\useXMLCreateObject\\src\\applicationContext.xml");
    }
}
```

2. 创建对象并获取

IoC 容器中的对象可以使用 getBean（String）方法来获取。

创建运行类，代码如下：

```
public class Test3 {
    public static void main(String[] args) {
        ApplicationContext context = new ClassPathXmlApplicationContext
("applicationContext.xml");
        Userinfo getUserinfo = (Userinfo) context.getBean("userinfo1");
        System.out.println("main end print " + getUserinfo);
    }
}
```

方法 getBean（String name）的参数是<bean>的 id 属性值，通过 id 值获得相应的对象。

程序运行后控制台输出的信息如下：

```
类 Userinfo 被实例化=entity.Userinfo@3c756e4d
main end print entity.Userinfo@3c756e4d
```

从控制台打印的信息来看，创建的 Userinfo 对象和获取的 Userinfo 对象是同一个。

3. 出现 NoUniqueBeanDefinitionException 及解决办法

在使用 getBean（Class）方法获取对象时，如果有多个对象属于同一个类型，那么在获取对象时会出现 NoUniqueBeanDefinitionException 异常。

创建配置文件 applicationContextError.xml，代码如下：

```
<bean id="userinfo1" class="entity.Userinfo"></bean>
<bean id="userinfo2" class="entity.Userinfo"></bean>
```

创建运行类，代码如下：

```
public class Test4 {
    public static void main(String[] args) {
        ApplicationContext context = new ClassPathXmlApplicationContext
("applicationContextError.xml");
        context.getBean(Userinfo.class);
    }
}
```

程序运行后控制台输出的异常信息如下：

```
类 Userinfo 被实例化=entity.Userinfo@3c756e4d
类 Userinfo 被实例化=entity.Userinfo@1d057a39
Exception in thread "main" org.springframework.beans.factory.NoUniqueBeanDefinitionException:
No qualifying bean of type 'entity.Userinfo' available: expected single matching bean but
found 2: userinfo1,userinfo2
```

控制台打印的信息提示，Spring 找到两个对象，分别是 userinfo1 和 userinfo2，两者都属于 Userinfo.class 类的对象，Spring 并不知道应该获取哪一个，因此出现了异常。

如果使用 getBean（Class）找到相同类型的对象，则出现异常，解决的办法是使用重载方法 getBean（String），代码如下：

```
public class Test5 {
    public static void main(String[] args) {
        ApplicationContext context = new ClassPathXmlApplicationContext
("applicationContextError.xml");
        Userinfo userinfo1 = (Userinfo) context.getBean("userinfo1");
        Userinfo userinfo2 = (Userinfo) context.getBean("userinfo2");
        System.out.println(userinfo1);
        System.out.println(userinfo2);
    }
}
```

程序运行的结果如下：

```
类 Userinfo 被实例化=entity.Userinfo@3c756e4d
类 Userinfo 被实例化=entity.Userinfo@1d057a39
entity.Userinfo@3c756e4d
entity.Userinfo@1d057a39
```

根据前面实验的结果可以总结出，getBean（Userinfo.class）和 getBean（"userinfo1"）的使用场景如下。

（1）getBean（Userinfo.class）：如果 IoC 容器中只有 1 个 Userinfo 类的实例，那么可以使用此方法。

（2）getBean（"userinfo1 "）：如果 IoC 容器中出现多个相同数据类型的对象，那么就要使用 getBean（String）的写法来根据对象的 id 找到指定的对象。

4. 创建 Inner bean

创建测试用的项目 createInnerBean。

创建类代码如下：

```
public class Out {
    private Inner inner;

    public Out() {
        System.out.println("public Out()=" + this);
    }

    public Inner getInner() {
        return inner;
    }

    public void setInner(Inner inner) {
        this.inner = inner;
    }
}
```

创建类代码如下：

```
public class Inner {
    public Inner() {
        System.out.println("public Inner()=" + this);
    }
}
```

配置文件 applicationContext.xml，代码如下：

```
<bean id="out" class="entity.Out">
    <property name="inner">
        <bean class="entity.Inner"></bean>
    </property>
</bean>
```

创建运行类，代码如下：

```
public class Test1 {
    public static void main(String[] args) {
        ApplicationContext context = new ClassPathXmlApplicationContext
("applicationContext.xml");
        Out out = (Out) context.getBean("out");
        System.out.println("out=" + out);
        System.out.println("out.getInner()" + out.getInner());
    }
}
```

程序运行后控制台输出的信息如下：

```
public Out()=entity.Out@402f32ff
public Inner()=entity.Inner@29774679
out=entity.Out@402f32ff
out.getInner()entity.Inner@29774679
```

5. 延迟 lazy Bean

测试延迟 lazy Bean 的主要目的是确定 JavaBean 实例化的时机。

创建测试用的项目 lazyBeanTest。

创建类代码如下：

```
public class Userinfo {
    public Userinfo() {
        System.out.println("类 Userinfo 被实例化=" + this);
    }
}
```

配置文件 applicationContext.xml，代码如下：

```
<bean id="userinfo" class="entity.Userinfo"></bean>
```

创建运行类，代码如下：

```
public class Test1 {
    public static void main(String[] args) {
        System.out.println("1");
        ApplicationContext context = new ClassPathXmlApplicationContext
("applicationContext.xml");
        System.out.println("2");
        context.getBean("userinfo");
        System.out.println("3");
    }
}
```

程序运行后控制台输出的信息如下：

```
1
类 Userinfo 被实例化=entity.Userinfo@3c756e4d
2
3
```

控制台输出的信息说明，id 为 userinfo 的对象在执行 getBean() 方法之前，也就是执行 new ClassPathXmlApplicationContext() 代码时就被创建了，属于立即加载。

修改配置，代码如下：

```
<bean id="userinfo" class="entity.Userinfo" lazy-init="true"></bean>
```

再次执行运行类，程序运行后在控制台输出的信息如下：

```
1
2
类 Userinfo 被实例化=entity.Userinfo@4b9e13df
3
```

控制台输出的信息说明，当执行 context.getBean（"userinfo"）方法时，userinfo 对象才被创建，属于延迟加载。

6. 创建内置类

使用 XML 声明法可以创建内置类。

创建测试用的项目 xmlCreateInnerClass。

创建实体类，代码如下：

```
public class UserinfoOut {
    public UserinfoOut() {
        System.out.println("类 UserinfoOut 被实例化=" + this);
    }

    public static class UserinfoInner {
        public UserinfoInner() {
            System.out.println("类 UserinfoInner 被实例化=" + this);
        }
    }
}
```

创建配置文件，代码如下：

```
<bean id="userinfoInner" class="entity.UserinfoOut$UserinfoInner"></bean>
```

创建运行类，代码如下：

```
public class Test1 {
    public static void main(String[] args) {
        ApplicationContext context = new ClassPathXmlApplicationContext
("applicationContext.xml");
        System.out.println(context.getBean("userinfoInner"));
    }
}
```

程序运行后控制台输出的异常信息如下：

```
类 UserinfoInner 被实例化=entity.UserinfoOut$UserinfoInner@3c756e4d
entity.UserinfoOut$UserinfoInner@3c756e4d
```

7. 使用 factory-method 属性创建对象

使用 factory-method 属性调用静态工厂方法来创建对象。

创建测试用的项目 xml_factory_method_property。

创建实体类，代码如下：

```
public class Userinfo {
    private Userinfo() {
    }

    public static Userinfo createUserinfo() {
        Userinfo userinfo = new Userinfo();
        System.out.println("执行了 public static Userinfo createUserinfo()方法 userinfo=" +
userinfo);
        return userinfo;
    }
}
```

创建配置文件 applicationContext.xml 的代码如下：

```
<bean id="userinfo1" class="entity.Userinfo"
    factory-method="createUserinfo"></bean>
```

创建运行类的代码如下：

```
public class Test1 {
    public static void main(String[] args) {
        ApplicationContext context = new ClassPathXmlApplicationContext
("applicationContext.xml");
        Userinfo getUserinfo = (Userinfo) context.getBean("userinfo1");
        System.out.println("main end print " + getUserinfo);
    }
}
```

程序运行后控制台输出的异常信息如下：

```
执行了 public static Userinfo createUserinfo()方法 userinfo=entity.Userinfo@685cb137
main end print entity.Userinfo@685cb137
```

使用普通的反射代码也可以创建构造方法为私有的对象，示例代码如下：

```
class Userinfo {
    private Userinfo() {
        System.out.println("private Userinfo() " + this);
    }
}

public class Test2 {
    public static void main(String[] args) throws NoSuchMethodException, SecurityException,
InstantiationException,
            IllegalAccessException, IllegalArgumentException, InvocationTargetException {
        Constructor constructor = Userinfo.class.getDeclaredConstructor(null);
        constructor.setAccessible(true);
        System.out.println(constructor.newInstance(null));
    }
}
```

程序运行后也能创建出构造方法为 private 的对象。

而构造方法为 private 的类也可以在 Spring 框架中通过反射创建类的对象。

8. 使用 factory-bean 和 factory-method 属性创建对象

可以调用 factory-bean 类中的非静态 factory-method 方法来创建对象。

创建测试用的项目 xml_factory_bean_method。

创建实体类，代码如下：

```
public class Userinfo {
    public Userinfo() {
        System.out.println("private Userinfo() " + this);
    }
}
```

创建工具类，代码如下：

```
public class Creator {
    public Userinfo createUserinfo() {
        Userinfo userinfo = new Userinfo();
```

```
        System.out.println("执行了 public static Userinfo createUserinfo()方法 userinfo=" +
userinfo);
        return userinfo;
    }
}
```

创建配置文件 applicationContext.xml 的代码如下：

```
<bean id="creator" class="tools.Creator"></bean>
<bean id="userinfo1" factory-bean="creator"
    factory-method="createUserinfo"></bean>
```

创建运行类的代码如下：

```
public class Test1 {
    public static void main(String[] args) {
        ApplicationContext context = new ClassPathXmlApplicationContext
("applicationContext.xml");
        Userinfo getUserinfo = (Userinfo) context.getBean("userinfo1");
        System.out.println("main end print " + getUserinfo);
    }
}
```

程序运行后控制台输出的异常信息如下：

```
private Userinfo() entity.Userinfo@5622fdf
执行了 public static Userinfo createUserinfo()方法 userinfo=entity.Userinfo@5622fdf
main end print entity.Userinfo@5622fdf
```

9. <bean>无 id 属性获取 JavaBean

创建测试用的项目 noId。
使用以下配置代码：

```
<bean class="entity.Userinfo"></bean>
```

创建 Userinfo 类的实例被放入 IoC 容器中，可以使用全路径名获取，示例代码如下：

```
public class Test {
    public static void main(String[] args) {
        ApplicationContext context = new ClassPathXmlApplicationContext
("applicationContext.xml");
        System.out.println(context.getBean("entity.Userinfo"));
        System.out.println(context.getBean(Userinfo.class));
    }
}
```

程序运行后控制台的输出信息如下：

```
类 Userinfo 被实例化=entity.Userinfo@3c756e4d
entity.Userinfo@3c756e4d
entity.Userinfo@3c756e4d
```

10. 出现 Overriding bean definition 及解决方法

当多个 applicationContext.xml 配置文件中<bean>的 id 属性值一样，并且数据类型也一样时，会出现 bean 覆盖的情况。

创建测试用的项目 beanOverriding1。

创建类的代码如下：

```
@Component
public class Userinfo {
    public Userinfo() {
        System.out.println("public Userinfo() " + this);
    }
}
```

创建配置文件 ac1.xml 的代码如下：

```
<bean id="useirnfo" class="root.Userinfo"></bean>
```

创建配置文件 ac2.xml 的代码如下：

```
<bean id="useirnfo" class="root.Userinfo"></bean>
```

创建类的代码如下：

```
public class MyTest {
    public static void main(String[] args) {
        ApplicationContext context = new ClassPathXmlApplicationContext("ac1.xml", "ac2.xml");
    }
}
```

运行程序后控制台输出的结果如下：

```
    信息: Overriding bean definition for bean 'useirnfo' with a different definition: replacing
[Generic bean: class [root.Userinfo]; scope=; abstract=false; lazyInit=false; autowireMode=0;
dependencyCheck=0; autowireCandidate=true; primary=false; factoryBeanName=null;
factoryMethodName=null; initMethodName=null; destroyMethodName=null; defined in class path
resource [ac1.xml]] with [Generic bean: class [root.Userinfo]; scope=; abstract=false;
lazyInit=false; autowireMode=0; dependencyCheck=0; autowireCandidate=true; primary=false;
factoryBeanName=null; factoryMethodName=null; initMethodName=null; destroyMethodName=null;
defined in class path resource [ac2.xml]]
    public Userinfo() root.Userinfo@573f2bb1
```

控制台输出 info 级别的日志信息，提示出现“Overriding bean definition”的情况，也就是 ac1.xml 和 ac2.xml 中<bean>的 id 值和数据类型都一样，造成 JavaBean 被覆盖。解决的办法就是将<bean>的 id 值设置为不同，解决此问题的项目在 beanOverriding2 中。

3.7.2　使用 Annotation 注解法创建对象

使用 XML 声明法创建对象容易造成 applicationContext.xml 文件中<bean>声明的配置代码过多，对代码的后期维护非常不利。因此可以使用 Spring 新版本提供的 Annotation 注解来解决这个问题。

1．使用<context:component-scan base-package="">创建对象

配置代码<context:component-scan base-package="">的作用是在指定的包中扫描创建对象的类。如果某些类需要被 Spring 实例化，那么类的上方必须使用相关的注解。

创建测试用的项目 annotationCreateBean。

实体类 Userinfo.java 的代码如下：

```
@Component
public class Userinfo {
    public Userinfo() {
        System.out.println("public Userinfo() " + this.hashCode());
    }
}
```

配置文件 applicationContext.xml 的代码如下：

```
<context:component-scan base-package="entity"></context:component-scan>
```

注解 @Component 的作用就是标识 Userinfo.java 类是 IoC 容器中的一个组件，能被<context:component-scan base-package="entity">扫描器所识别并自动实例化，最后将 Userinfo 对象放入 IoC 容器中。另外，也可以不使用@Component 注解，转而使用@Repository 来声明，虽然程序运行的效果没有变化，但两者还是有一些区别的。

（1）@Repository 主要用来声明 DAO 层。

（2）@Component 主要用来声明一些通用性的组件。

运行类代码如下：

```
public class Test1 {
    public static void main(String[] args) {
        ApplicationContext context = new ClassPathXmlApplicationContext("ac.xml");
    }
}
```

程序运行后的效果如下：

```
public Userinfo() 282432134
```

使用注解法成功地进行了 Userinfo.java 类的实例化。

2. 使用<context:component-scan base-package="">创建对象并获取

使用<context:component-scan base-package="">扫描器创建对象后，可以通过 getBean()方法获取对象，示例代码如下：

```
public class Test2 {
    public static void main(String[] args) {
        ApplicationContext context = new ClassPathXmlApplicationContext("ac.xml");
        System.out.println(context.getBean(Userinfo.class).hashCode());
    }
}
```

程序运行的结果如下：

```
public Userinfo() 282432134
282432134
```

3. 使用“全注解”创建对象

下面将使用“全注解”的方式创建对象，而不再使用 applicationContext.xml 配置文件。“全注解”配置法也称为 JavaConfig 配置法。

配置文件 applicationContext.xml 的根节点是<beans>，它起到全局配置（Configuration）定义的作用，它的子节点<bean>包含了创建 JavaBean 的细节信息。

在使用“全注解”法创建对象时，与<beans>标记起相同作用的注解就是@Configuration，与<bean>标记功能相同的注解是 @Bean。

创建测试用的项目 allAnnotationCreateObject1。

实体类代码如下：

```java
@Component
public class Userinfo {
    public Userinfo() {
        System.out.println("public Userinfo() " + this.hashCode());
    }
}
```

配置类代码如下：

```java
@Configuration
public class SpringConfig {
    @Bean(name = "u1")
    public Userinfo createUserinfo1() {
        Userinfo userinfo1 = new Userinfo();
        System.out.println("userinfo1 " + userinfo1.hashCode());
        return userinfo1;
    }

    @Bean(name = "u2")
    public Userinfo xxxxxxxxxxxxxxx() {
        Userinfo userinfo2 = new Userinfo();
        System.out.println("userinfo2 " + userinfo2.hashCode());
        return userinfo2;
    }
}
```

使用 @Bean 注解声明创建对象的方法是可以任意命名的，但必须要有返回值，不然会出现没有声明返回值的异常。

运行类代码如下：

```java
public class Test1 {
    public static void main(String[] args) {
        ApplicationContext context = new AnnotationConfigApplicationContext("createbean");
    }
}
```

实例化(new)AnnotationConfigApplicationContext() 类时传入参数 createbean 代表在 createbean 包中寻找哪个类带有 @Configuration 注解，再根据该类中的信息创建 JavaBean 对象。

程序运行的结果如下：

```
public Userinfo() 572191680
userinfo1 572191680
public Userinfo() 103536485
userinfo2 103536485
```

使用“全注解”法成功地创建了两个 Userinfo.java 类的对象。

注意：使用 new AnnotationConfigApplicationContext("createbean") 扫描配置类时，createbean 包中支持多个配置类。

4. 使用“全注解”获取对象时出现 NoUniqueBeanDefinitionException 及解决办法

使用“全注解”法获取对象时会出现获取相同类型对象的情况，并且也会出现 NoUniqueBeanDefinitionException 异常。

运行类代码如下：

```
public class Test2 {
    public static void main(String[] args) {
        ApplicationContext context = new AnnotationConfigApplicationContext("createbean");
        context.getBean(Userinfo.class);
    }
}
```

程序运行后出现以下信息表示异常，说明有多个对象的数据类型是相同的。

```
Exception in thread "main" org.springframework.beans.factory.NoUniqueBeanDefinitionEx
ception: No qualifying bean of type 'entity.Userinfo' available: expected single matching
bean but found 2: u1,u2
```

解决这个问题的办法是使用 beanId，示例代码如下：

```
public class Test3 {
    public static void main(String[] args) {
        ApplicationContext context = new AnnotationConfigApplicationContext("createbean");
        System.out.println();
        System.out.println(context.getBean("u2").hashCode());
    }
}
```

为 @Bean 中的 name 属性赋值，设置每个 Userinfo.java 类的对象拥有不同的 id 值，并且还要结合 getBean（String）方法。

程序运行后的效果如下：

```
public Userinfo() 572191680
userinfo1 572191680
public Userinfo() 103536485
userinfo2 103536485

103536485
```

5. 使用 @ComponentScan(basePackages = "")创建并获取对象

与命名空间<context:component-scan base-package="entity"></context:component-scan>作用相同的注解是 @ComponentScan（basePackages = ""），它也可以进行类扫描并将其实例化。

创建测试用的项目 ComponentScanTest。

实体类代码如下：

```
@Component
public class Userinfo {
    public Userinfo() {
        System.out.println("public Userinfo() " + this.hashCode());
```

```
    }
}
```

创建工厂类，代码如下：

```
@Configuration
public class SpringConfig {
    @Bean
    public Date createDate() {
        Date nowDate = new Date();
        System.out.println("createDate " + nowDate.hashCode());
        return nowDate;
    }
}
```

运行类代码如下：

```
public class Test1 {
    public static void main(String[] args) {
        ApplicationContext context = new AnnotationConfigApplicationContext("createbean");
    }
}
```

程序运行的结果如下：

```
createDate -312537331
```

从控制台输出的信息来看，只将 Date 对象进行了实例化，并没有创建 Userinfo.java 类的对象。如果想将其他包中的类进行实例化，那么就需要使用注解：

```
@ComponentScan(basePackages = "entity")
```

更改工厂类，代码如下：

```
@Configuration
@ComponentScan(basePackages = "entity")
public class SpringConfig {
    @Bean
    public Date createDate() {
        Date nowDate = new Date();
        System.out.println("createDate " + nowDate.hashCode());
        return nowDate;
    }
}
```

程序运行的效果如下：

```
public Userinfo() 1239548589
createDate -312501532
```

注意：在任何 IoC 容器的组件里，使用 @ComponentScan 注解可以扫描其他包中的类，那些类包含 JavaConfig 配置类。

6. 使用 @ComponentScan(basePackages = "")扫描多个包

注解 @ComponentScan 支持对多个包进行扫描。

创建测试用的项目 ComponentScanMorePackage。

创建实体类，代码如下：

```
@Component
public class Userinfo {
    public Userinfo() {
        System.out.println("public Userinfo() " + this.hashCode());
    }
}
```

创建实体类，代码如下：

```
@Component
public class Bookinfo {
    public Bookinfo() {
        System.out.println("public Bookinfo() " + this.hashCode());
    }
}
```

创建工厂类，代码如下：

```
@Configuration
@ComponentScan(basePackages = { "entity1", "entity2" })
public class SpringConfig {
    @Bean
    public Date createDate() {
        Date nowDate = new Date();
        System.out.println("createDate " + nowDate.hashCode());
        return nowDate;
    }
}
```

注解还可以使用如下写法：

```
@ComponentScan(basePackages = "entity1")
@ComponentScan(basePackages = "entity2")
```

作用也是相同的。

运行类代码如下：

```
public class Test1 {
    public static void main(String[] args) {
        ApplicationContext context = new AnnotationConfigApplicationContext("createbean");
    }
}
```

程序运行后的效果如下：

```
public Userinfo() 1795960102
public Bookinfo() 1027591600
createDate -312358510
```

7. 使用 @ComponentScan 的 basePackageClasses 属性进行扫描

注解 @ComponentScan 的 basePackageClasses 属性的作用是扫描指定*.class 文件所在的包的路径，然后创建该包以及子孙包下类的对象。

创建测试用的项目 basePackageClassesTest，项目结构如图 3-5 所示。

在 entity1.entity2.entity3 包中有 Bookinfo.java 类和 Userinfo.java 类，在 entity1.entity2 包中有 Entity2222222222222222222.java 类。下面要使用 basePackageClasses 属性扫描 Userinfo.java 类所在的包中所有的组件。

工厂类代码如下：

```
@Configuration
@ComponentScan(basePackageClasses = { Userinfo.class })
public class SpringConfig {
    @Bean
    public Date createDate() {
        Date nowDate = new Date();
        System.out.println("createDate " + nowDate.
hashCode());
        return nowDate;
    }
}
```

```
basePackageClassesTest
  JRE System Library [JavaSE-1.8]
  src
    createbean
      SpringConfig.java
    entity1.entity2
      Entity2222222222222222222.java
    entity1.entity2.entity3
      Bookinfo.java
      Userinfo.java
    test
      Test1.java
  SpringJAR
```

图 3-5　项目结构

运行类代码如下：

```
public class Test1 {
    public static void main(String[] args) {
        ApplicationContext context = new AnnotationConfigApplicationContext
(SpringConfig.class);
    }
}
```

程序运行的结果如下：

```
public Bookinfo() 1678854096
public Userinfo() 1849201180
createDate -312196541
```

Entity2222222222222222222.java 类并没有被实例化，只是将 Bookinfo.java 和 Userinfo.java 类实例化了，说明属性 basePackageClasses 实例化的对象为同级包以及子孙包下的类对象。

注解 @ComponentScan 还可以指定对多个 class 文件所在的路径进行扫描，示例代码如下：

```
@ComponentScan(basePackageClasses = { Entity2222222222222222222.class, Userinfo.class })
```

注解 @ComponentScan 允许拆分多行，下面的写法是有效的：

```
@ComponentScan(basePackageClasses = { Userinfo.class })
@ComponentScan(basePackageClasses = { Entity2222222222222222222.class })
```

8．使用 @ComponentScan 而不使用 basePackages 属性的效果

如果单独使用 @ComponentScan 注解而不使用任何的属性，则@ComponentScan 注解默认扫描的是使用 @Configuration 注解的配置类所在的包路径下的所有组件，包含子包中的组件。

创建项目 ComponentScanTest2，项目结构如图 3-6 所示。

实体类代码如下：

```
@Component
public class Userinfo {
    public Userinfo() {
        System.out.println("public Userinfo() " +
this.hashCode());
    }
}
```

ComponentScanTest2
JRE System Library [JavaSE-1.8]
src
createbean
entity1
Bookinfo.java
SpringConfig.java
entity
Userinfo.java
test
Test1.java
SpringJAR

图 3-6 项目结构

实体类代码如下：

```
@Component
public class Bookinfo {
    public Bookinfo() {
        System.out.println("public Bookinfo() " + this.hashCode());
    }
}
```

工厂类代码如下：

```
@Configuration
@ComponentScan
public class SpringConfig {
    @Bean
    public Date createDate() {
        Date nowDate = new Date();
        System.out.println("createDate " + nowDate.hashCode());
        return nowDate;
    }
}
```

运行类代码如下：

```
public class Test1 {
    public static void main(String[] args) {
        ApplicationContext context = new AnnotationConfigApplicationContext(SpringConfig.class);
    }
}
```

程序运行后控制台的输出如下：

```
public Bookinfo() 1027591600
createDate -311939949
```

9. 解决不同包中因有相同类名而出现异常的问题

如果在不同的包中有相同的类名，则在扫描时会发生异常：

```
org.springframework.context.annotation.ConflictingBeanDefinitionException
```

创建项目 packageDiffClassSame。
实体类在 entity1 包中，代码如下：

```
@Component
public class Userinfo {
    public Userinfo() {
        System.out.println("public entity1.Userinfo() " + this.hashCode());
    }
}
```

实体类在 entity2 包中，代码如下：

```
@Component
public class Userinfo {
    public Userinfo() {
        System.out.println("public entity2.Userinfo() " + this.hashCode());
    }
}
```

工厂类代码如下：

```
@Configuration
@ComponentScan(basePackages = "entity1")
@ComponentScan(basePackages = "entity2")
public class SpringConfig {
}
```

运行类代码如下：

```
public class Test1 {
    public static void main(String[] args) {
        ApplicationContext context = new AnnotationConfigApplicationContext("createbean");
    }
}
```

程序运行后控制台输出的异常信息如下：

```
    Caused by: org.springframework.context.annotation.ConflictingBeanDefinitionException:
 Annotation-specified bean name 'userinfo' for bean class [entity2.Userinfo] conflicts with
existing, non-compatible bean definition of same name and class [entity1.Userinfo]
```

解决异常的办法有两种。

（1）将类名设置为不相同。

（2）为组件定义一个别名。

下面实现第 2 种，修改两个实体类，代码如下：

```
@Component(value = "userinfo1")
public class Userinfo {

@Component(value = "userinfo2")
public class Userinfo {
```

修改运行类，代码如下：

```
public class Test1 {
    public static void main(String[] args) {
        ApplicationContext context = new AnnotationConfigApplicationContext("createbean");
        System.out.println(context.getBean("userinfo1").hashCode());
        System.out.println(context.getBean("userinfo2").hashCode());
    }
}
```

再次运行程序后输出的结果如下：

```
public entity1.Userinfo() 876926621
public entity2.Userinfo() 326298949
876926621
326298949
```

出现的异常问题被解决。

10．建议使用的代码结构

在使用 Spring 注解扫描包时，一个较好的实践方式是在包的 root 位置启用扫描，来对子包及子孙包中的类进行搜索。

创建测试用的项目 codeStruts。

代码结构的设计如下：

```
src
└─com
    └─ghy
        └─www
            │  Start.java
            │
            ├─entity1
            │      UserinfoA.java
            │
            └─entity2
                   UserinfoB.java
```

Start.java 类的主要作用就是在 root 位置启用扫描搜索，来对子包及其子孙包中的组件进行实例化，这样就不需要再使用如下注解进行扫描了：

```
@ComponentScan(value="packageName")
```

然后在 main 方法中使用如下代码进行启动：

```
    public static void main(String[] args) {
        ApplicationContext context = new AnnotationConfigApplicationContext("com.ghy.www");
    }
```

11．使用 @Lazy 注解实现延迟加载

创建测试用的项目 annotationLazyTest。

创建实体类，代码如下：

```
@Component
public class Userinfo {
    public Userinfo() {
        System.out.println("public Userinfo() " + this);
    }
}
```

创建 JavaConfig 类，代码如下：

```
@Configuration
public class JavaConfig {
    @Bean(name = "userinfo1")
    public Userinfo getUserinfo1() {
        Userinfo userinfo = new Userinfo();
        System.out.println("getUserinfo1 " + userinfo);
        return userinfo;
    }
```

```
    @Bean(name = "userinfo2")
    @Lazy(value = true)
    public Userinfo getUserinfo2() {
        Userinfo userinfo = new Userinfo();
        System.out.println("getUserinfo2 " + userinfo);
        return userinfo;
    }
}
```

创建运行类 1，代码如下：

```
public class Test1 {
    public static void main(String[] args) {
        ApplicationContext context = new AnnotationConfigApplicationContext("javaconfig");
    }
}
```

程序运行的结果如下：

```
public Userinfo() entity.Userinfo@59717824
getUserinfo1 entity.Userinfo@59717824
```

这里并没有执行 getUserinfo2 方法来创建 Userinfo 对象，说明使用 @Lazy（value = true）实现的效果是延迟加载。

创建运行类 2，代码如下：

```
public class Test2 {
    public static void main(String[] args) {
        ApplicationContext context = new AnnotationConfigApplicationContext("javaconfig");
        context.getBean("userinfo2");
    }
}
```

程序运行的结果如下：

```
public Userinfo() entity.Userinfo@59717824
getUserinfo1 entity.Userinfo@59717824
public Userinfo() entity.Userinfo@7bb58ca3
getUserinfo2 entity.Userinfo@7bb58ca3
```

这里执行 getUserinfo2 方法来创建 Userinfo 对象，说明执行 context.getBean（"userinfo2"）代码才会创建 Userinfo 对象，这属于延迟创建。

12．出现 Overriding bean definition 及其解决方法

当多个 applicationContext.xml 配置文件中<bean>的 id 属性值一样，并且数据类型也一样时，会出现 bean 覆盖的情况。

创建测试用的项目 beanOverriding3。

创建如下类代码：

```
@Component
public class Userinfo {
    public Userinfo() {
        System.out.println("public Userinfo() " + this);
    }
}
```

创建如下类代码：

```
@Configuration
public class JavaConfig {
    @Bean(name = "userinfo")
    public Userinfo getUserinfo() {
        System.out.println("@Bean creator getUser");
        return new Userinfo();
    }
}
```

创建如下类代码：

```
public class MyTest {
    public static void main(String[] args) {
        ApplicationContext context = new AnnotationConfigApplicationContext("root");
    }
}
```

运行程序后控制台输出的结果如下：

```
    信息: Overriding bean definition for bean 'userinfo' with a different definition: replacing
[Generic bean: class [root.Userinfo]; scope=singleton; abstract=false; lazyInit=false;
autowireMode=0; dependencyCheck=0; autowireCandidate=true; primary=false; factoryBeanName=
null; factoryMethodName=null; initMethodName=null; destroyMethodName=null; defined in file
[C:\spring-tool-suite-3.9.5.RELEASE-workspace\beanOverriding3\bin\root\ Userinfo.class]]
with [Root bean: class [null]; scope=; abstract=false; lazyInit=false; autowireMode=3;
dependencyCheck=0; autowireCandidate=true; primary=false; factoryBeanName=javaConfig;
factoryMethodName=getUserinfo; initMethodName=null; destroyMethodName=(inferred); defined in
root.JavaConfig]
    @Bean creator getUser
    public Userinfo() root.Userinfo@6305cb20
```

控制台输出 info 级别的日志信息，提示出现“Overriding bean definition”的情况，也就是在工厂方法中使用注解 @Bean（name = "userinfo"）创建的 Userinfo 的别名是 userinfo，而通过扫描创建的 Userinfo 的别名也是 userinfo，出现了 JavaBean 覆盖的情况。解决的办法是将 @Bean（name = "xxxxxxxxxx"）的 id 值设置为不同，解决此问题的项目在 beanOverriding4 中。

3.7.3　处理 JavaBean 的生命周期

IoC 容器中的 JavaBean 也具有生命周期，处理生命周期可以使用 XML 和注解这两种方式。

1．XML 方式——init-method 和 destroy-method 属性处理指定 JavaBean 的生命周期

创建测试用的项目 begin_end_scope。

创建业务类 UserinfoService1，代码如下：

```
package service;

public class UserinfoService1 {
    public UserinfoService1() {
        System.out.println("public UserinfoService1() " + this.hashCode());
    }
```

```
    public void initMethod() {
        System.out.println("public void initMethod() 1");
    }

    public void destroyMethod() {
        System.out.println("public void destroyMethod() 1");
    }

    public void save() {
        System.out.println("将数据保存到 A 数据库中 1");
    }
}
```

创建配置文件 ac1.xml，代码如下：

```
<!-- init-method: javaBean 生命周期初始化方法，对象创建后就进行调用 -->
<!-- destroy-method: 容器被销毁时，如果 bean 被容器管理，会调用该方法。 -->
<bean id="userinfoService1" class="service.UserinfoService1"
    init-method="initMethod" destroy-method="destroyMethod"></bean>
```

在<bean>标签中使用 init-method 和 destroy-method 属性处理生命周期，属性值就是方法名称。

创建运行类，代码如下：

```
package test;

import org.springframework.context.support.ClassPathXmlApplicationContext;

import service.UserinfoService1;

public class Test1 {
    public static void main(String[] args) {
        ClassPathXmlApplicationContext context = new ClassPathXmlApplicationContext
("ac1.xml");
        UserinfoService1 service = (UserinfoService1) context.getBean("userinfoService1");
        service.save();
        context.close();
    }
}
```

程序运行的结果如下：

```
public UserinfoService1() 1781256139
public void initMethod() 1
将数据保存到 A 数据库中 1
public void destroyMethod() 1
```

2．注解方式——@PostConstruct 和 @PreDestroy 注解处理指定 JavaBean 的生命周期

创建业务类 UserinfoService，代码如下：

```
package test1;

import javax.annotation.PostConstruct;
import javax.annotation.PreDestroy;

import org.springframework.stereotype.Service;
```

```
@Service
public class UserinfoService {
    public UserinfoService() {
        System.out.println("public UserinfoService() " + this.hashCode());
    }

    @PostConstruct
    public void initMethod() {
        System.out.println("public void initMethod()");
    }

    @PreDestroy
    public void destroyMethod() {
        System.out.println("public void destroyMethod()");
    }

    public void save() {
        System.out.println("将数据保存到 A 数据库中");
    }
}
```

创建 JavaConfig 配置类，代码如下：

```
package test1;

import org.springframework.context.annotation.ComponentScan;
import org.springframework.context.annotation.Configuration;

@Configuration
@ComponentScan
public class JavaConfig {

}
```

创建运行类，代码如下：

```
package test1;

import org.springframework.context.annotation.AnnotationConfigApplicationContext;

public class Test1 {
    public static void main(String[] args) {
        AnnotationConfigApplicationContext context = new AnnotationConfigApplicationContext
(JavaConfig.class);
        UserinfoService service = (UserinfoService) context.getBean(UserinfoService.class);
        service.save();
        context.close();
    }
}
```

程序运行后的效果如下：

```
public UserinfoService() 1786364562
public void initMethod()
将数据保存到 A 数据库中
public void destroyMethod()
```

3. 注解方式——@Bean 注解的 initMethod 和 destroyMethod 属性处理指定 JavaBean 的生命周期

创建业务类 UserinfoService，代码如下：

```
package test2;

public class UserinfoService {
    public UserinfoService() {
        System.out.println("public UserinfoService() " + this.hashCode());
    }

    public void initMethod() {
        System.out.println("public void initMethod()");
    }

    public void destroyMethod() {
        System.out.println("public void destroyMethod()");
    }

    public void save() {
        System.out.println("将数据保存到 A 数据库中");
    }
}
```

创建 JavaConfig 配置类，代码如下：

```
package test2;

import org.springframework.context.annotation.Bean;
import org.springframework.context.annotation.Configuration;

@Configuration
public class JavaConfig {
    @Bean(initMethod = "initMethod", destroyMethod = "destroyMethod")
    public UserinfoService getUserinfoService() {
        return new UserinfoService();
    }
}
```

创建运行类，代码如下：

```
package test2;

import org.springframework.context.annotation.AnnotationConfigApplicationContext;

public class Test1 {
    public static void main(String[] args) {
        AnnotationConfigApplicationContext context = new AnnotationConfigApplicationContext
(JavaConfig.class);
        UserinfoService service = (UserinfoService) context.getBean(UserinfoService.class);
        service.save();
        context.close();
    }
}
```

程序运行后的效果如下：

```
public UserinfoService() 376416077
public void initMethod()
将数据保存到 A 数据库中
public void destroyMethod()
```

4．XML 方式——使用 default-init-method 和 default-destroy-method 属性处理全部 JavaBean 的生命周期

在<bean>标签中使用 init-method 和 destroy-method 属性可以设置指定 JavaBean 的生命周期。如果在<beans>标签中使用 default-init-method 和 default-destroy-method 属性，则可批量设置全部 JavaBean 的生命周期。

创建新的业务类 UserinfoService2，代码如下：

```java
package service;

public class UserinfoService2 {
    public UserinfoService2() {
        System.out.println("public UserinfoService2() " + this.hashCode());
    }

    public void initMethod() {
        System.out.println("public void initMethod() 2");
    }

    public void destroyMethod() {
        System.out.println("public void destroyMethod() 2");
    }

    public void save() {
        System.out.println("将数据保存到 A 数据库中 2");
    }
}
```

创建配置文件 ac2.xml，代码如下：

```xml
<?xml version="1.0" encoding="UTF-8"?>
<beans xmlns="http://www.springframework.org/schema/beans"
    xmlns:xsi="http://www.w3.org/2001/XMLSchema-instance"
    xmlns:context="http://www.springframework.org/schema/context"
    xsi:schemaLocation="http://www.springframework.org/schema/beans http://www.
springframework.org/schema/beans/spring-beans.xsd
        http://www.springframework.org/schema/context http://www.springframework.org/
schema/context/spring-context-4.3.xsd"
    default-init-method="initMethod" default-destroy-method="destroyMethod">
    <bean id="userinfoService1" class="service.UserinfoService1"></bean>
    <bean id="userinfoService2" class="service.UserinfoService2"></bean>
</beans>
```

运行类代码如下：

```java
package test;

import org.springframework.context.support.ClassPathXmlApplicationContext;

import service.UserinfoService1;
import service.UserinfoService2;
```

```
public class Test2 {
    public static void main(String[] args) {
        ClassPathXmlApplicationContext context = new ClassPathXmlApplicationContext("ac2.xml");
        UserinfoService1 service1 = (UserinfoService1) context.getBean("userinfoService1");
        UserinfoService2 service2 = (UserinfoService2) context.getBean("userinfoService2");
        service1.save();
        service2.save();
        context.close();
    }
}
```

程序运行后的效果如下：

```
public UserinfoService1() 1307096070
public void initMethod() 1
public UserinfoService2() 1196765369
public void initMethod() 2
将数据保存到A数据库中 1
将数据保存到A数据库中 2
public void destroyMethod() 2
public void destroyMethod() 1
```

5. 属性 scope="prototype"对生命周期的影响

如果 bean 标签中的 scope="prototype"，那么容器只负责实例化和初始化 init-method，容器并不执行对象的销毁方法，而是由程序员进行显式调用 destroy-method 销毁方法。

创建配置文件 ac3.xml，代码如下：

```
<bean id="userinfoService1" class="service.UserinfoService1"
    scope="prototype" init-method="initMethod"
    destroy-method="destroyMethod"></bean>
<bean id="userinfoService2" class="service.UserinfoService2"
    scope="prototype" init-method="initMethod"
    destroy-method="destroyMethod"></bean>
```

运行类代码如下：

```
package test;

import org.springframework.context.support.ClassPathXmlApplicationContext;

import service.UserinfoService1;
import service.UserinfoService2;

public class Test3 {
    public static void main(String[] args) {
        ClassPathXmlApplicationContext context = new ClassPathXmlApplicationContext("ac3.xml");
        UserinfoService1 service11 = (UserinfoService1) context.getBean("userinfoService1");
        UserinfoService1 service12 = (UserinfoService1) context.getBean("userinfoService1");
        UserinfoService2 service21 = (UserinfoService2) context.getBean("userinfoService2");
        UserinfoService2 service22 = (UserinfoService2) context.getBean("userinfoService2");
        service11.save();
        service12.save();
        service21.save();
        service22.save();
        context.close();
```

```
    }
}
```

程序运行的结果如下：

```
public UserinfoService1() 214074868
public void initMethod() 1
public UserinfoService1() 1860944798
public void initMethod() 1
public UserinfoService2() 1179381257
public void initMethod() 2
public UserinfoService2() 258754732
public void initMethod() 2
将数据保存到 A 数据库中 1
将数据保存到 A 数据库中 1
将数据保存到 A 数据库中 2
将数据保存到 A 数据库中 2
```

3.8 装配 Spring Bean

依赖注入（DI）会产生类与类之间的关联，产生类之间关联的行为叫作装配（wiring），本节将使用多种方式来装配 JavaBean 之间的关系。

在 Spring 框架中，一个 JavaBean 不需要创建另一个 JavaBean，JavaBean 之间的关系全部由 IoC 容器进行管理，也就是说在 A 类中虽然声明了 B 类的对象，但是不需要实例化 B 类，B 类的对象由容器进行创建，容器还能为 A 类中的 B 对象进行赋值，这一切都由 Spring 框架的 IoC 容器来完成。

把 A 类实例化 B 类的写法交由 IoC 容器来处理，这样就使 A 类和 B 类产生了松耦合，程序代码利于扩展与后期维护。

Spring 默认按类型（byType）进行匹配，然后实施注入。

3.8.1 使用 XML 声明法注入对象

XML 声明法能实现注入对象的功能，这也是 Spring 最先采用的注入方式。

创建测试用的项目 diTest1。

实体类代码如下：

```
public class Userinfo {
    public Userinfo() {
        System.out.println("public Userinfo() " + this);
    }
}
```

配置文件的代码如下：

```
<bean id="userinfo" class="entity.Userinfo"></bean>
<bean id="test1" class="test.Test1">
    <property name="userinfo1" ref="userinfo"></property>
</bean>
<bean id="test2" class="test.Test1">
    <property name="userinfo2">
        <ref bean="userinfo" />
```

```
    </property>
</bean>
```

为 test 包中 Test1.java 类的 userinfo1 和 userinfo2 属性注入 userinfo 对象。

运行类代码如下：

```java
public class Test1 {

    public Test1() {
        System.out.println("public Test() " + this.hashCode());
    }

    private Userinfo userinfo1;
    private Userinfo userinfo2;

    public Userinfo getUserinfo1() {
        return userinfo1;
    }

    public void setUserinfo1(Userinfo userinfo1) {
        this.userinfo1 = userinfo1;
        System.out.println("setUserinfo1 userinfo1=" + userinfo1);
    }

    public Userinfo getUserinfo2() {
        return userinfo2;
    }

    public void setUserinfo2(Userinfo userinfo2) {
        this.userinfo2 = userinfo2;
        System.out.println("setUserinfo2 userinfo1=" + userinfo2);
    }

    public static void main(String[] args) {
        ApplicationContext context = new ClassPathXmlApplicationContext("ac.xml");
        Test1 test1 = (Test1) context.getBean("test1");
        Test1 test2 = (Test1) context.getBean("test2");
        System.out.println(test1.getUserinfo1());
        System.out.println(test2.getUserinfo2());

    }

}
```

程序运行后的效果如下：

```
public Userinfo() entity.Userinfo@573f2bb1
public Test() 346861221
setUserinfo1 userinfo1=entity.Userinfo@573f2bb1
public Test() 1497973285
setUserinfo2 userinfo1=entity.Userinfo@573f2bb1
entity.Userinfo@573f2bb1
entity.Userinfo@573f2bb1
```

使用反射技术动态调用 setUserinfo（Userinfo userinfo）方法传入 Userinfo.java 类的对象，最后在 main() 方法中输出 Userinfo.java 类的对象信息。

3.8.2 使用注解声明法注入对象

使用注解声明法注入对象时还需要使用以下代码：

```
<context:component-scan base-package="entity"></context:component-scan>
```

创建测试用的项目 diTest2。
实体类代码如下：

```
@Component
public class Userinfo {
    public Userinfo() {
        System.out.println("public Userinfo() " + this.hashCode());
    }
}
```

配置文件代码如下：

```
<context:component-scan base-package="entity"></context:component-scan>
<context:component-scan base-package="test"></context:component-scan>
```

运行类代码如下：

```
@Component
public class Test1 {
    public Test1() {
        System.out.println("public Test() " + this.hashCode());
    }

    @Autowired
    private Userinfo userinfo;

    public Userinfo getUserinfo() {
        return userinfo;
    }

    public void setUserinfo(Userinfo userinfo) {
        this.userinfo = userinfo;
        System.out.println("public void setUserinfo(Userinfo userinfo) userinfo=" + userinfo);
    }

    public static void main(String[] args) {
        ApplicationContext context = new ClassPathXmlApplicationContext("ac.xml");
        Test1 test = (Test1) context.getBean(Test1.class);
        System.out.println("main test " + test.hashCode());
        System.out.println(test.getUserinfo().hashCode());
    }
}
```

程序运行后的结果如下：

```
public Userinfo() 266437232
public Test() 323326911
main test 323326911
266437232
```

方法 setUserinfo（Userinfo）并没有执行，因为使用反射技术直接对 userinfo 变量进行赋值了。如果将 @Autowired 注解放在 public void setUserinfo(Userinfo userinfo) 方法之上，那么 set 方法会被调用。

3.8.3 多实现类的歧义性

如果对接口类型的变量进行注入，当 IoC 容器发现有多个实现类时，会在注入前抛出异常，Spring 并不知道应该通过哪个实现类对接口进行注入。

创建测试用的项目 diTest3。

接口代码如下：

```
public interface IUserinfoService {
    public void save();
}
```

实现类 A 代码如下：

```
@Service
public class UserinfoServiceA implements IUserinfoService {
    public UserinfoServiceA() {
        System.out.println("public Userinfo() " + this.hashCode());
    }

    public void save() {
        System.out.println("将数据保存到 A 数据库中");
    }
}
```

实现类 B 代码如下：

```
@Service
public class UserinfoServiceB implements IUserinfoService {
    public UserinfoServiceB() {
        System.out.println("public Userinfo() " + this.hashCode());
    }

    public void save() {
        System.out.println("将数据保存到 B 数据库中");
    }
}
```

配置文件的代码如下：

```
<context:component-scan base-package="service"></context:component-scan>
<context:component-scan base-package="test"></context:component-scan>
```

运行类代码如下：

```
@Component
public class Test1 {
    public Test1() {
        System.out.println("public Test() " + this.hashCode());
    }

    @Autowired
```

```
    private IUserinfoService userinfoService;

    public IUserinfoService getUserinfoService() {
        return userinfoService;
    }

    public void setUserinfoService(IUserinfoService userinfoService) {
        this.userinfoService = userinfoService;
    }

    public static void main(String[] args) {
        ApplicationContext context = new ClassPathXmlApplicationContext("ac.xml");
        Test1 test = (Test1) context.getBean(Test1.class);
        test.getUserinfoService().save();
    }
}
```

程序运行后出现的异常信息如下：

```
Caused by: org.springframework.beans.factory.NoUniqueBeanDefinitionException: No
qualifying bean of type 'service.IUserinfoService' available: expected single matching
bean but found 2: userinfoServiceA,userinfoServiceB
```

再次出现 NoUniqueBeanDefinitionException 异常。因为 Spring 框架不能确定具体注入的是哪个实现类对象，所以出现异常。解决的办法有以下 3 种。

（1）使用@Primary 注解。

（2）使用@Autowired 结合 @Qualifier 注解。

（3）使用 @Resource 注解。

1. 使用 @Primary 注解

创建测试用的项目 diTest3_1。

修改业务类，代码如下：

```
@Service
@Primary
public class UserinfoServiceB implements IUserinfoService {
```

注解 @Primary 表示在遇到相同类型的注入时，当前的组件具有高优先级。

2. 使用 @Autowired 结合 @Qualifier 注解

对 @Service 组件设置一个别名，@Qualifier 注解注入指定别名的组件。

创建测试用的项目 diTest3_2。

修改业务类，代码如下：

```
@Service(value = "userinfoServiceA")
public class UserinfoServiceA implements IUserinfoService {

@Service(value = "userinfoServiceB")
public class UserinfoServiceB implements IUserinfoService {
```

修改运行类，代码如下：

```
@Autowired
@Qualifier(value = "userinfoServiceB")
private IUserinfoService userinfoService;
```

3. 使用 @Resource 注解

创建测试用的项目 diTest3_3。

修改业务类，代码如下：

```
@Service(value = "userinfoServiceA")
public class UserinfoServiceA implements IUserinfoService {

@Service(value = "userinfoServiceB")
public class UserinfoServiceB implements IUserinfoService {
```

修改运行类，代码如下：

```
    @Resource(name = "userinfoServiceB")
    private IUserinfoService userinfoService;
```

3.8.4 使用 @Autowired 注解向构造方法进行注入

使用@Autowired 注解可以向构造方法的参数进行注入。

创建测试用的项目 diTest5。

实体类代码如下：

```
@Component
public class Userinfo {
    public Userinfo() {
        System.out.println("public Userinfo() " + this.hashCode());
    }
}
```

实体类代码如下：

```
@Component
public class Bookinfo {
    @Autowired
    public Bookinfo(Userinfo userinfo) {
        System.out.println("public Bookinfo(Userinfo userinfo) userinfo=" + userinfo.
hashCode());
    }
}
```

配置类代码如下：

```
@Configuration
@ComponentScan(basePackages = "entity")
public class SpringConfig {
}
```

运行类代码如下：

```
public class Test1 {
    public static void main(String[] args) {
```

```
            new AnnotationConfigApplicationContext(SpringConfig.class);
        }
    }
```

程序运行后的结果如下：

```
public Userinfo() 366590980
public Bookinfo(Userinfo userinfo) userinfo=366590980
```

如果类仅有 1 个有参构造方法，那么可以省略 @Autowire 注解，Spring 会自动进行注入。

3.8.5 使用 @Autowired 注解向 set 方法进行注入

可以向 set 方法注入参数。

创建测试用的项目 diTest6。

实体类代码如下：

```
@Component
public class Bookinfo {
    @Autowired
    public void setUserinfo(Userinfo userinfo) {
        System.out.println("public void setUserinfo(Userinfo userinfo) userinfo=" +
userinfo.hashCode());
    }
}
```

3.8.6 使用 @Autowired 注解向 Field 进行注入

可以向 Field 字段进行注入。

创建测试用的项目 diTestField。

实体类代码如下：

```
@Component
public class Bookinfo {
    @Autowired
    private Userinfo userinfo;

    public Userinfo getUserinfo() {
        return userinfo;
    }

    public void setUserinfo(Userinfo userinfo) {
        this.userinfo = userinfo;
    }
}
```

运行类代码如下：

```
public class Test1 {
    public static void main(String[] args) {
        ApplicationContext context = new AnnotationConfigApplicationContext
(SpringConfig.class);
        Bookinfo bookinfo = (Bookinfo) context.getBean(Bookinfo.class);
        System.out.println(bookinfo.getUserinfo());
    }
}
```

程序运行后的结果如下：

```
public Userinfo() entity.Userinfo@1c72da34
entity.Userinfo@1c72da34
```

通过前面的实验可以得知，使用 @Autowired 注解可以对字段、方法与构造方法的参数进行注入。

3.8.7 使用 @Inject 向 Field-setMethod-Constructor 进行注入

@Inject 注解可以实现注入，但需要 javaee-api-8.0.jar 文件，因为@Inject 注解是 Oracle 提供的，而 @Autowired 是 Spring 官方提供的注解。

使用 @Inject 向 Field 字段注入的测试项目为 inject_test1。

核心代码如下：

```
@Inject
private Userinfo userinfo;
```

使用 @Inject 向 setMethod 方法注入的测试项目为 inject_test2。

核心代码如下：

```
@Inject
public void setUserinfo(Userinfo userinfo) {
    this.userinfo = userinfo;
}
```

使用 @Inject 向 Constructor 方法注入的测试项目为 inject_test3。

```
@Component
public class Test1 {

    @Inject
    public Test1(Userinfo userinfo) {
        System.out.println("public Test1(Userinfo userinfo) userinfo=" + userinfo);
}
```

如果出现多个数据类型相同的情况，那么可以使用如下代码注入指定的 JavaBean：

```
@Inject
@Qualifier(value = "serviceB")
private IUserinfoService service;
```

上述实验代码在项目 inject_test4 中。

3.8.8 向 @Bean 工厂方法注入参数

本节将对工厂方法进行参数注入。

创建测试用的项目 diTest7。

实体类代码如下：

```
@Component
public class Userinfo {
    public Userinfo() {
        System.out.println("public Userinfo() " + this.hashCode());
    }
}
```

实体类代码如下：

```
public class Bookinfo {
    public Bookinfo(Userinfo userinfo) {
        System.out.println("public Bookinfo(Userinfo userinfo) userinfo=" + userinfo.
hashCode());
    }
}
```

工厂类代码如下：

```
@Configuration
@ComponentScan(basePackages = "entity2")
public class SpringConfig {
    @Bean
    public Bookinfo getBookinfo(Userinfo userinfo) {
        System.out.println("public Bookinfo getBookinfo(Userinfo userinfo) userinfo="
 + userinfo.hashCode());
        return new Bookinfo(userinfo);
    }
}
```

@Component 注解声明在类上，而注解 @Bean 声明在方法上，两个注解的功能都是将 JavaBean 放入 IoC 容器中。

运行类代码如下：

```
public class Test1 {
    public static void main(String[] args) {
        new AnnotationConfigApplicationContext(SpringConfig.class);
    }
}
```

程序运行后的结果如下：

```
public Userinfo() 477289012
public Bookinfo getBookinfo(Userinfo userinfo) userinfo=477289012
public Bookinfo(Userinfo userinfo) userinfo=477289012
```

3.8.9 使用 @Autowired (required = false) 的写法

在注入时，如果找不到符合条件的 JavaBean 对象，那么控制台会出现 NoSuchBeanDefinitionException 异常。

创建项目 diTest8，项目结构如图 3-7 所示。

```
diTest8
    JRE System Library [JavaSE-1.8]
    src
        entity
            Userinfo.java
                Userinfo
        springconfig
            SpringConfig.java
        test
            Test1.java
    SpringJAR
```

图 3-7 项目结构

配置类代码如下：

```
@Configuration
@ComponentScan(basePackages = "test")
public class SpringConfig {
}
```

运行类代码如下：

```
@Component
public class Test1 {
    @Autowired
```

```
    private Userinfo userinfo;

    public Userinfo getUserinfo() {
        return userinfo;
    }

    public void setUserinfo(Userinfo userinfo) {
        this.userinfo = userinfo;
    }

    public static void main(String[] args) {
        ApplicationContext context = new AnnotationConfigApplicationContext
(SpringConfig.class);
        Test1 test = (Test1) context.getBean(Test1.class);
        System.out.println(test.getUserinfo());
    }
}
```

程序运行后出现的异常信息如下：

```
Caused by: org.springframework.beans.factory.NoSuchBeanDefinitionException: No qualifying
bean of type 'entity.Userinfo' available: expected at least 1 bean which qualifies as auto
wire candidate. Dependency annotations: {@org.springframework.beans.factory.annotation.
Autowired(required=true)}
```

异常信息提示没有找到符合条件的 JavaBean 对象，这时为了避免发生异常，可以加入 required = false 属性，修改后的代码如下：

```
    @Autowired(required = false)
    private Userinfo userinfo;
```

程序运行后并没有出现异常，因为没有找到符合条件的记录，所以对象的值为 null，输出内容如下：

```
null
```

3.8.10 使用 @Bean 为 JavaBean 的 id 重命名

使用 @Bean 创建的 JavaBean 的 id 值默认就是方法的名称，还可以自定义 id 值。

创建测试用的项目 diTest10。

示例代码如下：

```
@Configuration
@ComponentScan(basePackages = "test")
public class SpringConfig {
    @Bean(name = "xxxxxxxxxxxxxxxxxxxxxxxx")
    public Date createDate() {
        Date date = new Date();
        System.out.println("public Date createDate() xxxxxxxxxxxxxxxxxxxxxxxx=" +
date.hashCode());
        return date;
    }

    @Bean(name = "zzzzzzzzzzzzzzzzzzzz")
    public Date getDate() {
        Date date = new Date();
```

```
        System.out.println("public Date getDate() zzzzzzzzzzzzzzzzzzzzz=" + date.hashCode());
        return date;
    }
}
```

在运行类中注入指定的 JavaBean 对象，代码如下：

```
@Component
public class Test1 {
    @Resource(name = "xxxxxxxxxxxxxxxxxxxxxxx")
    private Date date;

    public Date getDate() {
        return date;
    }

    public void setDate(Date date) {
        this.date = date;
    }

    public static void main(String[] args) {
        ApplicationContext context = new AnnotationConfigApplicationContext
(SpringConfig.class);
        Test1 test1 = (Test1) context.getBean(Test1.class);
        System.out.println("main date=" + test1.getDate().hashCode());

    }
}
```

3.8.11　为构造方法进行注入

使用 @Autowired 可以为构造方法进行注入，在 XML 配置文件中也可以为构造方法进行注入。

1. 使用<constructor-arg>为构造方法注入基本类型

创建测试用的项目 diTest11。

创建实体类，代码如下：

```
public class Userinfo {

    public Userinfo(String username, String password) {
        System.out.println(
                "public Userinfo(String username, String password) username=" +
username + " password=" + password);
    }

    public Userinfo(String username, int age) {
        System.out.println("public Userinfo(String username, int age) username=" +
username + " age=" + age);
    }

    public Userinfo(String username, Integer age) {
        System.out.println("public Userinfo(String username, Integer age) username="
+ username + " age=" + age);
    }
}
```

配置文件 applicationContext.xml 的内容如下：

```
<bean id="userinfo1" class="entity.Userinfo">
    <constructor-arg type="java.lang.String" value="中国"></constructor-arg>
    <constructor-arg type="java.lang.String" value="中国人"></constructor-arg>
</bean>

<bean id="userinfo2" class="entity.Userinfo">
    <constructor-arg type="java.lang.String" value="中国"></constructor-arg>
    <constructor-arg type="int" value="123"></constructor-arg>
</bean>

<bean id="userinfo3" class="entity.Userinfo">
    <constructor-arg type="java.lang.String" value="中国"></constructor-arg>
    <constructor-arg type="java.lang.Integer" value="456"></constructor-arg>
</bean>
```

注入基本数据类型，使用 value 属性。

运行类代码如下：

```
public class Test1 {
    public static void main(String[] args) {
        ApplicationContext context = new ClassPathXmlApplicationContext("ac.xml");
        context.getBean("userinfo1");
        context.getBean("userinfo2");
        context.getBean("userinfo3");
    }
}
```

通过 getBean 方法来取得 IoC 容器中的对象，再执行对象中的构造方法。

程序运行后的结果如下：

```
public Userinfo(String username, String password) username=中国 password=中国人
public Userinfo(String username, int age) username=中国 age=123
public Userinfo(String username, Integer age) username=中国 age=456
```

2. 使用<constructor-arg>为构造方法注入复杂类型

使用<constructor-arg>标签可以为构造方法注入复杂数据类型。

创建测试用的项目 diTest12。

实体类代码如下：

```
public class Userinfo {
    public Userinfo() {
        System.out.println("public Userinfo() " + this.hashCode());
    }
}
```

实体类代码如下：

```
public class Bookinfo {
    public Bookinfo(Userinfo userinfo) {
        System.out.println("public Bookinfo(Userinfo userinfo) userinfo=" + userinfo.
hashCode());
    }

    public Bookinfo(Userinfo userinfo1, Userinfo userinfo2) {
```

```
            System.out.println("public Bookinfo(Userinfo userinfo1, Userinfo userinfo2)
userinfo1=" + userinfo1.hashCode()
                    + " userinfo2=" + userinfo2.hashCode());
        }
    }
```

配置文件的代码如下：

```
<bean id="userinfo1" class="entity.Userinfo">
</bean>
<bean id="userinfo2" class="entity.Userinfo">
</bean>

<bean id="bookinfo1" class="entity.Bookinfo">
    <constructor-arg ref="userinfo1"></constructor-arg>
</bean>

<bean id="bookinfo2" class="entity.Bookinfo">
    <constructor-arg ref="userinfo1"></constructor-arg>
    <constructor-arg ref="userinfo2"></constructor-arg>
</bean>
```

注入复杂数据类型要使用 ref 属性。

运行类代码如下：

```
public class Test1 {
    public static void main(String[] args) {
        ApplicationContext context = new ClassPathXmlApplicationContext("ac.xml");
        context.getBean("bookinfo1");
        context.getBean("bookinfo2");
    }
}
```

程序运行的结果如下：

```
public Userinfo() 1566502717
public Userinfo() 1757676444
public Bookinfo(Userinfo userinfo) userinfo=1566502717
public Bookinfo(Userinfo userinfo1, Userinfo userinfo2) userinfo1=1566502717 userinfo2=
1757676444
```

3. 使用<constructor-arg>的 index 属性进行注入

可以使用<constructor-arg>的 index 属性对指定索引的构造方法参数进行注入。

创建测试用的项目 diTestIndex。

实体类代码如下：

```
public class Userinfo {
    public Userinfo() {
        System.out.println("public Userinfo() " + this);
    }
}
```

实体类代码如下：

```
public class Bookinfo {
    public Bookinfo(Userinfo userinfo, String address) {
        System.out.println("public Bookinfo(Userinfo userinfo, String address)");
```

```
        System.out.println("userinfo=" + userinfo);
        System.out.println("address=" + address);
    }
}
```

配置文件的代码如下：

```
<bean id="userinfo1" class="entity.Userinfo">
</bean>
<bean id="bookinfo1" class="entity.Bookinfo">
    <constructor-arg index="0" ref="userinfo1"></constructor-arg>
    <constructor-arg index="1" value="北京市"></constructor-arg>
</bean>
```

运行类代码如下：

```
public class Test1 {
    public static void main(String[] args) {
        ApplicationContext context = new ClassPathXmlApplicationContext("ac.xml");
    }
}
```

程序运行后的结果如下：

```
public Userinfo() entity.Userinfo@6cc4c815
public Bookinfo(Userinfo userinfo, String address)
userinfo=entity.Userinfo@6cc4c815
address=北京市
```

4．使用<constructor-arg>的 name 属性进行注入

可以使用<constructor-arg>的 name 属性对指定构造方法的参数进行注入。

创建测试用的项目 diTestName。

实体类代码如下：

```
public class Userinfo {
    public Userinfo() {
        System.out.println("public Userinfo() " + this);
    }
}
```

实体类代码如下：

```
public class Bookinfo {
    public Bookinfo(Userinfo userinfo, String address) {
        System.out.println("public Bookinfo(Userinfo userinfo, String address)");
        System.out.println("userinfo=" + userinfo);
        System.out.println("address=" + address);
    }
}
```

配置文件的代码如下：

```
<bean id="userinfo1" class="entity.Userinfo">
</bean>
<bean id="bookinfo1" class="entity.Bookinfo">
    <constructor-arg name="address" value="北京市"></constructor-arg>
    <constructor-arg name="userinfo" ref="userinfo1"></constructor-arg>
</bean>
```

运行类代码如下：

```
public class Test1 {
    public static void main(String[] args) {
        ApplicationContext context = new ClassPathXmlApplicationContext("ac.xml");
    }
}
```

程序运行后的结果如下：

```
public Userinfo() entity.Userinfo@6cc4c815
public Bookinfo(Userinfo userinfo, String address)
userinfo=entity.Userinfo@6cc4c815
address=北京市
```

5. 使用<constructor-arg>向 factory-method 方法传入参数

创建测试用的项目 xml_factory_method_hasParam。
创建实体类，代码如下：

```
public class Userinfo {
    public Userinfo() {
        System.out.println("private Userinfo() " + this);
    }
}
```

创建工具类，代码如下：

```
public class Creator {
    public Userinfo createUserinfo(Date nowDate, String address) {
        Userinfo userinfo = new Userinfo();
        System.out.println("执行了 public static Userinfo createUserinfo()方法");
        System.out.println("userinfo=" + userinfo);
        System.out.println("nowDate=" + nowDate);
        System.out.println("address=" + address);
        return userinfo;
    }
}
```

创建配置文件，代码如下：

```
<bean id="creator" class="tools.Creator"></bean>

<bean id="nowDate" class="java.util.Date"></bean>

<bean id="userinfo1" factory-bean="creator"
    factory-method="createUserinfo">
    <constructor-arg name="nowDate" ref="nowDate"></constructor-arg>
    <constructor-arg name="address" value="上海市"></constructor-arg>
</bean>
```

创建运行类，代码如下：

```
public class Test1 {
    public static void main(String[] args) {
        ApplicationContext context = new ClassPathXmlApplicationContext
("applicationContext.xml");
    }
}
```

程序运行后控制台输出的异常信息如下：

```
private Userinfo() entity.Userinfo@6adede5
执行了public static Userinfo createUserinfo()方法
userinfo=entity.Userinfo@6adede5
nowDate=Mon Jun 11 15:23:37 CST 2018
address=上海市
```

6. 使用 c 命名空间的参数名对构造方法进行注入

前面使用如下配置代码对构造方法进行注入：

```
<constructor-arg ref="userinfo1"></constructor-arg>
```

这段配置代码比较冗长，可以使用 c 命名空间进行简化。

创建测试用的项目 diTest13。

实体类代码如下：

```
package entity;

public class Userinfo {
    public Userinfo() {
        System.out.println("public Userinfo() " + this);
    }
}
```

实体类代码如下：

```
public class Bookinfo {

    public Bookinfo(Userinfo xxxxxx) {
        System.out.println();
        System.out.println("public Bookinfo(Userinfo xxxxxx)");
        System.out.println("xxxxxx=" + xxxxxx);
        System.out.println();
    }

    public Bookinfo(long maxValue) {
        System.out.println();
        System.out.println("public Bookinfo(long maxValue)");
        System.out.println("maxValue=" + maxValue);
        System.out.println();
    }

    public Bookinfo(Userinfo xxxxxx, String address) {
        System.out.println("public Bookinfo(Userinfo xxxxxx, String address)");
        System.out.println("xxxxxx=" + xxxxxx);
        System.out.println("address=" + address);
        System.out.println();
    }

    public Bookinfo(Userinfo userinfo1, Userinfo userinfo2, int age) {
        System.out.println("public Bookinfo(Userinfo userinfo1, Userinfo userinfo2, int age)");
        System.out.println("userinfo1=" + userinfo1);
        System.out.println("userinfo2=" + userinfo2);
        System.out.println("age=" + age);
        System.out.println();
```

```
    }
}
```

配置文件的代码如下：

```
<bean id="userinfo1" class="entity.Userinfo">
</bean>
<bean id="userinfo2" class="entity.Userinfo">
</bean>

<bean id="bookinfo1_1" class="entity.Bookinfo"
    c:xxxxxx-ref="userinfo1">
</bean>
<bean id="bookinfo1_2" class="entity.Bookinfo" c:_-ref="userinfo1">
</bean>

<bean id="bookinfo2_1" class="entity.Bookinfo" c:maxValue="123">
</bean>
<bean id="bookinfo2_2" class="entity.Bookinfo" c:_="456">
</bean>

<bean id="bookinfo3" class="entity.Bookinfo"
    c:xxxxxx-ref="userinfo1" c:address="北京">
</bean>

<bean id="bookinfo4" class="entity.Bookinfo"
    c:userinfo1-ref="userinfo1" c:userinfo2-ref="userinfo2" c:age="123">
</bean>
```

运行类代码如下：

```
public class Test1 {
    public static void main(String[] args) {
        ApplicationContext context = new ClassPathXmlApplicationContext("ac.xml");
        context.getBean("bookinfo1_1");
        context.getBean("bookinfo1_2");
        context.getBean("bookinfo2_1");
        context.getBean("bookinfo2_2");
        context.getBean("bookinfo3");
        context.getBean("bookinfo4");
    }
}
```

程序运行后的结果如下：

```
public Userinfo() entity.Userinfo@6767c1fc
public Userinfo() entity.Userinfo@464bee09

public Bookinfo(Userinfo xxxxxx)
xxxxxx=entity.Userinfo@6767c1fc


public Bookinfo(Userinfo xxxxxx)
xxxxxx=entity.Userinfo@6767c1fc


public Bookinfo(long maxValue)
maxValue=123
```

```
public Bookinfo(long maxValue)
maxValue=456

public Bookinfo(Userinfo xxxxxx, String address)
xxxxxx=entity.Userinfo@6767c1fc
address=北京

public Bookinfo(Userinfo userinfo1, Userinfo userinfo2, int age)
userinfo1=entity.Userinfo@6767c1fc
userinfo2=entity.Userinfo@464bee09
age=123
```

7. 使用 c 命名空间的索引对构造方法进行注入

可以根据构造方法参数的索引位置进行注入。

创建测试用的项目 diTest14。

实体类代码如下：

```java
public class Userinfo {
    public Userinfo() {
        System.out.println("public Userinfo() " + this.hashCode());
    }
}
```

实体类代码如下：

```java
public class Bookinfo {

    public Bookinfo(Userinfo xxxxxx) {
        System.out.println();
        System.out.println("public Bookinfo(Userinfo xxxxxx)");
        System.out.println("xxxxxx=" + xxxxxx);
        System.out.println();
    }

    public Bookinfo(long maxValue) {
        System.out.println();
        System.out.println("public Bookinfo(long maxValue)");
        System.out.println("maxValue=" + maxValue);
        System.out.println();
    }

    public Bookinfo(Userinfo xxxxxx, String address) {
        System.out.println("public Bookinfo(Userinfo xxxxxx, String address)");
        System.out.println("xxxxxx=" + xxxxxx);
        System.out.println("address=" + address);
        System.out.println();
    }

    public Bookinfo(Userinfo userinfo1, Userinfo userinfo2, int age) {
        System.out.println("public Bookinfo(Userinfo userinfo1, Userinfo userinfo2, int age)");
        System.out.println("userinfo1=" + userinfo1);
        System.out.println("userinfo2=" + userinfo2);
        System.out.println("age=" + age);
        System.out.println();
```

```
    }
}
```

配置文件的代码如下：

```
<bean id="userinfo1" class="entity.Userinfo">
</bean>
<bean id="userinfo2" class="entity.Userinfo">
</bean>

<bean id="bookinfo1_1" class="entity.Bookinfo"
    c:_0-ref="userinfo1">
</bean>
<bean id="bookinfo1_2" class="entity.Bookinfo" c:_-ref="userinfo1">
</bean>

<bean id="bookinfo2" class="entity.Bookinfo" c:_0="123">
</bean>

<bean id="bookinfo3" class="entity.Bookinfo" c:_0-ref="userinfo1"
    c:_1="北京">
</bean>

<bean id="bookinfo4" class="entity.Bookinfo" c:_0-ref="userinfo1"
    c:_1-ref="userinfo2" c:_2="123">
</bean>
```

运行类代码如下：

```
public class Test1 {
    public static void main(String[] args) {
        ApplicationContext context = new ClassPathXmlApplicationContext("ac.xml");
        context.getBean("bookinfo1_1");
        context.getBean("bookinfo1_2");
        context.getBean("bookinfo2");
        context.getBean("bookinfo3");
        context.getBean("bookinfo4");
    }
}
```

程序运行后的结果如下：

```
public Userinfo() entity.Userinfo@675d3402
public Userinfo() entity.Userinfo@53b32d7

public Bookinfo(Userinfo xxxxxx)
xxxxxx=entity.Userinfo@675d3402


public Bookinfo(Userinfo xxxxxx)
xxxxxx=entity.Userinfo@675d3402


public Bookinfo(long maxValue)
maxValue=123

public Bookinfo(Userinfo xxxxxx, String address)
xxxxxx=entity.Userinfo@675d3402
address=北京
```

```
public Bookinfo(Userinfo userinfo1, Userinfo userinfo2, int age)
userinfo1=entity.Userinfo@675d3402
userinfo2=entity.Userinfo@53b32d7
age=123
```

8．使用<constructor-arg>为构造方法注入 List/Set/Map/数组

可以使用<constructor-arg>为构造方法注入 List/Set/Map/数组。

创建测试用的项目 diTest16。

实体类代码如下：

```
public class Userinfo {
    private String username;

    public Userinfo() {
    }

    public Userinfo(String username) {
        super();
        this.username = username;
    }
    //省略 set 和 get 方法
}
```

实体类代码如下：

```
public class Bookinfo {
    public Bookinfo(String username, List<String> listString1) {
        for (int i = 0; i < listString1.size(); i++) {
            System.out.println("listString=" + listString1.get(i) + " username=" + username);
        }
        System.out.println();
    }

    public Bookinfo(int age, List<Userinfo> listString2) {
        for (int i = 0; i < listString2.size(); i++) {
            System.out.println("listUserinfo=" + listString2.get(i).getUsername() + "
age=" + age);
        }
        System.out.println();
    }

    public Bookinfo(String username, Set<String> set1) {
        Iterator<String> iterator = set1.iterator();
        while (iterator.hasNext()) {
            System.out.println("setString=" + iterator.next() + " username=" + username);
        }
        System.out.println();
    }

    public Bookinfo(int age, Set<Userinfo> set2) {
        Iterator<Userinfo> iterator = set2.iterator();
        while (iterator.hasNext()) {
            System.out.println("setUserinfo=" + iterator.next().getUsername() + " age
=" + age);
        }
```

```
            System.out.println();
        }

        public Bookinfo(String username, Map<String, String> map1) {
            Iterator<String> iterator = map1.keySet().iterator();
            while (iterator.hasNext()) {
                String key = iterator.next();
                System.out.println("map1<String, String> key=" + key + " value=" +
map1.get(key) + " username=" + username);
            }
            System.out.println();
        }

        public Bookinfo(int age, Map<String, Userinfo> map2) {
            Iterator<String> iterator = map2.keySet().iterator();
            while (iterator.hasNext()) {
                String key = iterator.next();
                System.out.println(
                        "map2<String, Userinfo> key=" + key + " value=" + map2.get(key).
getUsername() + " age=" + age);
            }
            System.out.println();
        }

        public Bookinfo(int age, int[] intArray) {
            for (int i = 0; i < intArray.length; i++) {
                System.out.println("int[]=" + intArray[i]);
            }
            System.out.println();
        }

        public Bookinfo(int age, Userinfo[] userinfoArray) {
            for (int i = 0; i < userinfoArray.length; i++) {
                System.out.println("Userinfo[]=" + userinfoArray[i].getUsername());
            }
        }
    }
```

配置文件的代码如下：

```
<bean id="userinfo1" class="entity.Userinfo">
    <property name="username" value="usernameValue1"></property>
</bean>
<bean id="userinfo2" class="entity.Userinfo">
    <property name="username" value="usernameValue2"></property>
</bean>
<bean id="userinfo3" class="entity.Userinfo">
    <property name="username" value="usernameValue3"></property>
</bean>

<bean id="bookinfo1" class="entity.Bookinfo">
    <constructor-arg type="java.lang.String" value="中国"></constructor-arg>
    <constructor-arg type="java.util.List">
        <list>
            <value>中国人 1</value>
            <value>中国人 2</value>
            <value>中国人 3</value>
            <value>中国人 4</value>
```

```
            </list>
        </constructor-arg>
    </bean>
    <bean id="bookinfo2" class="entity.Bookinfo">
        <constructor-arg type="int" value="1"></constructor-arg>
        <constructor-arg type="java.util.List">
            <list>
                <ref bean="userinfo1"></ref>
                <ref bean="userinfo2"></ref>
                <ref bean="userinfo3"></ref>
            </list>
        </constructor-arg>
    </bean>


    <bean id="bookinfo3" class="entity.Bookinfo">
        <constructor-arg type="java.lang.String" value="中国"></constructor-arg>
        <constructor-arg type="java.util.Set">
            <set>
                <value>中国人 1</value>
                <value>中国人 2</value>
                <value>中国人 3</value>
                <value>中国人 4</value>
            </set>
        </constructor-arg>
    </bean>
    <bean id="bookinfo4" class="entity.Bookinfo">
        <constructor-arg type="int" value="1"></constructor-arg>
        <constructor-arg type="java.util.Set">
            <set>
                <ref bean="userinfo1"></ref>
                <ref bean="userinfo2"></ref>
                <ref bean="userinfo3"></ref>
            </set>
        </constructor-arg>
    </bean>

    <bean id="bookinfo5" class="entity.Bookinfo">
        <constructor-arg type="java.lang.String" value="中国"></constructor-arg>
        <constructor-arg type="java.util.Map">
            <map>
                <entry key="a1" value="中国1" />
                <entry key="a2" value="中国2" />
                <entry key="a3" value="中国3" />
            </map>
        </constructor-arg>
    </bean>
    <bean id="bookinfo6" class="entity.Bookinfo">
        <constructor-arg type="int" value="1"></constructor-arg>
        <constructor-arg type="java.util.Map">
            <map>
                <entry key="a1" value-ref="userinfo1" />
                <entry key="a2" value-ref="userinfo2" />
                <entry key="a3" value-ref="userinfo3" />
            </map>
        </constructor-arg>
    </bean>
```

```
<bean id="bookinfo7" class="entity.Bookinfo">
    <constructor-arg type="int" value="123"></constructor-arg>
    <constructor-arg type="int[]">
        <array>
            <value>101</value>
            <value>102</value>
            <value>103</value>
            <value>104</value>
        </array>
    </constructor-arg>
</bean>

<bean id="bookinfo8" class="entity.Bookinfo">
    <constructor-arg type="int" value="456"></constructor-arg>
    <constructor-arg type="Userinfo[]">
        <array>
            <ref bean="userinfo1" />
            <ref bean="userinfo2" />
            <ref bean="userinfo3" />
        </array>
    </constructor-arg>
</bean>
```

运行类代码如下：

```
public class Test1 {
    public static void main(String[] args) {
        ApplicationContext context = new ClassPathXmlApplicationContext("ac.xml");
        context.getBean("bookinfo1");
        context.getBean("bookinfo2");
        context.getBean("bookinfo3");
        context.getBean("bookinfo4");
        context.getBean("bookinfo5");
        context.getBean("bookinfo6");
        context.getBean("bookinfo7");
        context.getBean("bookinfo8");
    }
}
```

程序运行后的结果如下：

```
listString=中国人 1 username=中国
listString=中国人 2 username=中国
listString=中国人 3 username=中国
listString=中国人 4 username=中国

listUserinfo=usernameValue1 age=1
listUserinfo=usernameValue2 age=1
listUserinfo=usernameValue3 age=1

setString=中国人 1 username=中国
setString=中国人 2 username=中国
setString=中国人 3 username=中国
setString=中国人 4 username=中国

setUserinfo=usernameValue1 age=1
setUserinfo=usernameValue2 age=1
setUserinfo=usernameValue3 age=1
```

```
map1<String, String> key=a1 value=中国 1 username=中国
map1<String, String> key=a2 value=中国 2 username=中国
map1<String, String> key=a3 value=中国 3 username=中国

map2<String, Userinfo> key=a1 value=usernameValue1 age=1
map2<String, Userinfo> key=a2 value=usernameValue2 age=1
map2<String, Userinfo> key=a3 value=usernameValue3 age=1

int[]=101
int[]=102
int[]=103
int[]=104

Userinfo[]=usernameValue1
Userinfo[]=usernameValue2
Userinfo[]=usernameValue3
```

3.8.12 使用 p 命名空间对属性值进行注入

前面使用<bean>的子标签<property>来对属性值进行注入，还可以使用 p 命名空间进行值的注入。

创建测试用的项目 diTest17。

实体类代码如下：

```java
public class Userinfo {
    private String username;
    public Userinfo() {
    }
    //省略 set 和 get 方法
}
```

实体类代码如下：

```java
public class Bookinfo {
    private String username;
    private Userinfo userinfo;
    private List<String> listString;
    private List<Userinfo> listUserinfo;
    private Set<String> setString;
    private Set<Userinfo> setUserinfo;
    private Map<String, String> mapString;
    private Map<String, Userinfo> mapUserinfo;
    private int[] intArray;
    private Userinfo[] userinfoArray;

    //省略 set 和 get 方法
}
```

配置文件的代码如下：

```xml
<bean id="userinfo1" class="entity.Userinfo">
    <property name="username" value="usernameValue1"></property>
</bean>
<bean id="userinfo2" class="entity.Userinfo">
    <property name="username" value="usernameValue2"></property>
```

```
</bean>
<bean id="userinfo3" class="entity.Userinfo">
    <property name="username" value="usernameValue3"></property>
</bean>

<util:list id="listString">
    <value>List 中国 1</value>
    <value>List 中国 2</value>
    <value>List 中国 3</value>
</util:list>

<util:list id="listUserinfo">
    <ref bean="userinfo1" />
    <ref bean="userinfo2" />
    <ref bean="userinfo3" />
</util:list>

<util:set id="setString">
    <value>Set 中国 1</value>
    <value>Set 中国 2</value>
    <value>Set 中国 3</value>
</util:set>

<util:set id="setUserinfo">
    <ref bean="userinfo1" />
    <ref bean="userinfo2" />
    <ref bean="userinfo3" />
</util:set>

<util:map id="mapString">
    <entry key="key1" value="Map 中国 1"></entry>
    <entry key="key2" value="Map 中国 2"></entry>
    <entry key="key3" value="Map 中国 3"></entry>
</util:map>

<util:map id="mapUserinfo">
    <entry key="key1" value-ref="userinfo1"></entry>
    <entry key="key2" value-ref="userinfo2"></entry>
    <entry key="key3" value-ref="userinfo3"></entry>
</util:map>

<util:list id="intArray">
    <value>1</value>
    <value>2</value>
    <value>3</value>
</util:list>

<util:list id="userinfoArray">
    <ref bean="userinfo1" />
    <ref bean="userinfo2" />
    <ref bean="userinfo3" />
</util:list>

<bean id="bookinfo" class="entity.Bookinfo" p:username="中国"
    p:userinfo-ref="userinfo1" p:listString-ref="listString"
    p:listUserinfo-ref="listUserinfo" p:setString-ref="setString"
    p:setUserinfo-ref="setUserinfo" p:mapString-ref="mapString"
    p:mapUserinfo-ref="mapUserinfo" p:intArray-ref="intArray"
```

```
    p:userinfoArray-ref="userinfoArray">
</bean>
```

运行类代码如下：

```
public class Test1 {
    public static void main(String[] args) {
        ApplicationContext context = new ClassPathXmlApplicationContext("ac.xml");
        Bookinfo bookinfo = (Bookinfo) context.getBean("bookinfo");
        System.out.println(bookinfo.getUsername());
        System.out.println(bookinfo.getUserinfo());
        System.out.println();
        List<String> listString = bookinfo.getListString();
        for (int i = 0; i < listString.size(); i++) {
            System.out.println("listString=" + listString.get(i));
        }
        System.out.println();
        List<Userinfo> listUserinfo = bookinfo.getListUserinfo();
        for (int i = 0; i < listUserinfo.size(); i++) {
            Userinfo userinfo = listUserinfo.get(i);
            System.out.println("listUserinfo=" + userinfo.getUsername());
        }
        System.out.println();
        Set<String> setString = bookinfo.getSetString();
        Iterator iterator = setString.iterator();
        while (iterator.hasNext()) {
            System.out.println("setString=" + iterator.next());
        }
        System.out.println();
        Set<Userinfo> setUserinfo = bookinfo.getSetUserinfo();
        iterator = setUserinfo.iterator();
        while (iterator.hasNext()) {
            Userinfo userinfo = (Userinfo) iterator.next();
            System.out.println("setUserinfo=" + userinfo.getUsername());
        }
        System.out.println();
        Map<String, String> mapString = bookinfo.getMapString();
        iterator = mapString.keySet().iterator();
        while (iterator.hasNext()) {
            String key = "" + iterator.next();
            String value = "" + mapString.get(key);
            System.out.println("mapString=" + key + " " + value);
        }
        System.out.println();
        Map<String, Userinfo> mapUserinfo = bookinfo.getMapUserinfo();
        iterator = mapUserinfo.keySet().iterator();
        while (iterator.hasNext()) {
            String key = "" + iterator.next();
            Userinfo userinfo = (Userinfo) mapUserinfo.get(key);
            System.out.println("mapUserinfo=" + key + " " + userinfo.getUsername());
        }
        System.out.println();
        int[] intArray = bookinfo.getIntArray();
        for (int i = 0; i < intArray.length; i++) {
            System.out.println("intArray=" + intArray[i]);
        }
        System.out.println();
        Userinfo[] userinfoArray = bookinfo.getUserinfoArray();
```

```
        for (int i = 0; i < userinfoArray.length; i++) {
            Userinfo userinfo = userinfoArray[i];
            System.out.println("userinfoArray=" + userinfo.getUsername());
        }
    }
}
```

程序运行后的结果如下：

```
中国
entity.Userinfo@2a17b7b6

listString=List 中国 1
listString=List 中国 2
listString=List 中国 3

listUserinfo=usernameValue1
listUserinfo=usernameValue2
listUserinfo=usernameValue3

setString=Set 中国 1
setString=Set 中国 2
setString=Set 中国 3

setUserinfo=usernameValue1
setUserinfo=usernameValue2
setUserinfo=usernameValue3

mapString=key1 Map 中国 1
mapString=key2 Map 中国 2
mapString=key3 Map 中国 3

mapUserinfo=key1 usernameValue1
mapUserinfo=key2 usernameValue2
mapUserinfo=key3 usernameValue3

intArray=1
intArray=2
intArray=3

userinfoArray=usernameValue1
userinfoArray=usernameValue2
userinfoArray=usernameValue3
```

3.8.13 Spring 上下文环境的相关知识

Spring 的上下文环境可以理解成是 Spring 运行的环境，可以创建多个 Spring 上下文环境。在默认的情况下，不同上下文环境中的 JavaBean 对象不可以共享。

1. 创建多个 Spring 上下文环境

创建多个 Spring 上下文环境就是创建多个 ApplicationContext 对象。

创建测试用的项目 springTest1。

实体类代码如下：

```
public class Userinfo {
    public Userinfo() {
        System.out.println("public Userinfo() " + this.hashCode());
    }
}
```

实体类代码如下：

```
public class Bookinfo {
    public Bookinfo() {
        System.out.println("public Bookinfo() " + this.hashCode());
    }
}
```

配置类代码如下：

```
@Configuration
public class SpringConfig1 {
    @Bean
    public Bookinfo getBookinfo() {
        Bookinfo bookinfo = new Bookinfo();
        System.out.println("getBookinfo bookinfo=" + bookinfo.hashCode());
        return bookinfo;
    }
}
```

配置类代码如下：

```
@Configuration
public class SpringConfig2 {
    @Bean
    public Userinfo getUserinfo() {
        Userinfo userinfo = new Userinfo();
        System.out.println("getUserinfo userinfo=" + userinfo.hashCode());
        return userinfo;
    }
}
```

运行类代码如下：

```
public class Test1 {
    public static void main(String[] args) {
        ApplicationContext context1 = new AnnotationConfigApplicationContext
(SpringConfig1.class);
        ApplicationContext context2 = new AnnotationConfigApplicationContext
(SpringConfig2.class);

        System.out.println("context1=" + context1.getBean(Bookinfo.class).hashCode());
        System.out.println("context2=" + context2.getBean(Userinfo.class).hashCode());
    }
}
```

程序运行后的结果如下：

```
public Bookinfo() 518522822
getBookinfo bookinfo=518522822
public Userinfo() 2008966511
getUserinfo userinfo=2008966511
context1=518522822
context2=2008966511
```

成功创建出不同的 Spring 上下文环境，在自己的上下文环境中可以获取自己的 JavaBean。

2. 不同的 Spring 上下文环境中的 JavaBean 对象是不共享的

这部分要测试在不同的 Spring 上下文环境中的 JavaBean 对象是不共享的。
创建测试用的项目 springTest2。
运行类代码如下：

```
public class Test1 {
    public static void main(String[] args) {
        ApplicationContext context1 = new AnnotationConfigApplicationContext
(SpringConfig1.class);
        System.out.println("context1=" + context1.getBean(Bookinfo.class).hashCode());
        context1.getBean(Userinfo.class);
    }
}
```

程序运行后出现的异常信息如下：

```
public Bookinfo() 518522822
getBookinfo bookinfo=518522822
context1=518522822
Exception in thread "main" org.springframework.beans.factory.NoSuchBeanDefinitionException:
No qualifying bean of type 'entity.Userinfo' available
```

提示 Userinfo.java 类并没有被找到。

3. 让多个配置类互相通信

使用配置类创建的不同上下文环境中的 JavaBean 如何互相共享呢？
创建测试用的项目 springTest3。
配置类代码如下：

```
@Configuration
@Import(value = SpringConfig2.class)
public class SpringConfig1 {
    @Bean
    public Bookinfo getBookinfo() {
        Bookinfo bookinfo = new Bookinfo();
        System.out.println("getBookinfo bookinfo=" + bookinfo.hashCode());
        return bookinfo;
    }
}
```

使用注解：

```
@Import(value = SpringConfig2.class)
```

导入其他上下文环境中的 JavaBean，还可以使用如下的注解写法一次性导入多个配置类，代码如下：

```
@Import({ MyContextX.class, MyContextY.class })
```

运行类代码如下：

```
public class Test1 {
    public static void main(String[] args) {
        ApplicationContext context1 = new AnnotationConfigApplicationContext
(SpringConfig1.class);
        System.out.println("context1=" + context1.getBean(Bookinfo.class).hashCode());
        System.out.println("context2=" + context1.getBean(Userinfo.class).hashCode());
    }
}
```

程序运行后的结果如下：

```
public Userinfo() 302155142
getUserinfo userinfo=302155142
public Bookinfo() 24606376
getBookinfo bookinfo=24606376
context1=24606376
context2=302155142
```

注意：如果多个上下文中的 JavaBean 的 id 值一样，并且数据类型也一样，那么会出现“Overriding bean definition”的效果。

4. 不同上下文环境中的工厂方法的名称不可以相同

注意：在不同配置类中创建的 JavaBean 工厂方法的名称不能一样。因为在默认的情况下，JavaBean 的 id 和方法名称一样。如果多个工厂方法名称一样，则不会创建其他的 JavaBean，这会导致找不到对象的异常发生。在项目 springTest4 中对此实验的源代码进行了测试，配置类代码如下：

```
@Configuration
@Import(SpringConfig2.class)
public class SpringConfig1 {
    @Bean
    public Bookinfo abc() {
        Bookinfo bookinfo = new Bookinfo();
        System.out.println("getBookinfo bookinfo=" + bookinfo.hashCode());
        return bookinfo;
    }
}
```

配置类代码如下：

```
@Configuration
public class SpringConfig2 {
    @Bean
    public Userinfo abc() {
        Userinfo userinfo = new Userinfo();
        System.out.println("getUserinfo userinfo=" + userinfo.hashCode());
        return userinfo;
    }
}
```

运行类代码如下：

```
public class Test1 {
    public static void main(String[] args) {
        ApplicationContext context1 = new AnnotationConfigApplicationContext
(SpringConfig1.class);
        System.out.println("context1=" + context1.getBean(Bookinfo.class).hashCode());
```

```
        System.out.println("context2=" + context1.getBean(Userinfo.class).hashCode());
    }
}
```

程序运行的结果如下：

```
public Bookinfo() 1237598030
getBookinfo bookinfo=1237598030
context1=1237598030
Exception in thread "main" org.springframework.beans.factory.NoSuchBeanDefinitionException: No qualifying bean of type 'entity.Userinfo' available
```

解决异常的办法是在 @Bean（name ="别名"）注解中对 name 属性设置不同的别名，或者修改工厂方法的名称。

5．创建 AllContext.java 全局配置类

对多个 Spring 上下文共享 JavaBean 的代码组织方式是创建 1 个全局的配置类，然后使用 @Import 注解在这个类中导入其他的配置类。

创建测试用的项目 springTest5。

全局配置代码如下：

```
@Configuration
@Import(value = { SpringConfig1.class, SpringConfig2.class })
public class AllConfig {
}
```

配置类代码如下：

```
@Configuration
public class SpringConfig1 {
    @Bean
    public Bookinfo getBookinfo() {
        Bookinfo bookinfo = new Bookinfo();
        System.out.println("getBookinfo bookinfo=" + bookinfo.hashCode());
        return bookinfo;
    }
}
```

配置类代码如下：

```
@Configuration
public class SpringConfig2 {
    @Bean
    public Userinfo getUserinfo() {
        Userinfo userinfo = new Userinfo();
        System.out.println("getUserinfo userinfo=" + userinfo.hashCode());
        return userinfo;
    }
}
```

运行类代码如下：

```
public class Test1 {
    public static void main(String[] args) {
        ApplicationContext context1 = new AnnotationConfigApplicationContext(AllConfig.class);
        System.out.println("context1=" + context1.getBean(Bookinfo.class).hashCode());
```

```
        System.out.println("context2=" + context1.getBean(Userinfo.class).hashCode());
    }
}
```

程序运行后的结果如下：

```
public Bookinfo() 1772160903
getBookinfo bookinfo=1772160903
public Userinfo() 756185697
getUserinfo userinfo=756185697
context1=1772160903
context2=756185697
```

6. 使用 @ImportResource 导入 XML 文件中的配置

如果使用 XML 文件创建另一个 Spring 上下文环境，那么在配置类中如何引用这个上下文环境呢？

创建测试用的项目 springTest6。

配置文件代码如下：

```
<bean class="entity.Userinfo"></bean>
```

配置类代码如下：

```
@Configuration
@ImportResource(value = "ac.xml")
public class SpringConfig1 {
    @Bean
    public Bookinfo getBookinfo() {
        Bookinfo bookinfo = new Bookinfo();
        System.out.println("getBookinfo bookinfo=" + bookinfo.hashCode());
        return bookinfo;
    }
}
```

在配置类中使用 @ImportResource 注解导入 XML 配置文件中的 JavaBean。

此实验也证明了可以使用不同的途径来创建 Spring 上下文环境，不同 Spring 上下文环境中的 JavaBean 是可以互相共享的。

运行类代码如下：

```
public class Test1 {
    public static void main(String[] args) {
        ApplicationContext context1 = new AnnotationConfigApplicationContext
(SpringConfig1.class);
        System.out.println("context1=" + context1.getBean(Bookinfo.class).hashCode());
        System.out.println("context2=" + context1.getBean(Userinfo.class).hashCode());
    }
}
```

程序运行后的结果如下：

```
public Bookinfo() 874217650
getBookinfo bookinfo=874217650
public Userinfo() 1436664465
context1=874217650
context2=1436664465
```

7. 使用 XML 文件创建多个上下文环境的总结

前面都是通过配置类的方式来创建多个 Spring 上下文，其实使用多个 XML 配置文件也可以达到相同的结果。

在 XML 文件中创建多个 Spring 上下文环境的总结如下。

（1）使用 XML 文件可以创建多个 Spring 上下文环境，但每个 Spring 上下文环境都有自己私有的 JavaBean 对象，在默认的情况下不可以共享这些 JavaBean 对象。测试的源代码在项目 springTest7 中。

（2）为了令多个使用 XML 文件创建的 Spring 上下文中的 JavaBean 能相互共享，要在 XML 配置文件中使用<import resource="ac.xml" />标签导入其他的 XML 配置文件，也就是导入其他的 Spring 上下文环境。测试的源代码在项目 springTest8 中。

（3）可以创建一个全局的 allContext.xml 配置文件，然后使用多个<import resource="ac.xml" />标签导入其他的 XML 配置文件。测试的源代码在项目 springTest9 中。

（4）在 XML 中导入 JavaConfig 配置类 Spring 上下文环境的用法是在 XML 文件中使用如下代码将配置类导入到当前上下文环境中：

```
<context:annotation-config />
<bean class="config.MyContext1">
</bean>
```

测试的源代码在项目 springTest10 中。

3.8.14 BeanFactory 与 ApplicationContext

包 org.springframework.beans 和包 org.springframework.context 是 Spring 框架 IoC 容器的基础。

BeanFactory 接口提供了一种能够管理任何类型对象的高级配置机制。ApplicationContext 是 BeanFactory 的一个子接口，它增加了 AOP 功能，更容易集成，支持消息资源处理（用于国际化）、事件发布和应用程序层特定的上下文和用于 Web 应用程序的 WebApplicationContext。

其实 Spring 的 IoC 容器就是一个实现了 BeanFactory 接口的实现类，因为 ApplicationContext 接口继承自 BeanFactory 接口。可以通过工厂模式来获得 JavaBean 对象。

看一下 Spring 的 API 帮助文档，如图 3-8 所示。

org.springframework.beans.factory

Interface BeanFactory

All Known Subinterfaces:

ApplicationContext, AutowireCapableBeanFactory, ConfigurableApplicationContext, ConfigurableBeanFactory, ConfigurableListableBeanFactory, ConfigurablePortletApplicationContext, ConfigurableWebApplicationContext, HierarchicalBeanFactory, ListableBeanFactory, WebApplicationContext

All Known Implementing Classes:

AbstractApplicationContext, AbstractAutowireCapableBeanFactory, AbstractBeanFactory, AbstractRefreshableApplicationContext, AbstractRefreshableConfigApplicationContext, AbstractRefreshablePortletApplicationContext, AbstractRefreshableWebApplicationContext, AbstractXmlApplicationContext, AnnotationConfigApplicationContext, AnnotationConfigWebApplicationContext, ClassPathXmlApplicationContext, DefaultListableBeanFactory, FileSystemXmlApplicationContext, GenericApplicationContext, GenericGroovyApplicationContext, GenericWebApplicationContext, GenericXmlApplicationContext, GroovyWebApplicationContext, ResourceAdapterApplicationContext, SimpleJndiBeanFactory, StaticApplicationContext, StaticListableBeanFactory, StaticPortletApplicationContext, StaticWebApplicationContext, XmlBeanFactory, XmlPortletApplicationContext, XmlWebApplicationContext

图 3-8 BeanFactory 接口结构

从图 3-8 中可以发现，ApplicationContext 是 BeanFactory 的子接口，BeanFactory 接口提供了基本的对象管理功能，而子接口 ApplicationContext 提供了更多的附加功能，比如与 Web 整合，支持国际化、事件发布和通知等。

3.8.15 注入 null 类型

创建测试用的项目 springTest12。

关键的配置文件 applicationContext.xml 的代码如下：

```
<bean id="test" class="test.Test">
    <property name="username">
        <null></null>
    </property>
</bean>
```

运行类代码如下：

```
public class Test {
    private String username = "我是中国";

    public String getUsername() {
        return username;
    }

    public void setUsername(String username) {
        this.username = username;
    }

    public static void main(String[] args) {
        ApplicationContext context = new ClassPathXmlApplicationContext("ac1.xml");
        Test test = context.getBean(Test.class);
        System.out.println(test.getUsername());
        System.out.println(test.getUsername() == null);
    }
}
```

程序运行后的结果如下：

```
null
true
```

3.8.16 注入 Properties 类型

创建测试用的项目 springTest13。

配置文件的内容如下：

```
<bean id="test" class="test.Test">
    <property name="prop1">
        <props>
            <prop key="a1">aa1</prop>
            <prop key="a2">aa2</prop>
            <prop key="a3">aa3</prop>
            <prop key="a4">aa4</prop>
        </props>
    </property>
    <property name="prop2">
        <value>
            a5=aa5
            a6=aa6
```

```
        </value>
    </property>
</bean>
```

工具类代码如下：

```
public class Test {
    private Properties prop1;
    private Properties prop2;

    //省略 get 和 set 方法

    public static void main(String[] args) {
        ApplicationContext context = new ClassPathXmlApplicationContext("ac1.xml");
        Test test = context.getBean(Test.class);
        Properties prop1 = test.getProp1();
        Properties prop2 = test.getProp2();
        System.out.println("a1" + " " + test.getProp1().get("a1"));
        System.out.println("a2" + " " + test.getProp1().get("a2"));
        System.out.println("a3" + " " + test.getProp1().get("a3"));
        System.out.println("a4" + " " + test.getProp1().get("a4"));
        System.out.println("a5" + " " + test.getProp2().get("a5"));
        System.out.println("a6" + " " + test.getProp2().get("a6"));
    }
}
```

程序运行后的结果如下：

```
a1 aa1
a2 aa2
a3 aa3
a4 aa4
a5 aa5
a6 aa6
```

3.8.17 在 Spring 中注入外部属性文件的属性值

在开发针对数据库软件的项目时，通常将数据库的连接信息放入属性文件中，但如何通过 Spring 将属性文件中的属性值提取出来并且注入类的属性中呢？

Spring 提供了 PropertyPlaceholderConfigurer 类来操作属性文件。

创建测试用的项目 springTest15。

属性文件 dbinfo.properties 的内容如下：

```
driver=drivervalue
url=urlvalue
username=usernamevalue
password=passwordvalue
```

配置文件的内容设计如下：

```
<bean
    class="org.springframework.beans.factory.config.PropertyPlaceholderConfigurer">
    <property name="locations">
        <list>
            <value>dbinfo.properties</value>
        </list>
```

```
    </property>
</bean>
<bean id="test" class="test.Test1">
    <property name="username" value="${username}"></property>
    <property name="password" value="${password}"></property>
    <property name="url" value="${url}"></property>
    <property name="driver" value="${driver}"></property>
</bean>
```

代码段如下：

```
<bean
    class="org.springframework.beans.factory.config.PropertyPlaceholderConfigurer">
    <property name="locations">
        <list>
            <value>dbinfo.properties</value>
        </list>
    </property>
</bean>
```

使用 Spring 框架中的类 PropertyPlaceholderConfigurer 来加载指定的属性文件。在本示例中，将属性文件 dbinfo.properties 从 classpath 路径中加载到内存，然后通过指定属性文件中的 key 来将属性文件中的 value 注入类的属性中，代码如下：

```
<bean id="test" class="test.Test1">
    <property name="username" value="${username}"></property>
    <property name="password" value="${password}"></property>
    <property name="url" value="${url}"></property>
    <property name="driver" value="${driver}"></property>
</bean>
```

通过使用类似 EL（Expression Language）表达式的形式${}，指定属性文件中的 key 来将对应的 value 注入类的属性中。

运行类代码如下：

```
public class Test1 {
    private String driver;
    private String url;
    private String username;
    private String password;

//省略 set 和 get 方法

    public static void main(String[] args) throws InterruptedException {
        ApplicationContext context = new ClassPathXmlApplicationContext("ac.xml");
        Test1 test = (Test1) context.getBean("test");
        System.out.println(test.getDriver());
        System.out.println(test.getUrl());
        System.out.println(test.getUsername());
        System.out.println(test.getPassword());
    }
}
```

程序输出的结果如下：

```
drivervalue
urlvalue
```

```
usernamevalue
passwordvalue
```

3.8.18 在 IoC 容器中创建单例和多例的对象——XML 配置文件法

IoC 容器对 JavaBean 的管理也有作用域，在 Spring 中一共有 7 种作用域：singleton、prototype、request、session、globalSession、application 和 websocket。

singleton 代表在 IoC 容器中只有一个 JavaBean 的对象，而 prototype 代表当使用 getBean 方法获得一个 JavaBean 时，IoC 容器会新建一个指定 JavaBean 的实例并且返回给程序员。它们的区别仅仅是 singleton 永远是一个实例，而 prototype 是多个实例。

singleton 一定要注意 JavaBean 的线程安全问题。非线程安全是指多个线程访问同一个对象的同一个实例变量，此变量值有可能被覆盖。

设计一个示例来具体看一下 singleton 和 prototype 这两种作用域的区别。

创建测试用的项目 springTest14。

配置文件的设计如下：

```
<bean id="mydate" class="java.util.Date" scope="prototype">
</bean>
```

运行类代码如下：

```
public class Test1 {
    public static void main(String[] args) throws InterruptedException {
        ApplicationContext context = new ClassPathXmlApplicationContext("ac1.xml");
        System.out.println(context.getBean("mydate"));
        Thread.sleep(3000);
        System.out.println(context.getBean("mydate"));
        Thread.sleep(3000);
        System.out.println(context.getBean("mydate"));
        Thread.sleep(3000);
        System.out.println(context.getBean("mydate"));
        context.getBean("mydate");
    }

}
```

在运行类中使用 getBean 方法来获得 java.util.Date 类的实例，由于在配置文件中使用了 prototype 作用域，因此每次执行 getBean 时 IoC 容器都要新创建一个 java.util.Date 类的实例，运行结果如下：

```
Mon Jun 11 16:02:24 CST 2018
Mon Jun 11 16:02:27 CST 2018
Mon Jun 11 16:02:30 CST 2018
Mon Jun 11 16:02:33 CST 2018
```

修改下面的代码：

```
<bean id="mydate" class="java.util.Date" scope="prototype">
</bean>
```

修改后的代码如下：

```
<bean id="mydate" class="java.util.Date" scope="singleton">
</bean>
```

根据上面介绍的知识，singleton 是单实例，因此，获得的 JavaBean 永远是一个，Date 的日期也是一样，输出的结果如下：

```
Mon Jun 11 16:03:19 CST 2018
Mon Jun 11 16:03:19 CST 2018
Mon Jun 11 16:03:19 CST 2018
Mon Jun 11 16:03:19 CST 2018
```

当有多个线程对同一对象的同一个实例变量进行写操作时，要避免出现“非线程安全”问题，因此要使用 prototype 多例。Struts2 的 Action 必须使用多例，因为 Action 中存在实例变量。

当没有出现多个线程对同一对象的同一个实例变量进行写操作时，为了降低内存的使用率，可以使用 singleton 单例模式。在开发中，Service 类和 DAO 类都使用单例模式，因为这两个类中的实例变量在大多数情况下都是只读的。

3.8.19　在 IoC 容器中创建单例和多例的对象——注解法

使用注解实现 JavaBean 单例或多例模式的测试项目为 scopeTest。

单例代码如下：

```
@Component
@Scope(scopeName = "singleton")
public class Userinfo1 {
    public Userinfo1() {
        System.out.println("public Userinfo1() " + this);
    }
}
```

多例代码如下：

```
@Component
@Scope(scopeName = "prototype")
public class Userinfo2 {
    public Userinfo2() {
        System.out.println("public Userinfo2() " + this);
    }
}
```

3.8.20　父子容器

通过 HierarchicalBeanFactory 接口，Spring 的 IoC 容器可以建立父子层级关联的容器体系，子容器可以访问父容器中的 Bean，但父容器不能访问子容器的 Bean。在容器内，Bean 的 id 必须是唯一的，但子容器可以拥有一个和父容器 id 相同的 Bean。父子容器层级体系增强了 Spring 容器架构的扩展性和灵活性，因为第三方可以通过编程的方式，为一个已经存在的容器添加一个或多个特殊用途的子容器，以提供额外的功能。

Spring 使用父子容器实现了很多功能，比如在 Spring MVC 中，控制层 Bean 位于一个子容器中，而业务层和持久层的 Bean 位于父容器中。这样，控制层 Bean 就可以引用业务层和持久层的 Bean，而业务层和持久层的 Bean 则不能访问控制层的 Bean。

1. 创建父子容器

创建测试用的项目 parent_sub_container1。

创建类代码如下：

```
public class Userinfo1 {
    public Userinfo1() {
        System.out.println("public Userinfo1() " + this);
    }
}
```

创建类代码如下：

```
public class Userinfo2 {
    public Userinfo2() {
        System.out.println("public Userinfo2() " + this);
    }
}
```

配置文件 ac1.xml 代码如下：

```
<bean id="userinfo1" class="entity.Userinfo1"></bean>
```

配置文件 ac2.xml 代码如下：

```
<bean id="userinfo2" class="entity.Userinfo2"></bean>
```

运行类代码如下：

```
public class Test {
    public static void main(String[] args) {
        ApplicationContext context1 = new ClassPathXmlApplicationContext("ac1.xml");
        ApplicationContext context2 = new ClassPathXmlApplicationContext(new String[]
{ "ac2.xml" }, context1);
        System.out.println("context1=" + context1);
        System.out.println("context2=" + context2);
        System.out.println("context2.getParent()=" + context2.getParent());
    }

}
```

程序运行后的结果如下：

```
    context1=org.springframework.context.support.ClassPathXmlApplicationContext@18769467:
startup date; root of context hierarchy
    context2=org.springframework.context.support.ClassPathXmlApplicationContext@5f341870:
startup date; parent: org.springframework.context.support.ClassPathXmlApplicationContext@18769467
    context2.getParent()=org.springframework.context.support.ClassPathXmlApplicationContext
@18769467: startup date; root of context hierarchy
```

2. 父容器不可以获取子容器的对象

创建测试用的项目 parent_sub_container2。

运行类代码如下：

```
public class Test {
    public static void main(String[] args) {
        ApplicationContext context1 = new ClassPathXmlApplicationContext("ac1.xml");
```

```
        ApplicationContext context2 = new ClassPathXmlApplicationContext(new String[]
{ "ac2.xml" }, context1);
        System.out.println(context1.getBean("userinfo2"));
    }
}
```

程序运行后的结果如下：

```
public Userinfo2() entity.Userinfo2@25bbe1b6
Exception in thread "main" org.springframework.beans.factory.NoSuchBeanDefinitionException:
No bean named 'userinfo2' available
```

3. 子容器可以获取父容器的对象

创建测试用的项目 parent_sub_container3。

运行类代码如下：

```
public class Test {
    public static void main(String[] args) {
        ApplicationContext context1 = new ClassPathXmlApplicationContext("ac1.xml");
        ApplicationContext context2 = new ClassPathXmlApplicationContext(new String[]
{ "ac2.xml" }, context1);
        System.out.println(context2.getBean("userinfo1"));
    }
}
```

程序运行后的结果如下：

```
public Userinfo1() entity.Userinfo1@3c756e4d
public Userinfo2() entity.Userinfo2@25bbe1b6
entity.Userinfo1@3c756e4d
```

4. 子容器注入的对象来自父容器

创建测试用的项目 parent_sub_container4。

在子容器中创建类的代码如下：

```
public class MyObject {
    private Userinfo1 userinfo1;

    public Userinfo1 getUserinfo1() {
        return userinfo1;
    }

    public void setUserinfo1(Userinfo1 userinfo1) {
        this.userinfo1 = userinfo1;
    }

}
```

父容器的配置文件代码如下：

```
<bean id="userinfo" class="entity.Userinfo1"></bean>
```

子容器的配置文件代码如下：

```
<bean id="userinfo" class="entity.Userinfo2"></bean>
<bean id="myObject" class="entity.MyObject">
```

```
    <property name="userinfo1">
        <ref parent="userinfo" />
    </property>
</bean>
```

父容器和子容器中都有 id 为 userinfo 的 JavaBean，在子容器中使用配置代码<ref parent= "userinfo" />将父容器中 id 为 userinfo 的对象注入子容器中 MyObject 的 userinfo1 属性。

运行类代码如下：

```
public class Test {
    public static void main(String[] args) {
        ApplicationContext context1 = new ClassPathXmlApplicationContext("ac1.xml");
        ApplicationContext context2 = new ClassPathXmlApplicationContext(new String[]
{ "ac2.xml" }, context1);
        MyObject myObject = (MyObject) context2.getBean("myObject");
        System.out.println("main end print " + myObject.getUserinfo1());
    }
}
```

程序运行后的结果如下：

```
public Userinfo1() entity.Userinfo1@3c756e4d
public Userinfo2() entity.Userinfo2@73846619
main end print entity.Userinfo1@3c756e4d
```

3.8.21　注入特殊字符

如果为属性注入特殊字符，那么会出现异常，效果如图 3-9 所示。

```
<bean id="test" class="test.Test">
    <property name="username" value="<"></property>
</bean>
```

图 3-9　出现异常

正确的注入代码如下：

```
<bean id="test" class="test.Test">
    <property name="username">
        <value><![CDATA[<>~!@#$%^&*()]]></value>
    </property>
</bean>
```

此实验在名称为 injectOtherChar 的项目中。

3.8.22　使用@Value 注解进行注入

@Value 注解也可以实现注入。

创建测试用的项目 valueTest。

创建实体类 Userinfo，代码如下：

```
package entity;

public class Userinfo {
```

```
    public Userinfo() {
        System.out.println("public Userinfo() " + this.hashCode());
    }
}
```

创建属性文件 db.properties，代码如下：

```
driver=oracleDriver
url=oracleURL
username=oracleUsername
password=oraclePassword
```

创建业务类 UserinfoService，代码如下：

```
package service;

import org.springframework.beans.factory.annotation.Value;
import org.springframework.stereotype.Service;

import entity.Userinfo;

@Service(value = "userinfoService")
public class UserinfoService {

    @Value("UserinfoService 常<量>值&")
    private String username;

    private Userinfo userinfo = new Userinfo();

    public String getUsername() {
        return username;
    }

    public void setUsername(String username) {
        this.username = username;
    }

    public Userinfo getUserinfo() {
        return userinfo;
    }

    public void setUserinfo(Userinfo userinfo) {
        this.userinfo = userinfo;
    }

}
```

运行类代码如下：

```
package test;

import org.springframework.beans.factory.annotation.Value;
import org.springframework.context.ApplicationContext;
import org.springframework.context.annotation.AnnotationConfigApplicationContext;
import org.springframework.context.annotation.ComponentScan;
import org.springframework.context.annotation.Configuration;
import org.springframework.context.annotation.PropertySource;
import org.springframework.stereotype.Component;
```

```
import entity.Userinfo;

@Component
@ComponentScan(basePackages = { "controller", "service" })
@PropertySource(value = { "classpath:db.properties" })
@Configuration
public class Test {

    @Value("#{userinfoService.userinfo}")
    public Userinfo userinfo;

    @Value("#{userinfoService.username}")
    public String injectStringValue;

    @Value("${driver}")
    public String a;
    @Value("${url}")
    public String b;
    @Value("${username}")
    public String c;
    @Value("${password}")
    public String d;

    public static void main(String[] args) {
        ApplicationContext context = new AnnotationConfigApplicationContext(Test.class);
        Test test = context.getBean(Test.class);
        System.out.println("userinfo=" + test.userinfo.hashCode());
        System.out.println("injectStringValue=" + test.injectStringValue);
        System.out.println("a=" + test.a);
        System.out.println("b=" + test.b);
        System.out.println("c=" + test.c);
        System.out.println("d=" + test.d);
    }
}
```

程序运行后的结果如下：

```
public Userinfo() 1150284200
userinfo=1150284200
injectStringValue=UserinfoService 常<量>值&
a=oracleDriver
b=oracleURL
c=ghy
d=oraclePassword
```

第 4 章　Spring 5 核心技术之 AOP

本章目标：

- 什么是 AOP
- 代理设计模式与 AOP 的关系
- 实现 AOP
- Spring 5 和 MyBatis 3 的整合

4.1　AOP 的使用

在以往的程序设计中，当处理记录日志这样的功能时，需要将与其相关的程序代码嵌入业务代码中，这样就造成了模块之间严重的紧耦合，不利于代码的后期维护。这种情况可以使用 AOP（面向切面编程）来进行分离。

AOP 给程序员最直观的感受就是，它可以在不改动程序代码的基础上做功能上的增强，比如在原来代码的基础上加入数据库事务处理和记录日志的功能。这种增强方式是通过 Proxy（代理）原理实现的。

那么代理具体的概念是什么呢？先来看一看生活中的例子，比如你在某网站买书，该网站只需要将书放在邮递包中，并且写上你的地址就可以了，剩下的任务由快递公司来完成。在这个过程中，快递公司就相当于网站送书服务的代理公司，网站并不负责送书，送书的任务由快递公司代理。这就是代理模式的通俗解释。因此，如果想对某一个类进行功能上的增强而又不改变原始代码，只需要让代理类对其进行增强就可以了，增强什么以及在哪方面增强由代理类决定。

4.1.1　AOP 的原理之代理设计模式

在 Java 中实现动态代理可以使用 4 种常见方式。

（1）JDK 提供的动态代理。

（2）使用 cglib 框架。

（3）使用 javaassist 框架。

（4）使用 Spring 框架。

本节将使用这 4 种方式实现动态代理。

在学习 AOP 技术之前，一定要先了解代理设计模式，Spring 的 AOP 技术原理就是基于代理设计模式的，下面来看一下代理模式的定义。

为其他对象提供一种代理以控制对这个对象的访问。在某些情况下，一个对象不适合或者不能直接引用另一个对象，而代理对象可以在客户端和目标对象之间起到中介的作用。

上面这段话正是“房东”—“中介”—“租房者”的映射，“租房者”联系不到“房东”，必须要通过“房产中介”，“房产中介”就是“代理人”，帮“租房者”寻找房子。

代理设计模式可以在不改变原始代码的基础上进行功能上的增强，使原始对象中的代码与增强代码得到充分解耦。代理模式分为静态代理与动态代理，为了更好地理解 AOP 技术，先来学习一下代理设计模式。

1. 静态代理的实现

在静态代理中，代理对象与被代理对象必须实现同一个接口，完整保留被代理对象的接口样式，并将一直保持接口不变的原则。

查看一个静态代理的代码实例是学习静态代理知识的简便方式。

创建测试用的项目 aopTest1。

新建接口，代码如下：

```
public interface ISendBook {
    public void sendBook();
}
```

新建接口的实现类，代码如下：

```
public class DangDangBook implements ISendBook {
    @Override
    public void sendBook() {
        System.out.println("网站图书部门知道你的地址，电话，备注，要给你送书了！");
    }
}
```

新建快递送书代理类，代码如下：

```
//快递就是代理类，代理类就是帮别人做一些事情的类
//被代理类就是 DangDangBook 或者 JDBook
//SFBookProxy.java 就是代理类
public class SFBookProxy implements ISendBook {
    private ISendBook sendBook;// JDBook or DangDangBook

    public SFBookProxy(ISendBook sendBook) {
        super();
        this.sendBook = sendBook;
    }

    @Override
    public void sendBook() {
        System.out.println("快递收件-事务开启-连接开启-增强开始");
        sendBook.sendBook();
        System.out.println("快递送达-事务提交-连接关闭-增强结束");
```

```
    }
}
```

代理类 SFBookProxy.java 也要实现 ISendBook.java 接口，来达到行为的一致。

新建 Test.java 运行类，代码如下：

```
public class Test {
    public static void main(String[] args) {
        JDBook jdBook = new JDBook();
        DangDangBook dangdangBook = new DangDangBook();

        ISendBook sf1 = new SFBookProxy(jdBook);
        sf1.sendBook();

        System.out.println();

        ISendBook sf2 = new SFBookProxy(dangdangBook);
        sf2.sendBook();
    }
}
```

程序运行的结果如下：

```
快递收件-事务开启-连接开启-增强开始
网站图书部门知道你的地址、电话、备注，要给你送书了！
快递送达-事务提交-连接关闭-增强结束

快递收件-事务开启-连接开启-增强开始
网站图书部门知道你的地址、电话、备注，要给你送书了！
快递送达-事务提交-连接关闭-增强结束
```

在不改变原有 DangDangBook 类代码的基础上进行功能的增强，在控制台输出的“网站书籍部门知道你的地址、电话、备注，要给你送书了！”信息之前和之后分别输出“快递收件—事务开启—连接开启—增强开始”和“快递送达—事务提交—连接关闭—增强结束”的信息，这就是一个典型的日志或事务的功能增强模型，不改变原始的代码就实现了事务处理，不改变原始的代码就实现了日志处理。

但从上面这个示例代码中可以发现，静态代理类有自身的缺点，即扩展性不好，因为 SFBookProxy 类绑定了 ISendBook 接口，如果快递想代理更多的送货任务，则需要创建更多的代理类，比如以下几种。

（1）送鼠标代理类：public class SFSendMouseProxy implements ISendMouse。

（2）送电视代理类：public class SFSendTVProxy implements ISendTV。

（3）送电话代理类：public class SFSendPhoneProxy implements ISendPhone。

静态代理这个致命的缺点导致静态代理不会被应用到实际的软件项目中，那么有没有一个更好的解决办法使代理类不再绑定接口呢？有！这就是动态代理。

2．使用 JDK 实现动态代理

静态代理的缺点是代理类绑定了固定的接口，不利于扩展，动态代理则不然，通过动态代理，可以对任何实现某一接口的类进行功能上的增强，而不会出现代理类绑定接口的情况。

在 Java 中动态代理类的对象由 Proxy 类的 newProxyInstance()方法创建，这就说明 Java 实现动态代理中的代理类并不像静态代理一样是由程序员自己创建的，而是由 JVM 创建出来的，增强的算法需要由 InvocationHandler 接口来实现。

创建测试用的项目 aopTest2。

新建增强算法类 LogInvocationHandler，代码如下：

```
public class LogInvocationHandler implements InvocationHandler {
    private Object anyObject;

    public LogInvocationHandler(Object anyObject) {
        super();
        this.anyObject = anyObject;
    }

    @Override
    public Object invoke(Object proxy, Method method, Object[] args) throws Throwable {
        System.out.println("log begin time=" + System.currentTimeMillis());
        method.invoke(anyObject);
        System.out.println("log   end time=" + System.currentTimeMillis());
        return null;
    }
}
```

运行类代码如下：

```
public class Test1 {
    public static void main(String[] args) {
        DangDangBook dangdangBook = new DangDangBook();
        JDBook jdBook = new JDBook();
        LogInvocationHandler handler = new LogInvocationHandler(dangdangBook);
        ISendBook sendBook = (ISendBook) Proxy.newProxyInstance(Test1.class.getClassLoader(),
                dangdangBook.getClass().getInterfaces(), handler);
        sendBook.sendBook();
    }
}
```

程序运行的结果如下：

```
log begin time=1528706044884
网站图书部门知道你的地址、电话、备注，要给你送书了！
log   end time=1528706044885
```

程序正确运行，总结有以下 3 点。

（1）代理类由 JVM 创建，程序员不需要自己创建代理类，代理类*.java 文件的数量急剧下降。

（2）代理类不再绑定固定的接口，以达到代理类与接口的解耦。

（3）在 InvocationHandler 中通过反射技术能更灵活地处理增强算法。

静态代理和动态代理都是针对 public void sendBook()方法进行增强，虽然在动态代理中使用了反射技术，但是 Spring 只支持对 Method 方法进行增强，不支持 Field 字段级的增强，Spring 认为那样做违反了面向对象编程（OOP）的思想，因此，支持 Method 方法的增强是极为合适的，而且与 Spring 的其他模块进行整合开发时更有标准性。

3. 使用 Spring 实现动态代理

要实现动态代理的功能可以使用 JDK 中 Proxy 类的 newProxyInstance()方法结合 InvocationHandler 接口，还可以使用第三方的 cglib（Code Generation Library）框架。

实现 AOP 功能时可以使用 Spring AOP 框架以及 AspectJ 框架，这两个框架是对动态代理模式的封装。AspectJ 框架是业界 AOP 的标准与规范，Spring AOP 框架支持 AspectJ 框架的部分特性，这些特性已经足够支撑在 Java EE 环境中进行使用。

虽然使用 JDK 中原始的动态代理能实现 AOP 功能，但还是不太方便，并且每个程序员的写法不同，不利于代码风格的一致性与后期维护，因此，Spring 框架对动态代理进行了封装，产生了自己的 AOP 框架，称为 Spring AOP。Spring AOP 框架中常见的接口如下。

（1）方法前通知要实现 org.springframework.aop.MethodBeforeAdvice 接口。

（2）方法后通知要实现 org.springframework.aop.AfterReturningAdvice 接口。

（3）方法环绕通知要实现 org.aopalliance.intercept.MethodInterceptor 接口。

（4）异常处理通知要实现 org.springframework.aop.ThrowsAdvice 接口。

上面 4 个接口是 Spring 实现 AOP 的早期实现，现在已经被废弃，不建议使用，但通过这 4 个接口完全可以认识与了解 AOP 的思想。

创建测试用的项目 aopTest3。

创建方法执行前通知类，代码如下：

```
public class MyMethodBeforeAdvice implements MethodBeforeAdvice {
    @Override
    public void before(Method arg0, Object[] arg1, Object arg2) throws Throwable {
        System.out.println("MyMethodBeforeAdvice");
    }
}
```

创建方法执行后通知类，代码如下：

```
public class MyAfterReturningAdvice implements AfterReturningAdvice {
    @Override
    public void afterReturning(Object arg0, Method arg1, Object[] arg2, Object arg3)
throws Throwable {
        System.out.println("AfterReturningAdvice");
    }
}
```

创建方法环绕通知类，代码如下：

```
public class MyMethodInterceptor implements MethodInterceptor {
    @Override
    public Object invoke(MethodInvocation arg0) throws Throwable {
        System.out.println("MethodInterceptor begin");
        Object value = arg0.proceed();
        System.out.println("MethodInterceptor   end");
        return value;
    }
}
```

创建异常通知类，代码如下：

```
public class MyThrowsAdvice implements ThrowsAdvice {
    public void afterThrowing(Method method, Object[] args, Object target, Exception ex) {
        System.out.println("ThrowsAdvice 信息，方法名称=" + method.getName() + " 参数个数:
" + args.length + " 原始对象: " + target
                + " 异常信息: " + ex.getMessage());
    }
}
```

业务接口 ISendBook 代码如下：

```
public interface ISendBook {
    public void sendBook();
    public void sendBookError();
}
```

业务接口 ISendBook 的实现类代码如下：

```
public class DangDangBook implements ISendBook {
    @Override
    public void sendBook() {
        System.out.println("网站图书部门知道你的地址、电话、备注，要给你送书了！");
    }

    @Override
    public void sendBookError() {
        System.out.println("网站图书部门知道你的地址、电话、备注，要给你送书了！");
        Integer.parseInt("a");
    }
}
```

配置文件 applicationContext.xml 代码如下：

```
<bean id="myAfterReturningAdvice"
    class="advice.MyAfterReturningAdvice"></bean>
<bean id="myMethodBeforeAdvice"
    class="advice.MyMethodBeforeAdvice"></bean>
<bean id="myMethodInterceptor" class="advice.MyMethodInterceptor"></bean>
<bean id="myThrowsAdvice" class="advice.MyThrowsAdvice"></bean>

<bean id="dangdangbook" class="bookservice.DangDangBook"></bean>

<bean id="proxy"
    class="org.springframework.aop.framework.ProxyFactoryBean">
    <property name="interfaces">
        <value>bookservice.ISendBook</value>
    </property>
    <property name="target" ref="dangdangbook"></property>
    <property name="interceptorNames">
        <list>
            <value>myMethodBeforeAdvice</value>
            <value>myAfterReturningAdvice</value>
            <value>myMethodInterceptor</value>
            <value>myThrowsAdvice</value>
        </list>
    </property>
</bean>
```

运行类 Test1.java 代码如下：

```
public class Test {
    public static void main(String[] args) {
        ApplicationContext context = new ClassPathXmlApplicationContext("ac1.xml");
        ISendBook sendBook = (ISendBook) context.getBean("proxy");
        sendBook.sendBook();
        System.out.println();
        System.out.println();
        System.out.println();
        sendBook.sendBookError();
    }
}
```

程序运行的结果如下：

```
MyMethodBeforeAdvice
MethodInterceptor begin
网站图书部门知道你的地址、电话、备注，要给你送书了！
MethodInterceptor   end
AfterReturningAdvice



MyMethodBeforeAdvice
MethodInterceptor begin
网站图书部门知道你的地址、电话、备注，要给你送书了！
ThrowsAdvice 信息，方法名称=sendBookError 参数个数：0 原始对象：
bookservice.DangDangBook@27808f31 异常信息：For input string: "a"
Exception in thread "main" java.lang.NumberFormatException: For input string: "a"
```

在控制台中输出了全部通知类中的功能增强方法，所有通知都参与了动态代理的算法增强。

4．使用 cglib 实现动态代理

创建测试用的项目 proxyType_cglib，并引用 asm-6.2.1.jar 和 cglib-3.2.8.jar 文件。

创建业务类，代码如下：

```
package service;

public class UserinfoService {
    public void save() {
        System.out.println("将数据保存到数据库！");
    }
}
```

创建 cglib 强化类，代码如下：

```
package myinterceptor;

import java.lang.reflect.Method;

import net.sf.cglib.proxy.MethodInterceptor;
import net.sf.cglib.proxy.MethodProxy;

public class MyMethodInterceptor implements MethodInterceptor {
    private Object object;

    public MyMethodInterceptor(Object object) {
        super();
```

```
        this.object = object;
    }

    @Override
    public Object intercept(Object arg0, Method arg1, Object[] arg2, MethodProxy arg3)
throws Throwable {
        System.out.println("begin");
        Object returnValue = arg1.invoke(object, arg2);
        System.out.println("  end");
        return returnValue;
    }

}
```

创建运行类，代码如下：

```
package test;

import myinterceptor.MyMethodInterceptor;
import net.sf.cglib.proxy.Enhancer;
import service.UserinfoService;

public class Test {
    public static void main(String[] args) {
        Enhancer enhancer = new Enhancer();
        enhancer.setSuperclass(UserinfoService.class);
        enhancer.setCallback(new MyMethodInterceptor(new UserinfoService()));
        UserinfoService service = (UserinfoService) enhancer.create();
        service.save();
    }
}
```

程序运行的结果如下：

```
begin
将数据保存到数据库!
  end
```

5. 使用 javaassist 实现动态代理

创建测试用的项目 proxyType_javaassist，并引用 javassist.jar 文件。
创建业务类，代码如下：

```
package service;

public class UserinfoService {
    public void save() {
        System.out.println("将数据保存到数据库！");
    }
}
```

创建 javaassist 强化类，代码如下：

```
package mymethodhandler;

import java.lang.reflect.Method;

import javassist.util.proxy.MethodHandler;

public class MyMethodHandler implements MethodHandler {
```

```
    private Object object;

    public MyMethodHandler(Object object) {
        super();
        this.object = object;
    }

    @Override
    public Object invoke(Object arg0, Method arg1, Method arg2, Object[] arg3) throws
Throwable {
        System.out.println("begin");
        Object returnValue = arg1.invoke(object, arg3);
        System.out.println("  end");
        return returnValue;
    }

}
```

创建运行类，代码如下：

```
package test;

import javassist.util.proxy.Proxy;
import javassist.util.proxy.ProxyFactory;
import mymethodhandler.MyMethodHandler;
import service.UserinfoService;

public class Test {
    public static void main(String[] args) throws InstantiationException,
IllegalAccessException {
        ProxyFactory proxyFactory = new ProxyFactory();
        proxyFactory.setSuperclass(UserinfoService.class);

        Class classRef = proxyFactory.createClass();
        UserinfoService service = (UserinfoService) classRef.newInstance();
        ((Proxy) service).setHandler(new MyMethodHandler(new UserinfoService()));
        service.save();
    }
}
```

程序运行的结果如下：

```
begin
将数据保存到数据库!
  end
```

4.1.2 与 AOP 相关的基本概念

前面介绍过，AOP 技术的原理是基于动态代理的，掌握动态代理就是掌握 AOP 技术的前提。AOP 现在已经发展成为一种独立的技术，有自己独有的术语，掌握理解这些术语有助于学习 AOP 的知识，它们是学习 AOP 技术的必备知识，其不只存在于 Spring 框架，也存在于其他的 AOP 框架中。需要说明的是，AOP 术语并不特别直观，如果不了解它们，那么将会对后续学习造成影响。

1．横切关注点（Cross-cutting Concerns）

在开发软件项目时，需要关注一些“通用性”的功能，比如需要计算出方法（method）的执行时间，有没有访问资源的权限，做一些常规的日志（日志内容包括哪位用户正在登录、在什么时间做了哪些操作、操作的结果是正常还是出现异常等需要记录的信息），另外还有系统中分散在多处的数据库连接的开启与关闭等功能。这些“通用性”功能的代码大多数交织在Service 业务对象中，代码结构如图 4-1 所示。

在大多数 Service 业务类的代码中会重复出现这些“通用性”的功能代码，这些代码与业务代码进行混合，两者之间产生了密不可分的紧耦合，不利于软件的后期维护与扩展。为了解决这个问题，程序员就要关注这些“通用性”的功能，它们在 Spring 的 AOP 中称为“横切关注点”。把“通用性”的功能代码从 Service 业务代码中进行分离正是 AOP 所要达到的核心目的：解耦合。

依赖注入中的“解耦”主要是对象之间解耦，而 AOP 中的“解耦”主要是将“横切关注点”相关的代码与“业务代码”分离、解耦。虽然都是解耦，但是性质是不一样的。

常用的横切关注点如图 4-2 所示。

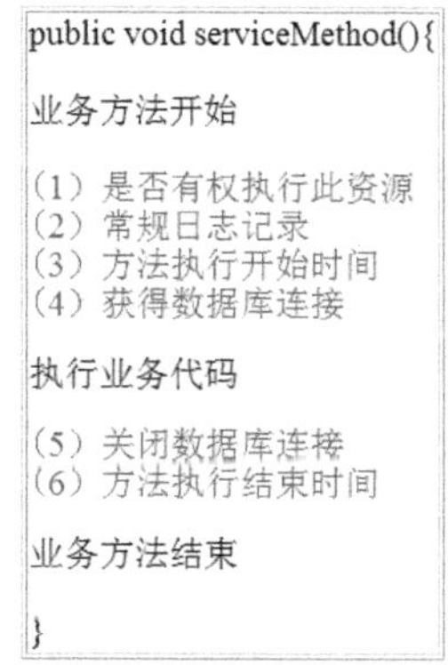

图 4-1 交织在一起的面条式代码

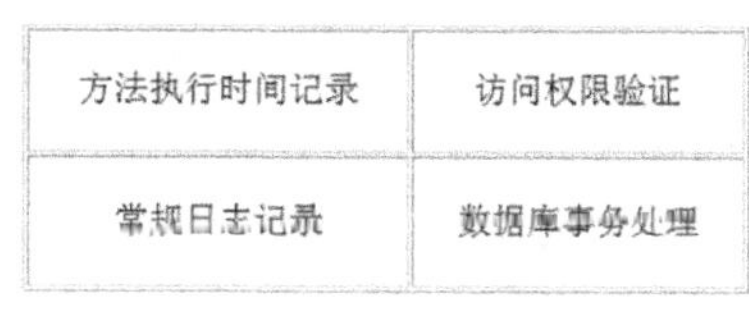

图 4-2 常用的横切关注点

横切关注点就是通用性的功能。

2．切面（Aspect）以及 AOP

在研究什么是切面之前，先来研究一下汉字中的“切”字，百度百科解释如图 4-3 所示。

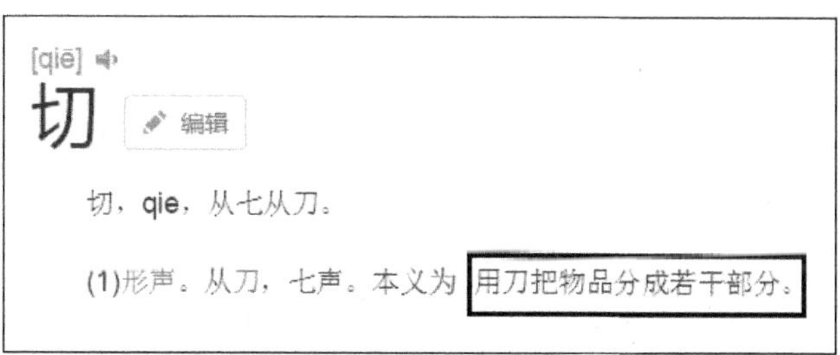

图 4-3 “切”的百度百科解释

汉字中的“切”是指将物品分成若干部分，这个特性和 AOP 是一致的，效果如图 4-4 所示。

在 AService.java、BService.java 和 CService.java 中都需要 3 个切面的功能，因此，要把这些功能从 Service.java 服务类中提取出来。当执行 Service 业务类中的代码时，动态地对切面中的功能代码进行调用，也就实现了对“横切关注点”通用性功能代码的复用，达到了解耦的目的。AOP 可以让一组类的对象拥有相同的行为。

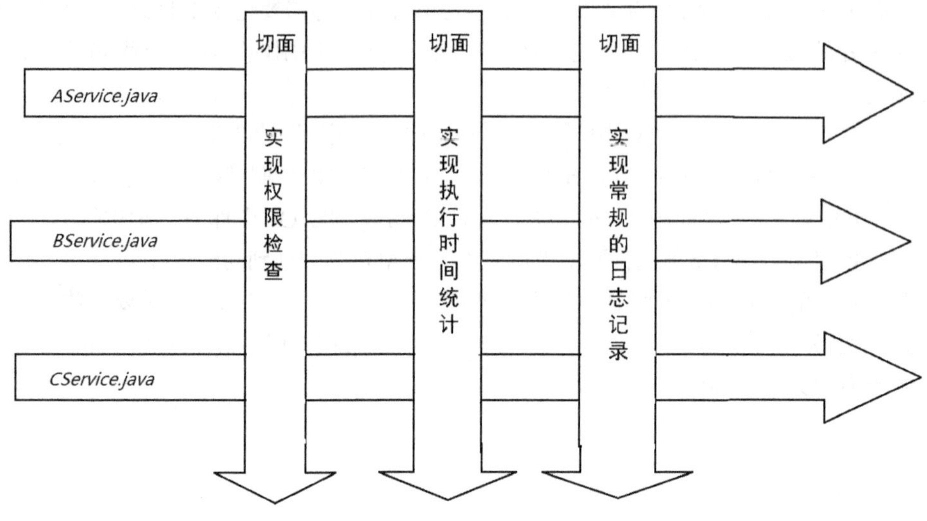

图 4-4　使用“切”的方法将流程分成若干部分

虽然使用“切”的手段能将流程分成若干部分，但是主流程的执行过程是不会被中断的，比如 AService.java 的流程代码不会因为有切面的存在而被中断，会一直运行到最后。在这个过程中，切面就有些类似于过滤器（Filter）：拦截下来，进行处理，然后放行。只不过过滤器拦截的是请求（request），而切面拦截的是方法的调用，运行的特性基本一致，但针对的对象不一样，一个是请求，另一个是方法的调用。

汉字中的“切”以及在 Spring 中如何应用“切”对执行的方法进行拦截已经解释完毕，那么什么是“切面”呢？“切面”就是对“横切关注点”的模块化，也就是将“横切关注点”的功能代码提取出来放入一个单独的类中进行统一处理，这个类就是“切面类”，也可称为切面。在 AOP 中，主要就是针对“切面”设计代码，因此 AOP 的全称就是面向切面编程（Aspect Oriented Programming）。

“切面”对“横切关注点”模块化的示例代码如图 4-5 所示。

将横切关注点中的功能代码放入切面类里的方法中，实现了横切关注点的模块化，切面中的方法可以被很多 Java 类以共享的方式进行访问。

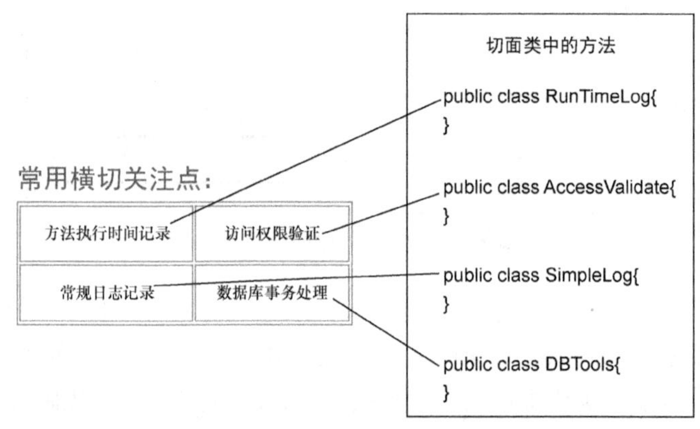

图 4-5　对横切关注点的代码模块化

3. 连接点（Joinpoint）

前面介绍了切面，那么什么是连接点呢？连接点是在软件执行过程中能够插入切面的一个点。连接点可以是在调用方法前、调用方法后、方法抛出异常时和方法返回值之后等位置。在这些连接点中，可以通过插入切面中定义的通用性功能来添加新的软件行为。

连接点示意图如图 4-6 所示。

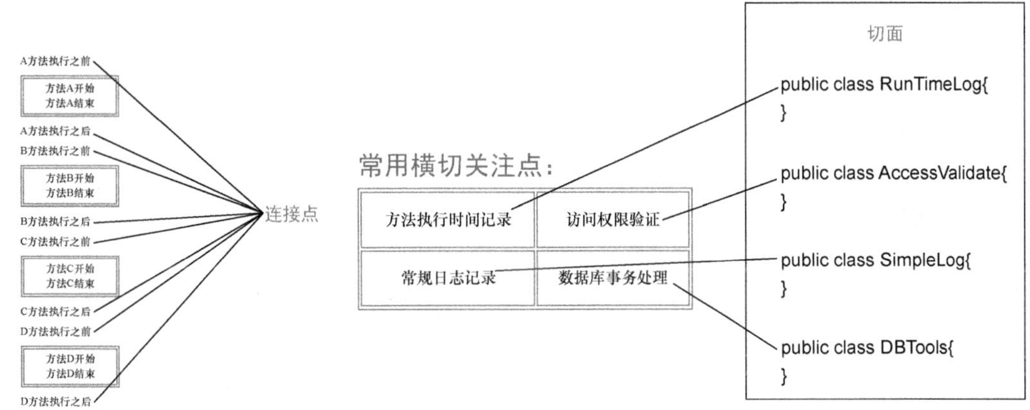

图 4-6　连接点示意图

在软件系统中，存在很多连接点，在这些连接点处可以插入切面，比如在执行 B 方法之前插入 SimpleLog()（日志）的功能，在执行 C 方法之前和之后插入 RunTimeLog()（记录执行时间）的功能等。

4. 切点（Pointcut）

考虑到软件运行的效率，对所有的连接点应用切面是不现实的，在大多数的情况下只想针对“部分的连接点”应用切面，这些“部分的连接点”被称为“切点”，在切点上应用切面。

切点示意图如图 4-7 所示。

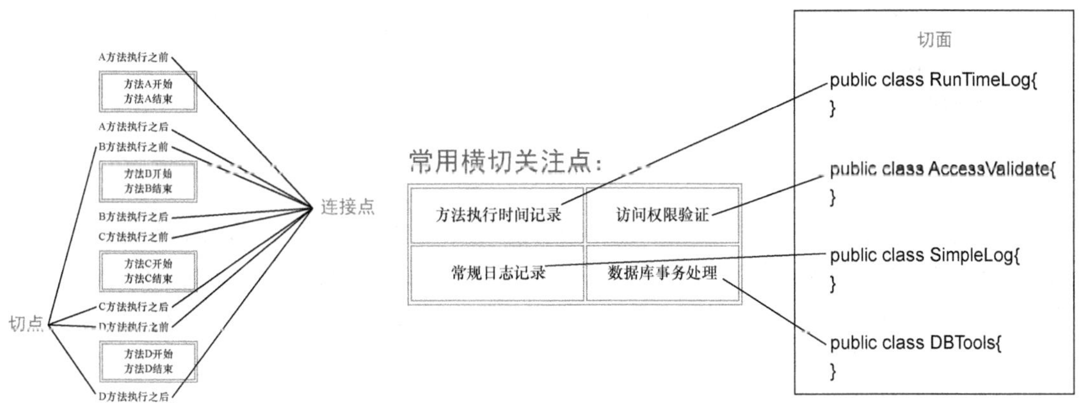

图 4-7　切点示意图

切点是缩小连接点数量后的范围，只针对某几个切点进行切面的应用，它可以精确地定义在什么位置放置切面。

5. 通知（Advice）

切点定义了应用切面的精准位置，但什么时候应用切面由通知来决定，比如可以在执行方法之前、执行方法之后、出现异常等情况下应用切面，这个时机可称为通知。

通知示意图如图 4-8 所示。

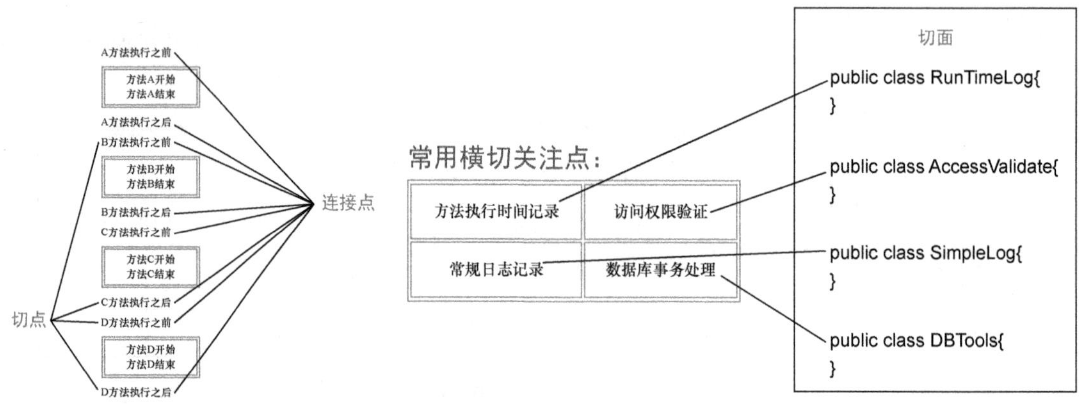

图 4-8 通知示意图

在 Spring 的 AOP 中，通知分为以下 5 种。

（1）前置通知（Before）：方法被调用之前。

（2）后置通知（After）：方法被调用之后。

（3）环绕通知（Around）：方法被调用之前与之后。

（4）返回通知（After-returning）：方法返回了值。

（5）异常通知（After-throwing）：方法出现了异常。

这 5 种通知类型都可以应用切面中的功能代码，在后面的章节会专门介绍这 5 种通知的使用。

切面包含了通知和切点，是两者的结合。通知和切点是切面的基本元素。

（1）通知：定义了在什么时机应用切面。

（2）切点：定义了可以在哪些连接点上放置切面。

6. 织入（Weaving）

织入是把切面应用到指定的对象中，在 Spring 的 AOP 中织入的原理是由 JVM 创建出代理对象，在代理对象中调用原始对象中的方法，再结合增强算法实现。Spring 的 AOP 技术的原理就是代理设计模式，因为 Spring 的 AOP 技术是基于动态代理的，所以它只支持方法的连接点，而不支持字段级的连接点。

织入示意图如图 4-9 所示。

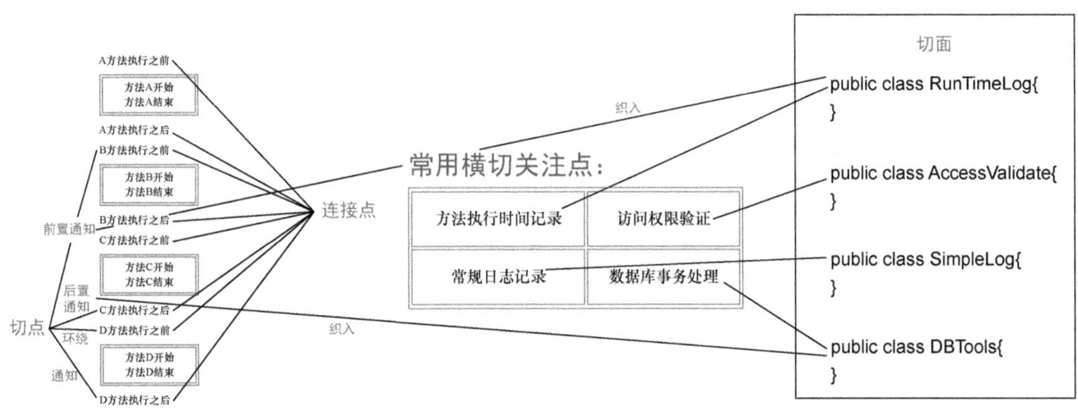

图 4-9　织入示意图

织入将切面应用到目标对象中。

4.1.3　AOP 核心案例

如果一个类实现了接口，那么 Spring 就使用 JDK 的动态代理完成 AOP（推荐）。

如果一个类没有实现接口，那么 Spring 就使用 cglib 完成 AOP。

AOP 技术不是 Spring 框架特有的，而 Spring 框架却高度依赖 AOP 技术实现功能。

前面用大量篇幅介绍了 AOP 技术的概念，并对术语进行了解释，下面开始用若干代码示例来学习 AOP 的程序设计。

1．使用注解实现前置通知、后置通知、返回通知和异常通知

本部分使用注解的方式来实现前置通知、后置通知、返回通知和异常通知的应用，从运行的结果来看，和代理设计模式一致。

创建测试用的项目 aopTest4。

创建切面类，代码如下：

```
@Component
@Aspect
public class AspectObject {
    @Before(value = "execution(* service.UserinfoService.*(..))")
    public void before() {
        System.out.println("public void before()");
    }

    @After(value = "execution(* service.UserinfoService.*(..))")
    public void after() {
        System.out.println("public void after()");
    }

    @AfterReturning(value = "execution(* service.UserinfoService.*(..))")
    public void afterReturning() {
        System.out.println("public void afterReturning()");
```

```
    }

    @AfterThrowing(value = "execution(* service.UserinfoService.*(..))")
    public void afterThrowing() {
        System.out.println("public void afterThrowing()");
    }
}
```

切面表达式（execution）的解释如图 4-10 所示。

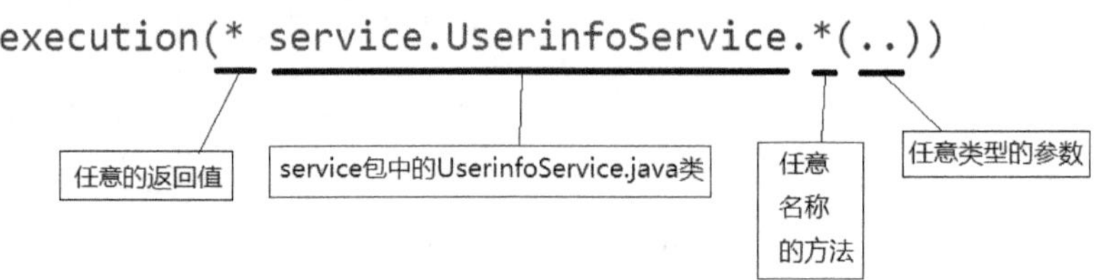

图 4-10 切面表达式解释

通知使用场景如表 4-1 所示。

表 4-1 通知使用场景

增 强 类 型	应 用 场 景
前置增强（Before advice）	权限控制、记录调用日志
后置增强（After returning advice）	统计分析结果数据
异常增强（After throwing advice）	通过日志记录方法异常信息
最终增强（After finally advice）	释放资源
环绕增强（Around advice）	缓存、性能日志、权限、事务管理

通知运行效果总结如图 4-11 所示。

5种不同的增强
aop:before（前置增强）：在方法执行之前执行增强操作。
aop:after-returning（后置增强）：在方法正常执行完成之后执行增强。
aop:throwing（异常增强）：在方法抛出异常退出时执行增强。
aop:after（最终增强）：在方法执行之后执行，相当于在finally中执行；可以通过配置throwing来获得拦截到的异常信息。
aop:around（环绕增强）：一种强大的增强类型。环绕增强可以在方法调用前后完成自定义的行为，环绕通知有两个要求。
（1）方法必须要返回一个Object（返回的结果）。
（2）方法的第一个参数必须是ProceedingJoinPoint（可以继续向下传递的切入点）。

图 4-11 通知运行效果总结

切面表达式（execution）的语法如下：

```
execution(modifiers-pattern?
          ret-type-pattern
          declaring-type-pattern?name-pattern(param-pattern)
          throws-pattern?)
```

（1）modifiers-pattern 代表修饰符，比如 public、private。

（2）ret-type-pattern 代表返回值类型。

（3）declaring-type-pattern 代表包路径。

（4）name-pattern 代表方法。

（5）param-pattern 代表参数。

（6）throws-pattern 代表异常。

带“?”（问号）表示可以忽略，也就是除了 ret-type-pattern，其他都是可选的。

常用的切点表达式示例如下。

（1）execution（public * *(..)）：匹配任何的 public 方法。

（2）execution（* set*(..)）：匹配任何以“set”开头的方法。

（3）execution（* com.xyz.service.AccountService.*(..)）：匹配 com.xyz.service.AccountService 类中所有的方法。

（4）execution（* com.xyz.service.*.*(..)）：匹配 com.xyz.service 包中的任何类中的任何方法。

（5）execution（* com.xyz.service..*.*(..)）：匹配 com.xyz.service 和子孙包中任何类中的任何方法。

（6）execution（public void addUser(com.senqi.entity.User)）：严格匹配。

（7）execution（public void *(com.entity.User)）：表示任意方法名。

（8）execution（public void save(..)）：表示任意参数。

（9）execution（public * save()）：表示任意返回值类型。

（10）execution（public * *(..)）：任意 public 方法。

（11）execution（* com.ghy.UserBiz.*(..)）：UserBiz 接口中的任意方法。

（12）execution（* com. ghy.*.*(..)）：com. ghy 包下任意类的任意方法。

（13）execution（* com. ghy..*.*(..)）：com. ghy 包及其子包下任意类的任意方法。

（14）execution（* set(..)）：任意 set 方法。

（15）execution（* set*(..)）：任意以“set”开头的方法。

切面类中的功能代码就是对横切关注点的模块化，把这些通用性的功能代码统一管理起来，便于复用。

注意：下面的实现类没有实现任何的接口。

创建业务类，代码如下：

```
@Service
public class UserinfoService {
    public void method1() {
        System.out.println("method1 run !");
    }

    public String method2() {
        System.out.println("method2 run !");
        return "我是返回值 A";
    }

    public String method3() {
        System.out.println("method3 run !");
```

```
        Integer.parseInt("a");
        return "我是返回值 B";
    }
}
```

Service 服务类中的方法包含无返回值和有返回值这两种无异常的情况，以及出现异常的情况。

创建配置类，代码如下：

```
@Configuration
@EnableAspectJAutoProxy
@ComponentScan(basePackages = { "aspect", "service" })
public class MyContext {
}
```

创建运行类，代码如下：

```
public class Test {
    public static void main(String[] args) {
        ApplicationContext context = new AnnotationConfigApplicationContext(MyContext.class);
        UserinfoService service = (UserinfoService) context.getBean(UserinfoService.class);
        service.method1();
        System.out.println();
        System.out.println();
        System.out.println("main get method2 returnValue=" + service.method2());
        System.out.println();
        System.out.println();
        System.out.println("main get method3 returnValue=" + service.method3());

    }
}
```

程序运行的结果如下：

```
public void before()
method1 run !
public void after()
public void afterReturning()


public void before()
method2 run !
public void after()
public void afterReturning()
main get method2 returnValue=我是返回值 A


public void before()
method3 run !
public void after()
public void afterThrowing()
Exception in thread "main" java.lang.NumberFormatException: For input string: "a"
```

在无异常的情况下，运行顺序如下。

（1）前置通知。

（2）业务类中的方法。

（3）后置通知。

（4）返回通知。

（5）返回值 A。

在有异常的情况下，运行顺序如下。

（1）前置通知。

（2）业务类中的方法。

（3）后置通知。

（4）异常通知。

在有异常情况出现时，返回通知@AfterReturning 的信息没有在控制台中输出，说明方法内部出现了程序运行流程的中断，程序不再继续执行，没有“返回通知”。也就是说，如果方法返回而不引发异常，通知@AfterReturning 就会被执行。

后置通知@After 和返回通知@AfterReturning 在输出效果上非常类似，这两者本质上的区别是以下两点。

（1）@After 着重点在于方法执行完毕，也就是方法执行完成的方式无论是正常的还是异常返回的，都将执行@After 通知。

（2）@AfterReturning 着重点在于方法无异常地执行完毕后进行返回并继续执行后面的方法，在返回的同时可以使用@AfterReturning 取得返回值，这个功能是@After 不具有的，使用@AfterReturning 取得返回值的代码如下：

```
@Aspect
public class AfterReturningExample {
    @AfterReturning(
        value = "execution(* service.UserinfoService.*(..))", returning="retVal")
    public void doAccessCheck(Object retVal) {
        // ...
    }
}
```

此实验在后面的章节有介绍。

注意： 在本节中的业务类并没有实现任何的接口，如果业务类实现了接口，则在获得业务类时，必须使用 context.getBean("beanId") 的写法，不能使用 context.getBean(Bean.class) 的写法，不然会出现找不到 JavaBean 的异常。

---

2. 对前置通知、后置通知、返回通知和异常通知传入 JoinPoint 参数

参数 JoinPoint 可以实现对连接点信息的获取。

创建测试用的项目 JoinPointParamTest。

创建切面类，代码如下：

```
package aspect;

import java.util.Arrays;

import org.aspectj.lang.JoinPoint;
import org.aspectj.lang.Signature;
import org.aspectj.lang.annotation.After;
import org.aspectj.lang.annotation.AfterReturning;
```

```
import org.aspectj.lang.annotation.AfterThrowing;
import org.aspectj.lang.annotation.Aspect;
import org.aspectj.lang.annotation.Before;
import org.springframework.stereotype.Component;

@Component
@Aspect
public class AspectObject {

    @Before(value = "execution(* service.UserinfoService.*(..))")
    public void before(JoinPoint point) {
        Signature signature = point.getSignature();
        System.out.println("public void before()" + " 对象: " + point.getThis() + " 方
法: " + signature.getName() + " 参数: "
                + Arrays.toString(point.getArgs()));
    }

    @After(value = "execution(* service.UserinfoService.*(..))")
    public void after(JoinPoint point) {
        Signature signature = point.getSignature();
        System.out.println("public void after()" + " 对象: " + point.getThis() + " 方法:
" + signature.getName() + " 参数: "
                + Arrays.toString(point.getArgs()));
    }

    @AfterReturning(value = "execution(* service.UserinfoService.*(..))")
    public void afterReturning(JoinPoint point) {
        Signature signature = point.getSignature();
        System.out.println("public void afterReturning()" + " 对象: " + point.getThis() + " 方
法: " + signature.getName()
                + " 参数: " + Arrays.toString(point.getArgs()));
    }

    @AfterThrowing(value = "execution(* service.UserinfoService.*(..))")
    public void afterThrowing(JoinPoint point) {
        Signature signature = point.getSignature();
        System.out.println("public void afterThrowing()" + " 对象: " + point.getThis() + " 方
法: " + signature.getName()
                + " 参数: " + Arrays.toString(point.getArgs()));
    }

}
```

程序运行后，通过 JoinPoint 对象可以输出对象、方法名称以及参数。

3. 使用 getBean(String) 和 getBean(class) 方法获得 JavaBean 的相关测试

在使用 getBean（String）和 getBean（class）获得容器中的 JavaBean 时有 4 种情况。

（1）实现了接口，使用 getBean（String）方法。

（2）实现了接口，使用 getBean（class）方法。

（3）未实现接口，使用 getBean（String）方法。

（4）未实现接口，使用 getBean（class）方法。

下面分别进行测试。

① 实现了接口

如果获得的 JavaBean 实现了接口，那么方法 getBean（String）和 getBean（class）在获得容器中的 JavaBean 时是有区别的。

创建测试用的项目 aop_hasInterface_hasImple1。

创建切面类，代码如下：

```
@Component
@Aspect
public class MyAspect {
    @Before(value = "execution(* service.*.*(..))")
    public void beforeMethod() {
        System.out.println("===beforeMethod");
    }

    @After(value = "execution(* service.*.*(..))")
    public void afterMethod() {
        System.out.println("===afterMethod");
    }
}
```

创建配置类，代码如下：

```
@Configuration
@EnableAspectJAutoProxy
@ComponentScan(basePackages = { "aspect", "service" })
public class MyContext {
}
```

创建业务接口，代码如下：

```
public interface ISendBook {
    public void sendBook();
}
```

创建实现类，代码如下：

```
@Service(value = "a")
public class DangDangBookA implements ISendBook {
    @Override
    public void sendBook() {
        System.out.println("DangDangBookA public void sendBook()");
    }
}
```

创建实现类，代码如下：

```
@Service(value = "b")
public class DangDangBookB implements ISendBook {
    @Override
    public void sendBook() {
        System.out.println("DangDangBookB public void sendBook()");
    }
}
```

使用 getBean（String）获得 JavaBean，创建运行类，代码如下：

```
public class Test1 {
    public static void main(String[] args) {
```

```
        ApplicationContext context = new AnnotationConfigApplicationContext(MyContext.class);
        ISendBook a = (ISendBook) context.getBean("a");
        ISendBook b = (ISendBook) context.getBean("b");
        a.sendBook();
        b.sendBook();
    }
}
```

程序运行的结果如下：

```
===beforeMethod
DangDangBookA public void sendBook()
===afterMethod
===beforeMethod
DangDangBookB public void sendBook()
===afterMethod
```

使用 getBean（class）获得 JavaBean，创建运行类，代码如下：

```
public class Test2 {
    public static void main(String[] args) {
        ApplicationContext context = new AnnotationConfigApplicationContext(MyContext.class);
        context.getBean(DangDangBookA.class);
    }
}
```

程序运行的结果如下：

```
    Exception in thread "main" org.springframework.beans.factory.NoSuchBeanDefinitionException:
No qualifying bean of type 'service.DangDangBookA' available
```

② 未实现接口

如果获得的 JavaBean 未实现接口，那么方法 getBean(String) 和 getBean(class) 在获得容器中的 JavaBean 时是没有区别的。

创建测试用的项目 aop_noInterface_hasImple1。

创建切面类，代码如下：

```
@Component
@Aspect
public class MyAspect {
    @Before(value = "execution(* service.*.*(..))")
    public void beforeMethod() {
        System.out.println("===beforeMethod");
    }

    @After(value = "execution(* service.*.*(..))")
    public void afterMethod() {
        System.out.println("===afterMethod");
    }
}
```

创建配置类，代码如下：

```
@Configuration
@EnableAspectJAutoProxy
@ComponentScan(basePackages = { "aspect", "service" })
public class MyContext {
}
```

创建实现类，代码如下：

```
@Service(value = "a")
public class DangDangBook {
    public void sendBook() {
        System.out.println("DangDangBook public void sendBook()");
    }
}
```

使用 getBean（String）获得 JavaBean，创建运行类，代码如下：

```
public class Test1 {
    public static void main(String[] args) {
        ApplicationContext context = new AnnotationConfigApplicationContext(MyContext.class);
        DangDangBook dangdangBook = (DangDangBook) context.getBean("a");
        dangdangBook.sendBook();
    }
}
```

程序运行的结果如下：

```
===beforeMethod
DangDangBook public void sendBook()
===afterMethod
```

使用 getBean（class）获得 JavaBean，创建运行类，代码如下：

```
public class Test2 {
    public static void main(String[] args) {
        ApplicationContext context = new AnnotationConfigApplicationContext(MyContext.class);
        DangDangBook dangdangBook = (DangDangBook) context.getBean(DangDangBook.class);
        dangdangBook.sendBook();
    }
}
```

程序运行的结果如下：

```
===beforeMethod
DangDangBook public void sendBook()
===afterMethod
```

4. 使用 XML 实现前置通知、后置通知、返回通知和异常通知

使用 XML 方式来实现 AOP 就不需要任何注解了，可以在 Java 类中删除全部的注解。

创建测试用的项目 aopTest5。

创建切面类，代码如下：

```
public class AspectObject {
    public void before() {
        System.out.println("public void before()");
    }

    public void after() {
        System.out.println("public void after()");
    }

    public void afterReturning() {
```

```
        System.out.println("public void afterReturning()");
    }

    public void afterThrowing() {
        System.out.println("public void afterThrowing()");
    }
}
```

创建业务类，代码如下：

```
public class UserinfoService {
    public void method1() {
        System.out.println("method1 run !");
    }

    public String method2() {
        System.out.println("method2 run !");
        return "我是返回值 A";
    }

    public String method3() {
        System.out.println("method3 run !");
        Integer.parseInt("a");
        return "我是返回值 B";
    }
}
```

创建配置文件 applicationContext.xml，代码如下：

```
<aop:aspectj-autoproxy></aop:aspectj-autoproxy>

<bean id="userinfoService" class="service.UserinfoService"></bean>
<bean id="myaspect" class="aspect.AspectObject"></bean>

<aop:config>
    <aop:aspect ref="myaspect">
        <aop:before method="before"
            pointcut="execution(* service.UserinfoService.*(..))" />
        <aop:after method="after"
            pointcut="execution(* service.UserinfoService.*(..))" />
        <aop:after-returning method="afterReturning"
            pointcut="execution(* service.UserinfoService.*(..))" />
        <aop:after-throwing method="afterThrowing"
            pointcut="execution(* service.UserinfoService.*(..))" />
    </aop:aspect>
</aop:config>
```

配置代码的作用如下。

（1）<aop:before>：方法执行前通知。

（2）<aop:after>：方法执行后通知。

（3）<aop:after-returning>：方法返回通知。

（4）<aop:after-throwing>：方法异常通知。

创建运行类 Test，代码如下：

```
public class Test {
    public static void main(String[] args) {
```

```
        ApplicationContext context = new ClassPathXmlApplicationContext("ac1.xml");
        UserinfoService service = (UserinfoService) context.getBean(UserinfoService.class);
        service.method1();
        System.out.println();
        System.out.println();
        System.out.println("main get method2 returnValue=" + service.method2());
        System.out.println();
        System.out.println();
        System.out.println("main get method3 returnValue=" + service.method3());
    }
}
```

程序运行的结果如下：

```
public void before()
method1 run !
public void after()
public void afterReturning()


public void before()
method2 run !
public void after()
public void afterReturning()
main get method2 returnValue=我是返回值 A


public void before()
method3 run !
public void after()
public void afterThrowing()
Exception in thread "main" java.lang.NumberFormatException: For input string: "a"
```

使用 XML 方式实现 AOP 与使用注解的方式实现 AOP 在运行结果上是一样的。

5. 使用注解实现环绕通知

所谓的环绕通知是在执行方法之前和执行方法之后都有切面的参与。

创建测试用的项目 aopTest6。

创建切面类，代码如下：

```
@Component
@Aspect
public class AspectObject {
    @Around(value = "execution(* service.UserinfoService.*(..))")
    public Object around(ProceedingJoinPoint point) {
        Object returnObject = null;
        try {
            System.out.println("开始");
            returnObject = point.proceed();
            System.out.println("结束");
        } catch (Throwable e) {
            e.printStackTrace();
        }
        return returnObject;
    }
}
```

必须要有参数 ProceedingJoinPoint point，数据类型不能写错。

创建业务类，代码如下：

```
@Service
public class UserinfoService {
    public void method1() {
        System.out.println("method1 run !");
    }

    public String method2() {
        System.out.println("method2 run !");
        return "我是返回值 A";
    }

    public String method3() {
        System.out.println("method3 run !");
        Integer.parseInt("a");
        return "我是返回值 B";
    }
}
```

创建配置类，代码如下：

```
@Configuration
@EnableAspectJAutoProxy
@ComponentScan(basePackages = { "aspect", "service" })
public class MyContext {
}
```

创建运行类，代码如下：

```
public class Test {
    public static void main(String[] args) {
        ApplicationContext context = new AnnotationConfigApplicationContext(MyContext.class);
        UserinfoService service = (UserinfoService) context.getBean(UserinfoService.class);
        service.method1();
        System.out.println();
        System.out.println();
        System.out.println("main get method2 returnValue=" + service.method2());
        System.out.println();
        System.out.println();
        System.out.println("main get method3 returnValue=" + service.method3());

    }
}
```

程序运行的结果如下：

```
开始
method1 run !
结束


开始
method2 run !
结束
main get method2 returnValue=我是返回值 A
```

```
开始
method3 run !
java.lang.NumberFormatException: For input string: "a"
    at java.lang.NumberFormatException.forInputString(NumberFormatException.java:65)
main get method3 returnValue=null
```

6. 使用 XML 实现环绕通知

创建测试用的项目 aopTest7。

创建切面类，代码如下：

```java
public class AspectObject {
    public Object around(ProceedingJoinPoint point) {
        Object returnObject = null;
        try {
            System.out.println("开始");
            returnObject = point.proceed();
            System.out.println("结束");
        } catch (Throwable e) {
            e.printStackTrace();
        }
        return returnObject;
    }
}
```

创建业务类，代码如下：

```java
public class UserinfoService {
    public void method1() {
        System.out.println("method1 run !");
    }

    public String method2() {
        System.out.println("method2 run !");
        return "我是返回值 A";
    }

    public String method3() {
        System.out.println("method3 run !");
        Integer.parseInt("a");
        return "我是返回值 B";
    }
}
```

创建配置文件，代码如下：

```xml
<aop:aspectj-autoproxy></aop:aspectj-autoproxy>

<bean id="userinfoService" class="service.UserinfoService"></bean>
<bean id="myaspect" class="aspect.AspectObject"></bean>

<aop:config>
    <aop:aspect ref="myaspect">
        <aop:around method="around"
            pointcut="execution(* service.UserinfoService.*(..))" />
    </aop:aspect>
</aop:config>
```

创建运行类，代码如下：

```
public class Test {
    public static void main(String[] args) {
        ApplicationContext context = new ClassPathXmlApplicationContext("ac1.xml");
        UserinfoService service = (UserinfoService) context.getBean(UserinfoService.class);
        service.method1();
        System.out.println();
        System.out.println();
        System.out.println("main get method2 returnValue=" + service.method2());
        System.out.println();
        System.out.println();
        System.out.println("main get method3 returnValue=" + service.method3());
    }
}
```

程序运行的结果如下：

```
开始
method1 run !
结束


开始
method2 run !
结束
main get method2 returnValue=我是返回值 A


开始
method3 run !
java.lang.NumberFormatException: For input string: "a"
    at java.lang.NumberFormatException.forInputString(NumberFormatException.java:65)
main get method3 returnValue=null
```

需要注意的是，@Before 和 @After 并不是真正的一对，在经过测试后发现，如果@Before、@After 和 @Around 联合使用，则输出的顺序如下：

```
aroundMethod begin
beforeMethod
a run !
aroundMethod   end
afterMethod
```

并不是想象中的：

```
aroundMethod begin
beforeMethod
a run !
afterMethod
aroundMethod   end
```

从上面若干实验的输出结果来看，@After 通知类似于 finally 块，无论有没有异常都会得到执行。

另外，如果在 XML 中也联合使用这 3 个通知，那么 @Before 与 @Around 的执行顺序取决于它们在 XML 文件中的配置顺序。

7. 在注解中使用 bean 表达式

在 Spring 中提供了 bean 表达式来限制切面可应用的目标对象。

创建测试用的项目 aopTest8。

创建切面类，代码如下：

```
@Component
@Aspect
public class AspectObject {
    @Around(value = "execution(* service..*(..)) and bean(service1)")
    public Object around(ProceedingJoinPoint point) {
        Object returnObject = null;
        try {
            System.out.println("开始");
            returnObject = point.proceed();
            System.out.println("结束");
        } catch (Throwable e) {
            e.printStackTrace();
        }
        return returnObject;
    }
}
```

切点表达式“service..”中的“..”代表任意子孙级的包。

表达式“and bean (service1)”表明切面必须要应用于 bean 的 ID 值为 service1 的对象。

创建业务类，代码如下：

```
@Service(value = "service1")
public class UserinfoServiceA {
    public void method1() {
        System.out.println("methodA run !");
    }
}
```

创建业务类，代码如下：

```
@Service(value = "service2")
public class UserinfoServiceB {
    public void method1() {
        System.out.println("methodB run !");
    }
}
```

创建配置类 MyContext，代码如下：

```
@Configuration
@EnableAspectJAutoProxy
@ComponentScan(basePackages = { "aspect", "service" })
public class MyContext {
}
```

创建运行类，代码如下：

```
public class Test {
    public static void main(String[] args) {
        ApplicationContext context = new AnnotationConfigApplicationContext(MyContext.class);
```

```
        UserinfoServiceA serviceA = (UserinfoServiceA) context.getBean(UserinfoServiceA.class);
        serviceA.method1();
        System.out.println();
        UserinfoServiceB serviceB = (UserinfoServiceB) context.getBean(UserinfoServiceB.class);
        serviceB.method1();
    }
}
```

程序运行的结果如下：

```
开始
methodA run !
结束

methodB run !
```

只对 bean 的 ID 是 service1 的对象应用了切面。

8. 在 XML 中使用 bean 表达式

创建测试用的项目 aopTest9。

创建切面类，代码如下：

```
public class AspectObject {
    public Object around(ProceedingJoinPoint point) {
        Object returnObject = null;
        try {
            System.out.println("开始");
            returnObject = point.proceed();
            System.out.println("结束");
        } catch (Throwable e) {
            e.printStackTrace();
        }
        return returnObject;
    }
}
```

创建业务类，代码如下：

```
public class UserinfoServiceA {
    public void method1() {
        System.out.println("methodA run !");
    }
}
```

创建业务类，代码如下：

```
public class UserinfoServiceB {
    public void method1() {
        System.out.println("methodB run !");
    }
}
```

创建配置文件，代码如下：

```
<bean id="userinfoService1" class="service.UserinfoServiceA"></bean>
<bean id="userinfoService2" class="service.UserinfoServiceB"></bean>

<bean id="myaspect" class="aspect.AspectObject"></bean>
```

```
<aop:config>
    <aop:aspect ref="myaspect">
        <aop:around method="around"
            pointcut="execution(* service..*(..)) and bean(userinfoService2)" />
    </aop:aspect>
</aop:config>
```

创建运行类，代码如下：

```
public class Test {
    public static void main(String[] args) {
        ApplicationContext context = new ClassPathXmlApplicationContext("ac1.xml");
        UserinfoServiceA serviceA = (UserinfoServiceA) context.getBean(UserinfoServiceA.class);
        serviceA.method1();
        System.out.println();
        UserinfoServiceB serviceB = (UserinfoServiceB) context.getBean(UserinfoServiceB.class);
        serviceB.method1();
    }
}
```

程序运行的结果如下：

```
methodA run !

开始
methodB run !
结束
```

9. 使用注解 @Pointcut 定义全局切点

前面在多处使用了一样的表达式，可以将表达式全局化，以减少冗余的配置。

创建测试用的项目 aopTest10。

创建切面类，代码如下：

```
@Component
@Aspect
public class AspectObject {
    @Pointcut(value = "execution(* service.UserinfoService.*(..))")
    public void publicPointcut() {

    }

    @Before(value = "publicPointcut()")
    public void before() {
        System.out.println("public void before()");
    }

    @After(value = "publicPointcut()")
    public void after() {
        System.out.println("public void after()");
    }

    @AfterReturning(value = "publicPointcut()")
    public void afterReturning() {
        System.out.println("public void afterReturning()");
    }
```

```
    @AfterThrowing(value = "publicPointcut()")
    public void afterThrowing() {
        System.out.println("public void afterThrowing()");
    }
}
```

使如下的全局配置依附于 public void publicPointcut() 方法，引用这个全局切点 Pointcut 时只需指出方法名称即可。

```
@Pointcut(value = "execution(* service.UserinfoService.*(..))")
```

创建业务类，代码如下：

```
@Service
public class UserinfoService {
    public void method1() {
        System.out.println("method1 run !");
    }

    public String method2() {
        System.out.println("method2 run !");
        return "我是返回值 A";
    }

    public String method3() {
        System.out.println("method3 run !");
        Integer.parseInt("a");
        return "我是返回值 B";
    }
}
```

创建配置类，代码如下：

```
@Configuration
@EnableAspectJAutoProxy
@ComponentScan(basePackages = { "aspect", "service" })
public class MyContext {
}
```

创建运行类，代码如下：

```
public class Test {
    public static void main(String[] args) {
        ApplicationContext context = new AnnotationConfigApplicationContext(MyContext.class);
        UserinfoService service = (UserinfoService) context.getBean(UserinfoService.class);
        service.method1();
        System.out.println();
        System.out.println();
        System.out.println("main get method2 returnValue=" + service.method2());
        System.out.println();
        System.out.println();
        System.out.println("main get method3 returnValue=" + service.method3());

    }
}
```

程序运行结果如下：

```
public void before()
method1 run !
public void after()
public void afterReturning()


public void before()
method2 run !
public void after()
public void afterReturning()
main get method2 returnValue=我是返回值 A


public void before()
method3 run !
public void after()
public void afterThrowing()
Exception in thread "main" java.lang.NumberFormatException: For input string: "a"
```

10．使用 xml<aop:pointcut> 定义全局切点

创建测试用的项目 aopTest11。

创建切面类，代码如下：

```
public class AspectObject {
    public void before() {
        System.out.println("public void before()");
    }

    public void after() {
        System.out.println("public void after()");
    }

    public void afterReturning() {
        System.out.println("public void afterReturning()");
    }

    public void afterThrowing() {
        System.out.println("public void afterThrowing()");
    }
}
```

创建业务类，代码如下：

```
public class UserinfoService {
    public void method1() {
        System.out.println("method1 run !");
    }

    public String method2() {
        System.out.println("method2 run !");
        return "我是返回值 A";
    }

    public String method3() {
        System.out.println("method3 run !");
```

```
        Integer.parseInt("a");
        return "我是返回值 B";
    }
}
```

创建配置文件，代码如下：

```
<aop:aspectj-autoproxy></aop:aspectj-autoproxy>

<bean id="userinfoService" class="service.UserinfoService"></bean>
<bean id="myAspect" class="aspect.AspectObject"></bean>

<aop:config>
    <aop:pointcut id="myExecution"
        expression="execution(* service..*(..))" />
    <aop:aspect ref="myAspect">
        <aop:before method="before" pointcut-ref="myExecution" />
        <aop:after method="after" pointcut-ref="myExecution" />
        <aop:after-returning method="afterReturning"
            pointcut-ref="myExecution" />
        <aop:after-throwing method="afterThrowing"
            pointcut-ref="myExecution" />
    </aop:aspect>
</aop:config>
```

创建运行类，代码如下：

```
public class Test {
    public static void main(String[] args) {
        ApplicationContext context = new ClassPathXmlApplicationContext("ac1.xml");
        UserinfoService service = (UserinfoService) context.getBean(UserinfoService.class);
        service.method1();
        System.out.println();
        System.out.println();
        System.out.println("main get method2 returnValue=" + service.method2());
        System.out.println();
        System.out.println();
        System.out.println("main get method3 returnValue=" + service.method3());

    }
}
```

程序运行的结果如下：

```
public void before()
method1 run !
public void after()
public void afterReturning()


public void before()
method2 run !
public void after()
public void afterReturning()
main get method2 returnValue=我是返回值 A


public void before()
method3 run !
```

```
public void after()
public void afterThrowing()
Exception in thread "main" java.lang.NumberFormatException: For input string: "a"
```

11. 使用注解向切面传入参数

只要 Service 业务方法有参数，切面类就能获得参数值。

创建测试用的项目 aopTest12。

创建切面类，代码如下：

```
@Component
@Aspect
public class AspectObject {
    // 下面是@Pointcut 注解的解释
    // (1)@Pointcut 注解的功能是声明 1 个切点表达式
    // (2)因为切面要取得调用方法时传入的参数，
    // 所以要使用 args 表达式：args(xxxxxx)来进行获取
    // (3)与@Pointcut 关联方法的参数名称必须和 args()表达式中的 xxxxxx 一样
    @Pointcut(value = "execution(* service.UserinfoService.method1(int)) && args(xxxxxx)")
    public void methodAspect1(int xxxxxx) {
    }

    // 下面是@Before 注解的解释
    // (1)属性 value = "methodAspect(ageabc)"是引用方法
    // public void methodAspect(int xxxxxx),
    // 引用时参数名称可以不一样，一个是 xxxxxx，另一个是 ageabc
    // (2)与@Before 关联的方法
    // public void method1Before(int ageabc)
    // 中的参数名称必须和@Before(value = "methodAspect(ageabc)")
    // 配置中方法的参数名称一样
    // (3)@Pointcut 和@Before 交换的关联点在于方法的名称 methodAspect,
    // 不包含参数的命名统一性
    @Before(value = "methodAspect1(ageabc)")
    public void method1Before(int ageabc) {
        System.out.println("切面: public void method1Before(int ageabc) ageabc=" + ageabc);
    }


    @Pointcut(value = "execution(* service.UserinfoService.method2(String,String,int,j
ava.util.Date)) && args(u,p,a,i)")
    public void methodAspect2(String u, String p, int a, Date i) {
    }

    @Before(value = "methodAspect2(uu,pp,aa,ii)")
    public void method2Before(String uu, String pp, int aa, Date ii) {
        System.out.println("切面: public void method2Before(String uu, String pp, int aa,
Date ii) uu=" + uu + " pp=" + pp
                + " aa=" + aa + " ii=" + ii);
    }
}
```

创建业务类，代码如下：

```
@Service
public class UserinfoService {
    public void method1(int ageage) {
```

```
            System.out.println("method1 age=" + ageage);
        }

        public String method2(String username, String password, int age, Date insertdate) {
            System.out.println(
                    "method2 username=" + username + " password=" + password + " age=" +
age + " insertdate=" + insertdate);
            return "我是返回值 method2";
        }
    }
```

创建配置类，代码如下：

```
@Configuration
@EnableAspectJAutoProxy
@ComponentScan(basePackages = { "aspect", "service" })
public class MyContext {
}
```

创建运行类，代码如下：

```
public class Test {
    public static void main(String[] args) {
        ApplicationContext context = new AnnotationConfigApplicationContext(MyContext.class);
        UserinfoService service = (UserinfoService) context.getBean(UserinfoService.class);
        service.method1(100);
        System.out.println();
        System.out.println();
        System.out.println("main get method2 returnValue=" + service.method2("中国", "
中国", 123, new Date()));
    }
}
```

程序运行的结果如下：

```
切面: public void method1Before(int ageabc) ageabc=100
method1 age=100


切面: public void method2Before(String uu, String pp, int aa, Date ii) uu=中国 pp=中
国 aa=123 ii=Mon Jun 11 16:58:17 CST 2018
method2 username=中国 password=中国 age=123 insertdate=Mon Jun 11 16:58:17 CST 2018
main get method2 returnValue=我是返回值 method2
```

12．使用 XML 向切面传入参数

创建测试用的项目 aopTest13。

创建切面类，代码如下：

```
public class AspectObject {
    public void method1Before(int xxxxxx) {
        System.out.println("切面: public void method1Before(int xxxxxx) xxxxxx=" + xxxxxx);
    }

    public void method2Before(String uu, String pp, int aa, Date ii) {
        System.out.println("切面: public void method2Before(String uu, String pp, int aa,
Date ii) uu=" + uu + " pp=" + pp
                + " aa=" + aa + " ii=" + ii);
```

```
    }
}
```

创建业务类，代码如下：

```
public class UserinfoService {
    public void method1(int ageage) {
        System.out.println("method1 age=" + ageage);
    }

    public String method2(String username, String password, int age, Date insertdate) {
        System.out.println(
                "method2 username=" + username + " password=" + password + " age=" +
age + " insertdate=" + insertdate);
        return "我是返回值 method2";
    }
}
```

创建配置文件，代码如下：

```
<bean id="userinfoService" class="service.UserinfoService"></bean>
<bean id="myaspect" class="aspect.AspectObject"></bean>

<aop:config>
    <aop:pointcut id="pointCut1"
        expression="execution (* service..method1(int)) and args(xxxxxx)" />
    <aop:pointcut id="pointCut2"
        expression="execution (* service..method2(String,String,int,java.util.Date))
and args(uu,pp,aa,ii)" />
    <aop:aspect ref="myaspect">
        <aop:before method="method1Before"
            pointcut-ref="pointCut1" />
        <aop:before method="method2Before"
            pointcut-ref="pointCut2" />
    </aop:aspect>
</aop:config>
```

以下配置代码中的 args()的参数名称必须要和切面类的通知方法的参数名称一致。

```
<aop:pointcut id="pointCut1"
    expression="execution (* service..method1(int)) and args(xxxxxx)" />
```

通知方法的代码如下：

```
public void method1Before(int xxxxxx) {
    System.out.println("切面: public void method1Before(int xxxxxx) xxxxxx=" + xxxxxx);
}
```

方法 public void method1Before（int xxxxxx）的参数名称必须是 xxxxxx。

创建运行类，代码如下：

```
public class Test {
    public static void main(String[] args) {
        ApplicationContext context = new ClassPathXmlApplicationContext("ac1.xml");
        UserinfoService service = (UserinfoService) context.getBean(UserinfoService.class);
        service.method1(100);
        System.out.println();
```

```
            System.out.println();
            System.out.println("main get method2 returnValue=" + service.method2("中国", "
中国", 123, new Date()));
        }
    }
```

程序运行的结果如下：

```
切面：public void method1Before(int xxxxxx) xxxxxx=100
method1 age=100


切面：public void method2Before(String uu, String pp, int aa, Date ii) uu=中国 pp=中
国 aa=123 ii=Mon Jun 11 17:01:13 CST 2018
method2 username=中国 password=中国 age=123 insertdate=Mon Jun 11 17:01:13 CST 2018
main get method2 returnValue=我是返回值 method2
```

13. 使用注解 @AfterReturning 和 @AfterThrowing 向切面传入参数

创建测试用的项目 aopTest14。

创建切面类，代码如下：

```
@Component
@Aspect
public class AspectObject {
    @Pointcut(value = "execution(* service.UserinfoService.method1(int)) && args(xxxxxx)")
    public void methodAspect1(int xxxxxx) {
    }

    @Pointcut(value = "execution(* service.UserinfoService.method2(String,String,int,
java.util.Date)) && args(u,p,a,i)")
    public void methodAspect2(String u, String p, int a, Date i) {
    }

    @Pointcut(value = "execution(* service.UserinfoService.*(..))")
    public void methodAspect3() {
    }

    @Before(value = "methodAspect1(xxxxxx)")
    public void method1Before(int xxxxxx) {
        System.out.println("切面：public void method1Before(int xxxxxx) xxxxxx=" + xxxxxx);
    }

    @Before(value = "methodAspect2(u, p, a, i)")
    public void method2Before(String u, String p, int a, Date i) {
        System.out.println("切面：public void method2Before(String u, String p, int a,
Date i) u=" + u + " p=" + p + " a="
                + a + " i=" + i);
    }

    @AfterReturning(value = "methodAspect3()", returning = "returnParam")
    public void method3AfterReturning(Object returnParam) {
        System.out.println("public void method3AfterReturning(Object returnParam)
returnParam=" + returnParam);
    }

    @AfterThrowing(value = "methodAspect3()", throwing = "t")
    public void method4AfterThrowing(Throwable t) {
```

```
        System.out.println("public void method4AfterThrowing(Throwable t) t=" + t);
    }
}
```

创建业务类，代码如下：

```
@Service
public class UserinfoService {
    public void method1(int ageage) {
        System.out.println("method1 age=" + ageage);
    }

    public String method2(String username, String password, int age, Date insertdate) {
        System.out.println(
                "method2 username=" + username + " password=" + password + " age=" +
age + " insertdate=" + insertdate);
        Integer.parseInt("a");
        return "我是返回值 method2";
    }
}
```

创建配置类，代码如下：

```
@Configuration
@EnableAspectJAutoProxy
@ComponentScan(basePackages = { "aspect", "service" })
public class MyContext {
}
```

创建运行类，代码如下：

```
public class Test {
    public static void main(String[] args) {
        ApplicationContext context = new AnnotationConfigApplicationContext(MyContext.class);
        UserinfoService service = (UserinfoService) context.getBean(UserinfoService.class);
        service.method1(100);
        System.out.println();
        System.out.println();
        System.out.println("main get method2 returnValue=" + service.method2("中国", "
中国", 123, new Date()));
    }
}
```

程序运行的结果如下：

```
切面: public void method1Before(int xxxxxx) xxxxxx=100
method1 age=100
public void method3AfterReturning(Object returnParam) returnParam=null


切面: public void method2Before(String u, String p, int a, Date i) u=中国 p=中国 a=123 i=
Mon Jun 11 17:04:02 CST 2018
method2 username=中国 password=中国 age=123 insertdate=Mon Jun 11 17:04:02 CST 2018
public void method4AfterThrowing(Throwable t) t=java.lang.NumberFormatException: For
input string: "a"
Exception in thread "main" java.lang.NumberFormatException: For input string: "a"
```

14. 使用 xml<aop:after-returning>和<aop:after-throwing> 向切面传入参数

创建测试用的项目 aopTest15。

创建切面类，代码如下：

```
public class AspectObject {
    public void method1Before(int xxxxxx) {
        System.out.println("切面: public void method1Before(int xxxxxx) xxxxxx=" + xxxxxx);
    }

    public void method2Before(String u, String p, int a, Date i) {
        System.out.println("切面: public void method2Before(String u, String p, int a,
Date i) u=" + u + " p=" + p + " a="
                + a + " i=" + i);
    }

    public void method3AfterReturning(Object returnParam) {
        System.out.println("public void method3AfterReturning(Object returnParam)
returnParam=" + returnParam);
    }

    public void method4AfterThrowing(Throwable t) {
        System.out.println("public void method4AfterThrowing(Throwable t) t=" + t);
    }
}
```

创建业务类，代码如下：

```
public class UserinfoService {
    public void method1(int ageage) {
        System.out.println("method1 age=" + ageage);
    }

    public String method2(String username, String password, int age, Date insertdate) {
        System.out.println(
                "method2 username=" + username + " password=" + password + " age=" +
age + " insertdate=" + insertdate);
        Integer.parseInt("a");
        return "我是返回值 method2";
    }
}
```

创建配置文件，代码如下：

```
<bean id="userinfoService" class="service.UserinfoService"></bean>
<bean id="myAspect" class="aspect.AspectObject"></bean>

<aop:aspectj-autoproxy></aop:aspectj-autoproxy>

<aop:config>
    <aop:pointcut id="pointCut1"
        expression="execution(* service.UserinfoService.method1(int)) and args(xxxxxx)" />
    <aop:pointcut id="pointCut2"
        expression="execution(* service.UserinfoService.method2(String,String,int,
java.util.Date)) and args(u,p,a,i)" />
    <aop:pointcut id="pointCut3"
        expression="execution(* service.UserinfoService.*(..)) " />
```

```
    <aop:aspect ref="myAspect">
        <aop:before method="method1Before"
            pointcut-ref="pointCut1" />
        <aop:before method="method2Before"
            pointcut-ref="pointCut2" />
        <aop:after-returning
            method="method3AfterReturning" pointcut-ref="pointCut3"
            returning="returnParam" />
        <aop:after-throwing method="method4AfterThrowing"
            pointcut-ref="pointCut3" throwing="t" />
    </aop:aspect>

</aop:config>
```

创建运行类，代码如下：

```
public class Test {
    public static void main(String[] args) {
        ApplicationContext context = new ClassPathXmlApplicationContext("ac1.xml");
        UserinfoService service = (UserinfoService) context.getBean(UserinfoService.class);
        service.method1(100);
        System.out.println();
        System.out.println();
        System.out.println("main get method2 returnValue=" + service.method2("中国", "
中国", 123, new Date()));
    }
}
```

程序运行的结果如下：

```
切面：public void method1Before(int xxxxxx) xxxxxx=100
method1 age=100
public void method3AfterReturning(Object returnParam) returnParam=null


切面：public void method2Before(String u, String p, int a, Date i) u=中国 p=中国 a=123 i=
Mon Jun 11 17:05:35 CST 2018
method2 username=中国 password=中国 age=123 insertdate=Mon Jun 11 17:05:35 CST 2018
public void method4AfterThrowing(Throwable t) t=java.lang.NumberFormatException: For
input string: "a"
Exception in thread "main" java.lang.NumberFormatException: For input string: "a"
```

15. 使用注解向环绕通知传入参数

本实验将实现向环绕通知传入参数。

创建测试用的项目 aopTest16。

创建切面类，代码如下：

```
@Component
@Aspect
public class AspectObject {
    @Pointcut(value = "execution(* service.UserinfoService.method1(int)) && args(xxxxxx)")
    public void methodAspect1(int xxxxxx) {
    }

    @Pointcut(value = "execution(* service.UserinfoService.method2(String,String,int,
java.util.Date)) && args(u,p,a,i)")
```

```
    public void methodAspect2(String u, String p, int a, Date i) {
    }

    @Pointcut(value = "execution(* service.UserinfoService.*(..))")
    public void methodAspect3() {
    }

    @Around(value = "methodAspect1(xxxxxx)")
    public void method1Around(ProceedingJoinPoint point, int xxxxxx) throws Throwable {
        System.out.println("切面开始:
public void method1Before(ProceedingJoinPoint point, int xxxxxx) xxxxxx=" + xxxxxx);
        point.proceed();
        System.out.println("切面结束:
public void method1Before(ProceedingJoinPoint point, int xxxxxx) xxxxxx=" + xxxxxx);
    }

    @Around(value = "methodAspect2(u, p, a, i)")
    public Object method2Around(ProceedingJoinPoint point, String u, String p, int a,
 Date i) throws Throwable {
        Object returnValue = null;
        System.out.println(
                "切面开始: public void method2Before(ProceedingJoinPoint point, String u,
String p, int a, Date i) u=" + u
                        + " p=" + p + " a=" + a + " i=" + i);
        returnValue = point.proceed();
        System.out.println(
                "切面开始: public void method2Before(ProceedingJoinPoint point, String u,
String p, int a, Date i) u=" + u
                        + " p=" + p + " a=" + a + " i=" + i);
        return returnValue;
    }

    @AfterReturning(value = "methodAspect3()", returning = "returnParam")
    public void method3AfterReturning(Object returnParam) {
        System.out.println("public void method3AfterReturning(Object returnParam)
returnParam=" + returnParam);
    }

    @AfterThrowing(value = "methodAspect3()", throwing = "t")
    public void method4AfterThrowing(Throwable t) {
        System.out.println("public void method4AfterThrowing(Throwable t) t=" + t);
    }
}
```

注意：不要在切面中捕获（catch）异常，而应将异常抛出（throws）给 Spring 框架进行后续处理，这样会使 @AfterThrowing 通知得到执行。

创建业务类，代码如下：

```
@Service
public class UserinfoService {
    public void method1(int ageage) {
        System.out.println("method1 age=" + ageage);
    }

    public String method2(String username, String password, int age, Date insertdate) {
        System.out.println(
```

```
                "method2 username=" + username + " password=" + password + " age=" +
age + " insertdate=" + insertdate);
            Integer.parseInt("a");
            return "我是返回值 method2";
        }
    }
```

创建配置类，代码如下：

```
@Configuration
@EnableAspectJAutoProxy
@ComponentScan(basePackages = { "aspect", "service" })
public class MyContext {
}
```

创建运行类，代码如下：

```
public class Test {
    public static void main(String[] args) {
        ApplicationContext context = new AnnotationConfigApplicationContext(MyContext.class);
        UserinfoService service = (UserinfoService) context.getBean(UserinfoService.class);
        service.method1(100);
        System.out.println();
        System.out.println();
        System.out.println("main get method2 returnValue=" + service.method2("中国", "
中国", 123, new Date()));
    }
}
```

程序运行的结果如下：

```
切面开始: public void method1Before(ProceedingJoinPoint point, int xxxxxx) xxxxxx=100
method1 age=100
切面结束: public void method1Before(ProceedingJoinPoint point, int xxxxxx) xxxxxx=100
public void method3AfterReturning(Object returnParam) returnParam=null


切面开始:
public void method2Before(ProceedingJoinPoint point, String u, String p, int a, Date i) u=
中国 p=中国 a=123 i=Mon Jun 11 17:07:19 CST 2018
method2 username=中国 password=中国 age=123 insertdate=Mon Jun 11 17:07:19 CST 2018
public void method4AfterThrowing(Throwable t) t=java.lang.NumberFormatException: For
input string: "a"
Exception in thread "main" java.lang.NumberFormatException: For input string: "a"
```

16. 使用 XML 向环绕通知传入参数

创建测试用的项目 aopTest17。

创建切面类，代码如下：

```
public class AspectObject {
    public void method1Before(ProceedingJoinPoint point, int xxxxxx) throws Throwable {
        System.out.println("切面开始:
public void method1Before(ProceedingJoinPoint point, int xxxxxx) xxxxxx=" + xxxxxx);
        point.proceed();
        System.out.println("切面结束: public void method1Before(ProceedingJoinPoint point,
int xxxxxx) xxxxxx=" + xxxxxx);
    }
```

```
    public Object method2Before(ProceedingJoinPoint point, String u, String p, int a,
 Date i) throws Throwable {
        Object returnValue = null;
        System.out.println(
                "切面开始: public void method2Before(ProceedingJoinPoint point, String u,
String p, int a, Date i) u=" + u
                        + " p=" + p + " a=" + a + " i=" + i);
        returnValue = point.proceed();
        System.out.println(
                "切面开始: public void method2Before(ProceedingJoinPoint point, String u,
String p, int a, Date i) u=" + u
                        + " p=" + p + " a=" + a + " i=" + i);
        return returnValue;
    }

    public void method3AfterReturning(Object returnParam) {
        System.out.println("public void method3AfterReturning(Object returnParam)
returnParam=" + returnParam);
    }

    public void method4AfterThrowing(Throwable t) {
        System.out.println("public void method4AfterThrowing(Throwable t) t=" + t);
    }
}
```

创建业务类，代码如下：

```
public class UserinfoService {
    public void method1(int ageage) {
        System.out.println("method1 age=" + ageage);
    }

    public String method2(String username, String password, int age, Date insertdate) {
        System.out.println(
                "method2 username=" + username + " password=" + password + " age=" +
age + " insertdate=" + insertdate);
        Integer.parseInt("a");
        return "我是返回值 method2";
    }
}
```

创建配置文件，代码如下：

```
<bean id="userinfoService" class="service.UserinfoService"></bean>
<bean id="myAspect" class="aspect.AspectObject"></bean>

<aop:aspectj-autoproxy></aop:aspectj-autoproxy>

<aop:config>
    <aop:pointcut id="pointCut1"
        expression="execution(* service.UserinfoService.method1(int)) and args(xxxxxx)" />
    <aop:pointcut id="pointCut2"
        expression="execution(* service.UserinfoService.method2(String,String,int,
java.util.Date)) and args(u,p,a,i)" />
    <aop:pointcut id="pointCut3"
        expression="execution(* service.UserinfoService.*(..))" />
```

```
        <aop:aspect ref="myAspect">
            <aop:around method="method1Before"
                pointcut-ref="pointCut1" />
            <aop:around method="method2Before"
                pointcut-ref="pointCut2" />
            <aop:after-returning
                method="method3AfterReturning" pointcut-ref="pointCut3"
                returning="returnParam" />
            <aop:after-throwing method="method4AfterThrowing"
                pointcut-ref="pointCut3" throwing="t" />
        </aop:aspect>
    </aop:config>
```

创建运行类，代码如下：

```
public class Test {
    public static void main(String[] args) {
        ApplicationContext context = new ClassPathXmlApplicationContext("ac1.xml");
        UserinfoService service = (UserinfoService) context.getBean(UserinfoService.class);
        service.method1(100);
        System.out.println();
        System.out.println();
        System.out.println("main get method2 returnValue=" + service.method2("中国", "
中国", 123, new Date()));
    }
}
```

程序运行的结果如下：

```
切面开始：public void method1Before(ProceedingJoinPoint point, int xxxxxx) xxxxxx=100
method1 age=100
切面结束：public void method1Before(ProceedingJoinPoint point, int xxxxxx) xxxxxx=100
public void method3AfterReturning(Object returnParam) returnParam=null


切面开始：public void method2Before(ProceedingJoinPoint point, String u, String p, int a,
Date i) u=中国 p=中国 a=123 i=Mon Jun 11 17:08:55 CST 2018
method2 username=中国 password=中国 age=123 insertdate=Mon Jun 11 17:08:55 CST 2018
public void method4AfterThrowing(Throwable t) t=java.lang.NumberFormatException: For
input string: "a"
Exception in thread "main" java.lang.NumberFormatException: For input string: "a"
```

17. xml<aop:aspectj-autoproxy></aop:aspectj-autoproxy>在 AOP 切面上的应用

创建测试用的项目 aopTest18。

创建切面类，代码如下：

```
@Component
@Aspect
public class AspectObject {
    @Pointcut(value = "execution(* service.UserinfoService.*(..))")
    public void methodAspect1() {
    }

    @Before(value = "methodAspect1()")
    public void method1Before() {
        System.out.println("切面开始：public void method1Before()");
```

```
    }

    @After(value = "methodAspect1()")
    public void method1After() {
        System.out.println("切面开始: public void method1After()");
    }

    @AfterReturning(value = "methodAspect1()")
    public void method3AfterReturning() {
        System.out.println("public void method3AfterReturning()");
    }

    @AfterThrowing(value = "methodAspect1()")
    public void method4AfterThrowing() {
        System.out.println("public void method4AfterThrowing()");
    }
}
```

配置类代码如下：

```
@Configuration
@ComponentScan(basePackages = { "aspect", "service" })
public class MyContext {
}
```

注意：在配置类中并没有使用 @EnableAspectJAutoProxy 注解，启动 AOP 切面的功能是由 XML 中的配置来开启的。

```
<aop:aspectj-autoproxy></aop:aspectj-autoproxy>
```

创建业务类，代码如下：

```
@Service
public class UserinfoService {
    public void method1() {
        System.out.println("public void method1()");
    }

    public String method2(int age) {
        System.out.println("public String method2(int age) age=" + age);
        return "我是中国 1";
    }

    public String method3(String username, String password, int age, Date insertdate) {
        System.out.println(
                "method2 username=" + username + " password=" + password + " age=" +
age + " insertdate=" + insertdate);
        Integer.parseInt("a");
        return "我是中国 2";
    }
}
```

创建配置文件，代码如下：

```
<aop:aspectj-autoproxy></aop:aspectj-autoproxy>
<context:component-scan
    base-package="javaconfig"></context:component-scan>
```

创建运行类，代码如下：

```
public class Test {
    public static void main(String[] args) {
        ApplicationContext context = new ClassPathXmlApplicationContext("ac1.xml");
        UserinfoService service = (UserinfoService) context.getBean(UserinfoService.class);
        service.method1();
        System.out.println();
        System.out.println();
        System.out.println("main get valueA =" + service.method2(100));
        System.out.println();
        System.out.println();
        System.out.println("main get valueB =" + service.method3("中国", "中国人", 100,
new Date()));
    }
}
```

程序运行的结果如下：

```
切面开始：public void method1Before()
public void method1()
切面开始：public void method1After()
public void method3AfterReturning()


切面开始：public void method1Before()
public String method2(int age) age=100
切面开始：public void method1After()
public void method3AfterReturning()
main get valueA =我是中国 1


切面开始：public void method1Before()
method2 username=中国 password=中国人 age=100 insertdate=Mon Jun 11 17:10:45 CST 2018
切面开始：public void method1After()
public void method4AfterThrowing()
Exception in thread "main" java.lang.NumberFormatException: For input string: "a"
```

18. 使用注解实现多切面的应用

在系统中多个切面可以同时运行。

创建测试用的项目 moreAOP。

创建切面类，代码如下：

```
@Component
@Aspect
public class AspectObject1 {
    @Around(value = "execution(* service.UserinfoService.*(..))")
    public Object around(ProceedingJoinPoint point) {
        Object returnObject = null;
        try {
            System.out.println("开始 1");
            returnObject = point.proceed();
            System.out.println("结束 1");
        } catch (Throwable e) {
            e.printStackTrace();
```

```
        }
        return returnObject;
    }
}

@Component
@Aspect
public class AspectObject2 {
    @Around(value = "execution(* service.UserinfoService.*(..))")
    public Object around(ProceedingJoinPoint point) {
        Object returnObject = null;
        try {
            System.out.println("开始 2");
            returnObject = point.proceed();
            System.out.println("结束 2");
        } catch (Throwable e) {
            e.printStackTrace();
        }
        return returnObject;
    }
}

@Component
@Aspect
public class AspectObject3 {
    @Around(value = "execution(* service.UserinfoService.*(..))")
    public Object around(ProceedingJoinPoint point) {
        Object returnObject = null;
        try {
            System.out.println("开始 3");
            returnObject = point.proceed();
            System.out.println("结束 3");
        } catch (Throwable e) {
            e.printStackTrace();
        }
        return returnObject;
    }
}
```

创建业务类，代码如下：

```
@Service
public class UserinfoService {
    public void method1() {
        System.out.println("method1 run !");
    }

    public String method2() {
        System.out.println("method2 run !");
        return "我是返回值 A";
    }

    public String method3() {
        System.out.println("method3 run !");
        Integer.parseInt("a");
        return "我是返回值 B";
    }
}
```

创建配置类，代码如下：

```
@Configuration
@EnableAspectJAutoProxy
@ComponentScan(basePackages = { "aspect", "service" })
public class MyContext {
}
```

创建运行类，代码如下：

```
public class Test {
    public static void main(String[] args) {
        ApplicationContext context = new AnnotationConfigApplicationContext(MyContext.class);
        UserinfoService service = (UserinfoService) context.getBean(UserinfoService.class);
        service.method1();
    }
}
```

程序运行的结果如下：

```
开始 1
开始 2
开始 3
method1 run !
结束 3
结束 2
结束 1
```

如果想制定切面运行的顺序，那么可以使用 @Order 注解。

19. 使用 @Order 注解制定切面的运行顺序

在系统中多个切面可以同时运行，并且可以制定执行的顺序。
创建测试用的项目 moreAOPOrder。
创建切面类，代码如下：

```
@Component
@Aspect
@Order(value = 3)
public class AspectObject1 {

@Component
@Aspect
@Order(value = 2)
public class AspectObject2 {

@Component
@Aspect
@Order(value = 1)
public class AspectObject3 {
```

创建运行类，代码如下：

```
public class Test {
    public static void main(String[] args) {
        ApplicationContext context = new AnnotationConfigApplicationContext(MyContext.class);
```

```
        UserinfoService service = (UserinfoService) context.getBean(UserinfoService.class);
        service.method1();
    }
}
```

程序运行的结果如下：

```
开始 3
开始 2
开始 1
method1 run !
结束 1
结束 2
结束 3
```

结果表明可以使用 @Order 注解制定切面运行的顺序。

20. 使用 XML 实现多切面的应用及运行顺序

创建测试用的项目 moreAOPOrderXML。

创建配置文件，代码如下：

```
<aop:aspectj-autoproxy></aop:aspectj-autoproxy>

<bean id="userinfoService" class="service.UserinfoService"></bean>
<bean id="myaspect1" class="aspect.AspectObject1"></bean>
<bean id="myaspect2" class="aspect.AspectObject2"></bean>
<bean id="myaspect3" class="aspect.AspectObject3"></bean>

<aop:config>
    <aop:aspect ref="myaspect1" order="3">
        <aop:around method="around"
            pointcut="execution(* service.UserinfoService.*(..))" />
    </aop:aspect>
</aop:config>

<aop:config>
    <aop:aspect ref="myaspect2" order="2">
        <aop:around method="around"
            pointcut="execution(* service.UserinfoService.*(..))" />
    </aop:aspect>
</aop:config>

<aop:config>
    <aop:aspect ref="myaspect3" order="1">
        <aop:around method="around"
            pointcut="execution(* service.UserinfoService.*(..))" />
    </aop:aspect>
</aop:config>
```

4.2 MyBatis 3 和 Spring 5 的整合

创建测试用的项目 MyBatis3_Spring5_AllOne。

在 Spring 5 框架中，因为默认不支持与 MyBatis 3 进行直接的整合，所以 MyBatis 官方发布了整合 MyBatis 与 Spring 的插件，此插件就是 mybatis-spring-1.3.2.jar，通过此 JAR 文件即可实现 MyBatis 3 与 Spring 5 的整合。

创建实体类，代码如下：

```
public class Userinfo implements java.io.Serializable {

    private Long id;
    private String username;
    private String password;
    private Long age;
    private Timestamp insertdate;

    public Userinfo() {
    }

    public Userinfo(String username, String password, Long age, Timestamp insertdate) {
        this.username = username;
        this.password = password;
        this.age = age;
        this.insertdate = insertdate;
    }

    //省略 get 和 set 方法
}
```

创建 SQL 映射接口，代码如下：

```
public interface IUserinfoMapping {
    public void save(Userinfo userinfo);
}
```

创建 SQL 映射 XML 文件，代码如下：

```
<mapper namespace="mapping.IUserinfoMapping">
    <insert id="save" parameterType="entity.Userinfo">
        <selectKey resultType="java.lang.Long" order="BEFORE"
            keyProperty="id">
            select idauto.nextval from dual
        </selectKey>
        insert into userinfo(id,username)
        values(#{id},#{username})
    </insert>
</mapper>
```

创建业务层，代码如下：

```
@Service
@Transactional
public class UserinfoService {
    @Autowired
    private IUserinfoMapping userinfoMapping;

    public void saveServiceMethod(Userinfo userinfo1, Userinfo userinfo2) {
        userinfoMapping.save(userinfo1);
        userinfoMapping.save(userinfo2);
    }
}
```

创建运行类，代码如下：

```
@Component
public class Test {
    @Autowired
    private UserinfoService userinfoService;

    public UserinfoService getUserinfoService() {
        return userinfoService;
    }

    public void setUserinfoService(UserinfoService userinfoService) {
        this.userinfoService = userinfoService;
    }

    public static void main(String[] args) {
        Userinfo userinfo1 = new Userinfo();
        userinfo1.setUsername("中国 1");

        Userinfo userinfo2 = new Userinfo();
        userinfo2.setUsername("中国 2");

        ApplicationContext context = new ClassPathXmlApplicationContext
("applicationContext.xml");
        Test test = context.getBean(Test.class);
        test.getUserinfoService().saveServiceMethod(userinfo1, userinfo2);

    }
}
```

创建 Spring 配置文件 applicationContext.xml，代码如下：

```
<context:component-scan
    base-package="controller"></context:component-scan>
<context:component-scan base-package="service"></context:component-scan>

<bean id="dataSource"
    class="org.springframework.jdbc.datasource.DriverManagerDataSource">
    <property name="url"
        value="jdbc:oracle:thin:@localhost:1521:orcl"></property>
    <property name="driverClassName"
        value="oracle.jdbc.OracleDriver"></property>
    <property name="username" value="y2"></property>
    <property name="password" value="123"></property>
</bean>

<bean id="sqlSessionFactory"
    class="org.mybatis.spring.SqlSessionFactoryBean">
    <property name="dataSource" ref="dataSource"></property>
</bean>

<bean class="org.mybatis.spring.mapper.MapperScannerConfigurer">
    <property name="basePackage" value="mapping"></property>
    <property name="sqlSessionFactoryBeanName" value="sqlSessionFactory"></property>
</bean>

<bean id="transactionManager"
    class="org.springframework.jdbc.datasource.DataSourceTransactionManager">
```

```
    <property name="dataSource" ref="dataSource"></property>
</bean>

<tx:annotation-driven
    transaction-manager="transactionManager" />
<aop:aspectj-autoproxy proxy-target-class="true"></aop:aspectj-autoproxy>
```

程序运行后，在数据表中增加了两条记录。

更改运行类，代码如下：

```
public static void main(String[] args) {
    Userinfo userinfo1 = new Userinfo();
    userinfo1.setUsername("中国 1");

    Userinfo userinfo2 = new Userinfo();
    userinfo2.setUsername(
           "中国 2 中国 2 中国 2 中国 2 中国 2 中国 2 中国 2 中国 2 中国 2 中国 2 中国 2 中国 2 中国 2 中国 2 中国 2
中国 2 中国 2 中国 2 中国 2 中国 2 中国 2 中国 2 中国 2 中国 2 中国 2 中国 2 中国 2 中国 2 中国 2 中国 2 中国 2");

    ApplicationContext context = new ClassPathXmlApplicationContext
("applicationContext.xml");
    Test test = context.getBean(Test.class);
    test.getUserinfoService().saveServiceMethod(userinfo1, userinfo2);

}
```

程序运行后出现异常，并未在数据表中增加两条记录，事务进行了回滚。

第 5 章 Spring 5 MVC 实战技术

本章目标：

- 掌握常用注解
- 结合 AJAX 和 JSON
- 转发与重定向
- 上传与下载文件
- 实现国际化
- 结合 AOP 切面

5.1 MVC、软件框架与 Spring 5 MVC 介绍

什么是 MVC 模式？MVC 模式是一种开发方式，它主要的用途是对组件之间进行隔离分层。M 代表模型，模型中包含传递的数据。在软件项目中，M 常常被定义为业务模型，也就是业务/服务层。V 代表视图层，也就是用什么组件显示数据，常用的就是 HTML 和 JSP 等文件。而 C 代表控制层，指出软件大体的执行流程以及用哪个视图对象将数据展示给客户。MVC 模式就是将不同功能的组件进行隔离和分层，有利于代码的后期维护。

什么是软件框架？软件框架就是软件功能的半成品，框架提供了针对某一个领域所写代码的基本模型，对大多数通用性的功能进行封装，程序员只需要在这些功能的半成品上继续开发，这样可以提高程序员开发的效率，缩短软件设计的整体周期，而且还统一以及规范了软件的整体架构，所有程序员使用一种方式进行开发，有利于新职员快速加入开发进程中。

Spring 5 MVC 框架是现在主流的 Java Web 服务端 MVC 分层框架，它在功能及代码执行效率上进行了优化和增强，现阶段越来越多的软件公司使用 Spring 5 MVC 框架开发软件项目，Spring 5 MVC 模块的使用率越来越高。软件公司中新的项目大多数是使用 Spring 5 MVC 作为分层框架，可见掌握该技术非常有必要。Spring 5 MVC 提供了很多功能性的注解，大大方便程序员设计与后期维护代码，提高组件的松耦合，有利于软件模块间的设计，更为重要的是减少了 XML 配置文件中的代码量。

5.2 Spring 5 MVC 核心控制器

Spring 5 MVC 使用基于 Servlet 的 DispatcherServlet 类作为核心控制器来处理请求（request），它的父类是真正的 HttpServlet 对象，继承关系如图 5-1 所示。

Object - java.lang
GenericServlet - javax.servlet
HttpServlet - javax.servlet.http
HttpServletBean - org.springframework.web.servlet
FrameworkServlet - org.springframework.web.servlet
DispatcherServlet - org.springframework.web.servlet
TestDispatcherServlet - org.springframework.test.web.servlet

图 5-1 DispatcherServlet 的继承关系

在使用 Spring 5 MVC 框架时，需要在 web.xml 文件中对 DispatcherServlet 类进行注册声明，代码如下：

```
<servlet>
    <servlet-name>springMVC</servlet-name>
    <servlet-class>org.springframework.web.servlet.DispatcherServlet</servlet-class>
    <load-on-startup>1</load-on-startup>
</servlet>

<servlet-mapping>
    <servlet-name>springMVC</servlet-name>
    <url-pattern>/</url-pattern>
</servlet-mapping>
```

上面配置代码整体的作用是在输入 URL 后访问服务器，并且把这个请求交给 DispatcherServlet 类来进行处理，处理的方式有可能是上传与下载文件、实现访问控制层、传递 JSON 与返回 JSON 等。

在配置代码中将 DispatcherServlet 类的别名设置为 springMVC，Spring MVC 框架还要在项目的 WEB-INF 文件夹中找出 springMVC-servlet.xml 配置文件，也就是将以下配置代码中的 springMVC 别名作为 XML 文件名的前缀，后面连接“-servlet.xml”作为文件名，所有 JavaBean 和相关的配置都要在这个 springMVC-servlet.xml 文件中声明。

```
<servlet-name>springMVC</servlet-name>
```

DispatcherServlet 类也被称为“前端控制器”，所有的请求都要经过它，DispatcherServlet 的任务是将请求交给控制层（Controller），因为系统中存在多个控制层，所以 DispatcherServlet 会根据 URL 的映射关系，找到目的控制层对象并执行控制层中的方法，进而执行 Service 业务层中的功能。

5.3 核心技术

本节中的实验技能都是必须要掌握的，因为在软件项目中经常使用它们。

5.3.1 执行控制层：无参数传递

在软件开发中，实践才是真正的学习方法！

本示例将实现在浏览器上访问控制层的 URL 后执行对应控制层中的代码，然后再转发到 JSP 文件的效果，“请求—响应”模型的全部过程都在此实验中得到了体现。

创建项目 springMVC_HelloWorld1。

（1）在 web.xml 中注册核心控制器，核心代码如下：

```
<servlet>
    <servlet-name>springMVC</servlet-name>
    <servlet-class>org.springframework.web.servlet.DispatcherServlet</servlet-class>
    <load-on-startup>1</load-on-startup>
</servlet>
<servlet-mapping>
    <servlet-name>springMVC</servlet-name>
    <url-pattern>/</url-pattern>
</servlet-mapping>
```

（2）创建 springMVC 上下文配置文件 springMVC-servlet.xml，核心代码如下：

```
<context:component-scan base-package="controller"></context:component-scan>
```

代码 <context:component-scan base-package="controller" /> 的作用是在指定包及子包中扫描带 @Controller 注解的控制层 Java 文件。如果找到，则说明该 Java 类是控制层，并不是一个普通的 JavaBean，它会参与处理 request 和 response 对象。

（3）控制层在 controller 包中，代码如下：

```
//@Controller 注解代表本 Java 类是控制层
@Controller
public class HelloController {
    // 通过@RequestMapping 注解可以用指定的 URL 路径
    // 访问本控制层中与 URL 关联的业务方法
    @RequestMapping(value = "helloWorld")
    public String helloWorldMethod() {
        System.out.println("run helloWorld Method!~");
        return "hello.jsp";
    }
}
```

（4）创建 hello.jsp 文件，核心代码如下：

```
<html>
    <head>
    </head>
    <body>
        this is hello.jsp
    </body>
</html>
```

（5）部署项目，输入网址：

```
http://localhost:8080/springMVC_HelloWorld1/helloWorld
```

执行了控制层，控制台输出的结果如下：

```
run helloWorld Method!~
```

并且在浏览器上显示了 JSP 文件，效果如图 5-2 所示。

图 5-2 成功转发到 JSP 文件

5.3.2 执行控制层：有参数传递

本示例将实现在浏览器上访问控制层的 URL 后执行对应控制层中的代码，并且向控制层传递参数，然后再转发到 JSP 文件的效果。

创建项目 springMVC_HelloWorld2。

（1）创建控制层，代码如下：

```
@Controller
public class HelloController {
    @RequestMapping(value = "helloWorld")
    public String helloWorldMethod(@RequestParam("username") String u) {
        System.out.println("hello " + u);
        return "hello.jsp";
    }
}
```

注解代码的作用是取得 URL 中参数名是 username 的值，再传递给参数 u。代码如下：

```
@RequestParam("username") String u
```

（2）部署项目，输入网址：

```
http://localhost:8080/springMVC_HelloWorld2/helloWorld?username=i_like_spring
```

执行了控制层，控制台输出的结果如下：

```
hello i_like_spring
```

控制层成功从 URL 中获得参数值。

5.3.3 执行控制层：有参数传递简化版

从请求中获得参数还可以更加简洁，注解 @RequestParam 是可以省略的。

创建项目 springMVC_HelloWorld3。

（1）创建控制层，代码如下：

```
@Controller
public class HelloController {
    @RequestMapping(value = "helloWorld")
    public String helloWorldMethod(String username, String password) {
```

```
        System.out.println("hello " + username + " " + password);
        return "hello.jsp";
    }
}
```

（2）部署项目，输入网址：

```
http://localhost:8080/springMVC_HelloWorld3/helloWorld?username=like&password=java
```

执行了控制层，控制台输出的结果如下：

```
hello like java
```

成功从 URL 中获得参数值。

基于前面实验的知识，可以使用 Spring MVC 实现一个经典的登录案例，来从整体的角度强化 Spring 5 MVC 使用的熟练度。

5.3.4 实现登录功能

创建实验用的项目 springMVC_login。

（1）本实验要创建 3 个 JSP 文件，显示登录的 login.jsp 核心代码如下：

```
<html>
    <head>
    </head>
    <body>
        post:
        <br/>
        <form action="login" method="post">
            username:<input type="text" name="username">
            <br/>
            username:<input type="text" name="password">
            <br/>
            <input type="submit" value="submit">
            <br/>
        </form>
        <br/>
        get:
        <br/>
        <form action="login" method="get">
            username:<input type="text" name="username">
            <br/>
            username:<input type="text" name="password">
            <br/>
            <input type="submit" value="submit">
            <br/>
        </form>
    </body>
</html>
```

（2）登录成功的 ok.jsp 核心代码如下：

```
<body>
    welcome:${param.username}
</body>
```

（3）登录失败的 no.jsp 核心代码如下：

```
<body>
    登录失败!
</body>
```

（4）下面开始设计最为关键的组件——控制层，核心代码如下：

```
@Controller
public class UserinfoController {
    @RequestMapping(value = "login")
    public String loginMethod(String username, String password) {
        if (username.equals("a") && password.equals("aa")) {
            return "ok.jsp";
        } else {
            return "no.jsp";
        }
    }
}
```

（5）项目中的代码设计完毕后，部署项目并输入网址：

```
http://localhost:8080/springMVC_login/login.jsp
```

显示登录界面，如图 5-3 所示。

可以分别在 post 或 get 提交的<form>中的 username 处填写“a”，在 password 处填写“aa”，以便使用 post 或 get 提交的请求实现成功登录的效果。单击登录界面中的两个“submit”按钮，显示成功登录，界面如图 5-4 所示。

如果不在 username 和 password 处填写信息，直接单击“submit”按钮，那么出现的效果如图 5-5 所示。

图 5-3 显示登录界面

图 5-4 登录成功界面

图 5-5 登录失败界面

（6）登录案例的总结。

要掌握 Spring 5 MVC 的开发与执行流程，应着重学习以下知识点。

- 在 web.xml 中配置 DispatcherServlet 核心控制器。
- WEB-INF 文件夹中要添加 XML 配置文件。
- @Controller、@RequestMapping 和 @RequestParam 注解的作用和使用。
- 控制层在默认的情况下可以处理 get 和 post 请求。
- 在 web.xml 中有如下配置代码：

```
<servlet>
    <servlet-name>springMVC</servlet-name>
    <servlet-class>org.springframework.web.servlet.DispatcherServlet</servlet-class>
    <load-on-startup>1</load-on-startup>
</servlet>
```

其中，<load-on-startup>1</load-on-startup>的执行效果就是在 Tomcat 容器启动时到 springMVC-servlet.xml 文件中进行<context: component-scan base-package="controller">扫描操作，找到控制层 JavaBean 并立即进行实例化，如果在 web.xml 中不添加<load-on-startup>1</load-on-startup>配置，那么在第一次访问 Controller 时，该 Controller 的实例才被创建。

5.3.5 将 URL 参数封装成实体类

可以将 URL 中的参数封装成实体类。

此实验在项目 springMVC_paramToEntity 中实现。

（1）创建封装 URL 参数的实体类，代码如下：

```
public class Userinfo {
    private String username;
    private String password;

    public Userinfo() {
    }

    public Userinfo(String username, String password) {
        super();
        this.username = username;
        this.password = password;
    }

    //省略 get 和 set 方法
}
```

（2）创建控制层，代码如下：

```
@Controller
public class UserinfoController {
    @RequestMapping(value = "login")
    public String loginMethod(Userinfo userinfo) {
        System.out.println("username=" + userinfo.getUsername());
        System.out.println("password=" + userinfo.getPassword());
        return "index.jsp";
    }
}
```

（3）输入网址：

```
http://localhost:8080/springMVC_paramToEntity/login.spring?username=123&password=456
```

控制台输出的内容如下：

```
username=123
password=456
```

成功将 URL 中的参数值封装到 Userinfo 实体类中。

5.3.6 限制提交 method 的方式

前面使用 Spring 5 MVC 实现了登录功能，控制层在默认的情况下允许以 post 和 get 方式进行提示，但标准的登录功能使用的是 post 提交，绝大多数不支持 get 方式，这就需要在控制层对提交方式进行限制。这个需求在项目 springMVC_login_method 中实现。

（1）创建控制层，代码如下：

```
@Controller
public class UserinfoController {
    @RequestMapping(value = "login", method = RequestMethod.POST)
    public String loginMethod(String username, String password) {
        if (username.equals("a") && password.equals("aa")) {
            return "ok.jsp";
        } else {
            return "no.jsp";
        }
    }
}
```

使用属性 method = RequestMethod.POST 限制提交的方式必须是 post 提交。

（2）部署项目并输入网址：

```
http://localhost:8080/springMVC_login_method/login.jsp
```

以 post 方式提交的表单能实现登录成功或失败的效果。

然而，在以 get 方式提交表单时，却出现了异常，效果如图 5-6 所示。

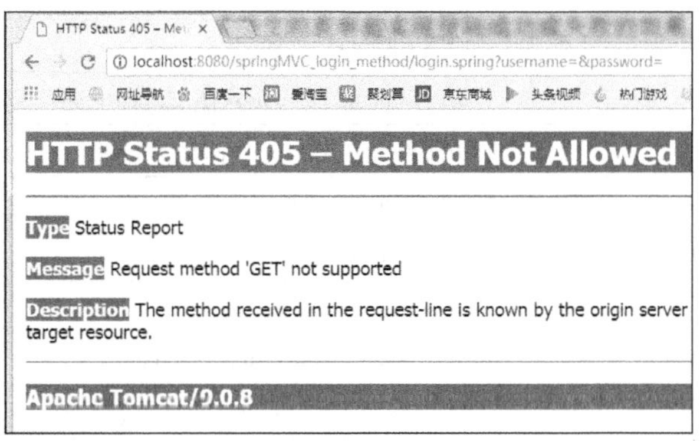

图 5-6 以 get 方式提交不被支持

使用注解 @RequestMapping（value = "login", method = RequestMethod.POST）可以限制接收提交的方式，有利于代码的规范。

5.3.7 控制层方法的参数类型

在前面，控制层方法的声明如下：

```
public String loginMethod(String username, String password)
```

或

```
public String loginMethod(Userinfo userinfo)
```

方法中的参数都是 URL 中的同名参数或实体类，目的是获得参数值，其实 Spring MVC 控制层方法的参数还可以是如下常见数据类型，如表 5-1 所示。

表 5-1 控制层方法的参数类型

控制层方法的参数类型	解　释
WebRequest NativeWebRequest	可以访问 request 的 parameters、request 和 session 的 attributes，而不需要使用 Servlet API
javax.servlet.ServletRequest javax.servlet.ServletResponse MultipartRequest MultipartHttpServletRequest	使用指定的 request 或 response 对象
javax.servlet.http.HttpSession	使用指定的 HttpSession 对象 注意：访问 HttpSession 不是线程安全的，如果有多个请求同时访问 HttpSession 对象，那么需要将 RequestMappingHandlerAdapter 类的 synchronizeOnSession 属性设置为 true
HttpMethod	request 请求的 method 方式
java.util.Locale	当前请求的区域
java.util.TimeZone + java.time.ZoneId	当前请求关联的 Zone
java.io.InputStream java.io.Reader	访问 request body 最原始的数据
java.io.OutputStream java.io.Writer	访问 response body 最原始的数据
@PathVariable	访问 URL 模板变量
@MatrixVariable	访问以 name-value 形式存在于 URL 路径中的片段
@RequestParam	访问 request 中的 parameter，该注解是可选的
@RequestHeader	访问 request header 中的数据
@CookieValue	访问 Cookie
@RequestBody	访问 request body
HttpEntity<B>	访问 request 中的 header 和 body
@RequestPart	处理"multipart/form-data"请求中的 part
java.util.Map org.springframework.ui.Model org.springframework.ui.ModelMap	用于与 View 层的交互
RedirectAttributes	在重定向时添加 Attributes 处理，有两种用法： （1）可以将数据放在 QueryString 中； （2）结合 FlashAttributes 将数据存储到临时的空间，当重定向结束后删除临时空间中的数据
@ModelAttribute	访问已存在的 Attribute

续表

控制层方法的参数类型	解　释
Errors BindingResult	访问的 Error 来自于数据验证和绑定，或者是对@RequestBody 或@RequestPart 的验证。一个 Error 或 BindingResult 参数要在验证方法参数之后声明
类级别的@SessionAttributes	定义 HttpSession 中的 Attributes 在处理完成后触发清理
@SessionAttribute	访问 Session 中的 Attribute
@RequestAttribute	访问 Request 中的 Attribute

5.3.8 控制层方法的返回值类型

控制层的方法可以返回如下常见数据类型（见表 5-2），而不仅仅是 String。

表 5-2 控制层方法的返回值类型

控制层方法的返回值类型	解　释
@ResponseBody	通过 HttpMessageConverter 转换返回值，并且写入 Response 中
HttpEntity<B> ResponseEntity<B>	response 包括完整的 header 和 body，并且通过 HttpMessageConverter 转换返回值，并且写入 response 中
HttpHeaders	返回 response header，但不包括 body
String	使用 ViewResolver 解析的视图名称
View	返回 View 实例
java.util.Map org.springframework.ui.Model	要添加到隐式 Model 中的属性，并通过 RequestToViewNameTranslator 确定视图名称
@ModelAttribute	要添加到隐式 Model 中的属性，并通过 RequestToViewNameTranslator 确定视图名称
ModelAndView object	用于确定 View 和 Attribute
void	如果具有 void 返回值或返回 null 值的方法具有 ServletResponse、OutputStream 参数或 @ResponseStatus 注解，那么视为已完全处理响应

5.3.9 取得 request、response 和 session 对象

有时需要在控制层取得 HttpServletRequest、HttpServletResponse 和 HttpSession 对象，以便调用这 3 个对象的方法。

新建名称为 springMVC_req_res_session 的项目。

（1）控制层代码如下：

```
@Controller
public class UserinfoController {
    @RequestMapping(value = "test")
    public String loginMethod(HttpServletRequest request, HttpServletResponse response,
HttpSession session) {
        System.out.println(request);
        System.out.println(response);
        System.out.println(session);
```

```
        request.setAttribute("requestKey", "request 中国");
        session.setAttribute("sessionKey", "session 中国");
        System.out.println(request.getSession().getServletContext().getRealPath("/"));
        return "index.jsp";
    }
}
```

（2）JSP 代码如下：

```
<body>
    ${requestKey}
    <br/>
    ${sessionKey}
</body>
```

（3）程序运行后在控制台及浏览器显示相关的数据信息，效果如图 5-7 所示。

```
org.apache.catalina.connector.RequestFacade@272350c2
org.apache.catalina.connector.ResponseFacade@dcc09d4
org.apache.catalina.session.StandardSessionFacade@4aa8d2d
C:\workspace-sts-3.9.4.RELEASE\.metadata\.plugins\org.eclipse.wst.server.core\tmp0\wtpwebapps\springMVC_req_res_session\
```

```
request中国
session 中国
```

图 5-7　运行后的效果

5.3.10　登录失败后显示错误信息

虽然 Spring 5 MVC 支持对前台参数的数据进行有效性验证，但是如果在操作性上存在业务型的验证，那么还需要程序员以手写代码的方式来处理。因此，本示例将演示如何用手动的方式来验证前台传递过来的登录信息，登录失败后在前台显示异常信息。

创建测试用的项目 springMVC_loginErrorMessage。

（1）创建登录界面，代码如下：

```
<html>
    <head>
    </head>
    <body>
        <form action="login" method="post">
            username:<input type="text" name="username">${message.usernameisnull}
            <br/>
            password:<input type="text" name="password">${message.passwordisnull}
            <br/>
            <input type="submit" value="submit">
            <br/>
        </form>
    </body>
</html>
```

（2）创建登录成功界面，代码如下：

```
<body>
    welcome:${param.username}
</body>
```

（3）创建控制层，核心代码如下：

```
@Controller
public class UserinfoController {
```

```
    public Map loginValidateMethod(String username, String password) {
        Map map = new HashMap();
        if (username == null || "".equals(username)) {
            map.put("usernameisnull", "账号为空！");
        }
        if (password == null || "".equals(password)) {
            map.put("passwordisnull", "密码为空！");
        }
        return map;
    }

    @RequestMapping(value = "login")
    public String loginMethod(String username, String password, HttpServletRequest request) {
        Map map = loginValidateMethod(username, password);
        if (map.size() > 0) {
            request.setAttribute("message", map);
            return "index.jsp";
        } else {
            return "ok.jsp";
        }
    }
}
```

（4）如果账号和密码不输入任何字符，那么提交表单后返回的 index.jsp 页面会显示错误信息，效果如图 5-8 所示。

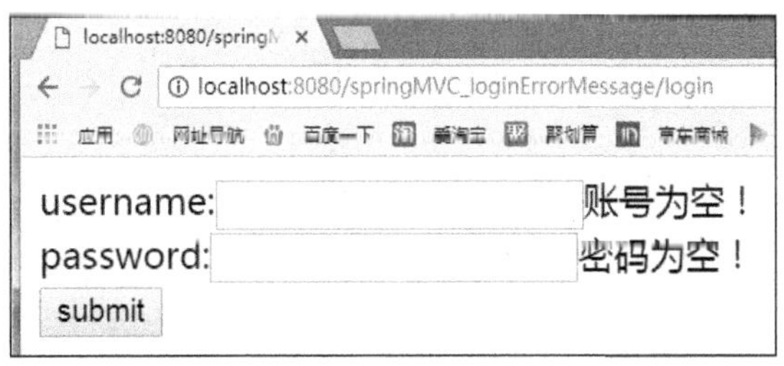

图 5-8　显示错误信息

5.3.11　向控制层注入 Service 业务逻辑层

前面的示例都是在控制层中进行业务的处理，下面向控制层注入 Service 业务层以实现严格的 MVC 分层，示例代码在项目 springMVC_setService 中。

（1）创建业务接口，代码如下：

```
public interface IUserinfoService {
    String getUsername();
}
```

（2）创建业务类，代码如下：

```
@Service(value = "serviceA")
public class UserinfoServiceA implements IUserinfoService {
    @Override
    public String getUsername() {
        return "业务逻辑层的中国 A";
    }
}
```

（3）创建业务类，代码如下：

```
@Service(value = "serviceB")
public class UserinfoServiceB implements IUserinfoService {
    public String getUsername() {
        return "业务逻辑层的中国 B";
    }
}
```

使用注解 @Service 代表本类是一个业务对象。

（4）控制层代码如下：

```
@Controller
public class UserinfoController {

    @Autowired
    @Qualifier(value = "serviceB")
    private IUserinfoService userinfoService;

    @RequestMapping(value = "test")
    public String test() {
        System.out.println(userinfoService.getUsername());
        return "index.jsp";
    }
}
```

（5）扫描两个包，代码如下：

```
<context:component-scan base-package="controller"></context:component-scan>
<context:component-scan base-package="service"></context:component-scan>
```

（6）运行项目后在控制台输出的结果如下：

```
业务逻辑层的中国 B
```

5.3.12 重定向：无参数传递

本节要实现在两个控制层中进行重定向操作，并且在重定向时不向目的控制层传递参数。

创建测试用的项目 springMVC_redirect1。

（1）控制层代码如下：

```
@Controller
public class UserinfoController {
    @RequestMapping(value = "login")
    public String loginMethod(String username) {
        System.out.println("loginMethod username=" + username);
        return "redirect:/listString.spring";// 重定向无传参
    }

    @RequestMapping(value = "listString")
    public String listStringMethod(HttpServletRequest request) {
        System.out.println("listStringMethod");
        List list = new ArrayList();
        list.add("中国 1");
        list.add("中国 2");
        list.add("中国 3");
```

```
        list.add("中国 4");
        request.setAttribute("list", list);
        return "listString.jsp";
    }
}
```

在 Spring 5 MVC 中重定向的关键代码就是在返回字符串中加入“redirect:/”前缀，代表这个操作是重定向的。

（2）文件 listString.jsp 核心代码如下：

```
<body>
    <c:forEach var="eachString" items="${list}">
        ${eachString}
        <br/>
    </c:forEach>
</body>
```

（3）部署项目，在浏览器输入网址：

```
http://localhost:8080/springMVC_redirect1/login?username=123
```

浏览器的地址栏发生变化，重定向到如下网址：

```
http://localhost:8080/springMVC_redirect1/listString.spring
```

证明成功重定向到其他的控制层，效果如图 5-9 所示。

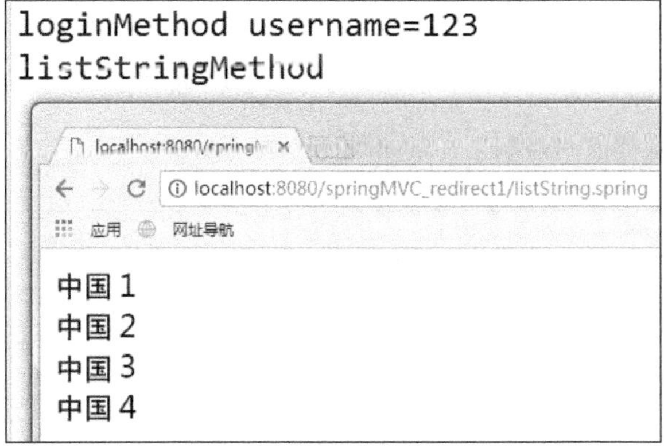

图 5-9　控制层重定向到控制层且无参数传递

5.3.13　重定向：有参数传递

本节要实现在两个控制层中进行重定向操作，并且在重定向时向目的控制层传递参数。

创建测试用的项目 springMVC_redirect2。

（1）控制层代码如下：

```
@Controller
public class UserinfoController {
    @RequestMapping(value = "login")
    public String loginMethod() throws UnsupportedEncodingException {
        System.out.println("loginMethod run !");
        String username = java.net.URLEncoder.encode("我是中文我是参数", "utf-8");
```

```
        username = username.replace("%", "_");
        return "redirect:/listString.spring?xxxxxxxxxx=" + username;
    }

    @RequestMapping(value = "listString")
    public String listStringMethod(String xxxxxxxxxx, HttpServletRequest request)
throws UnsupportedEncodingException {
        xxxxxxxxxx = xxxxxxxxxx.replace("_", "%");
        xxxxxxxxxx = java.net.URLDecoder.decode(xxxxxxxxxx, "utf-8");
        System.out.println("listStringMethod xxxxxxxxxx=" + xxxxxxxxxx);

        List list = new ArrayList();
        list.add("中国 1");
        list.add("中国 2");
        list.add("中国 3");
        list.add("中国 4");

        request.setAttribute("list", list);

        return "listString.jsp";
    }
}
```

（2）部署项目，在浏览器输入网址

```
http://localhost:8080/springMVC_redirect2/login
```

浏览器的地址栏发生变化，重定向到如下网址：

```
http://localhost:8080/springMVC_redirect2/listString.spring?xxxxxxxxxx=_E6_88_91_E6_9
8_AF_E4_B8_AD_E6_96_87_E6_88_91_E6_98_AF_E5_8F_82_E6_95_B0
```

证明成功重定向到其他的控制层，效果如图 5-10 所示。

```
loginMethod run !
listStringMethod xxxxxxxxxx=我是中文我是参数
```

localhost:8080/springMVC_redirect2/listString.spring?xxxxxxxxxx=_E6_88_91_E6_98_AF_E4_B8_AD_E6_96_87_E6_88_91_E6_98_AF_E5_8F_82_E6_95_B0

中国 1
中国 2
中国 3
中国 4

图 5-10 控制层重定向到控制层且传参（1）

5.3.14 重定向传递参数：RedirectAttributes.addAttribute() 方法

在重定向传递参数时，如果使用如下拼接 URL 的形式，则不太标准：

```
return "redirect:/listString.spring?xxxxxxxxxx=" + username;
```

可以使用 RedirectAttributes 来作为代替。

创建测试用的项目 springMVC_redirect3。

（1）控制层代码如下：

```
@Controller
public class UserinfoController {
    @RequestMapping(value = "a")
    public String a(RedirectAttributes attr) {
        System.out.println("into a method");
        attr.addAttribute("username", "abc");
        attr.addAttribute("age", 123);
        return "redirect:/b";
    }

    @RequestMapping(value = "b")
    public String b(String username, String age) {
        System.out.println("into b method");
        System.out.println("username:" + username);
        System.out.println("age:" + age);
        return "index.jsp";
    }
}
```

（2）部署项目，在浏览器输入网址

```
http://localhost:8080/springMVC_redirect3/a
```

浏览器的地址栏发生变化，重定向到如下网址：

```
http://localhost:8080/springMVC_redirect3/b?username=abc&age=123
```

证明成功重定向到其他的控制层，效果如图 5-11 所示。

图 5-11　控制层重定向到控制层且传参（2）

5.3.15　重定向传递参数：RedirectAttributes.addFlashAttribute() 方法

在正常的情况下，重定向的参数存在于 URL 中，但 Spring 5 MVC 框架还提供一种将重定向的参数放入 HttpSession 中的技术，URL 中由此不再出现参数值了。这样就可以通过透明的方式实现重定向传递参数，参数值暂存在 HttpSession 中，重定向结束后从 HttpSession 中清除数据。

创建测试用的项目 springMVC_redirect4。

（1）控制层代码如下：

```
@Controller
public class UserinfoController {
    @RequestMapping(value = "a")
    public String a(RedirectAttributes attr) {
        System.out.println("into a method");
        attr.addFlashAttribute("address", "地址");
```

```
        attr.addFlashAttribute("note", "备注");
        return "redirect:/b";
    }

    @RequestMapping(value = "b")
    public String b(@ModelAttribute("address") String address, @ModelAttribute("note")
String note) {
        System.out.println("into b method");
        System.out.println("address:" + address);
        System.out.println("note:" + note);
        return "index.jsp";
    }
}
```

（2）部署项目，在浏览器输入网址

```
http://localhost:8080/springMVC_redirect4/a
```

浏览器的地址栏发生变化，重定向到如下网址：

```
http://localhost:8080/springMVC_redirect4/b;jsessionid=7B90A4B33516066F11AD212EEB694C4E
```

证明成功重定向到其他的控制层，效果如图 5-12 所示。

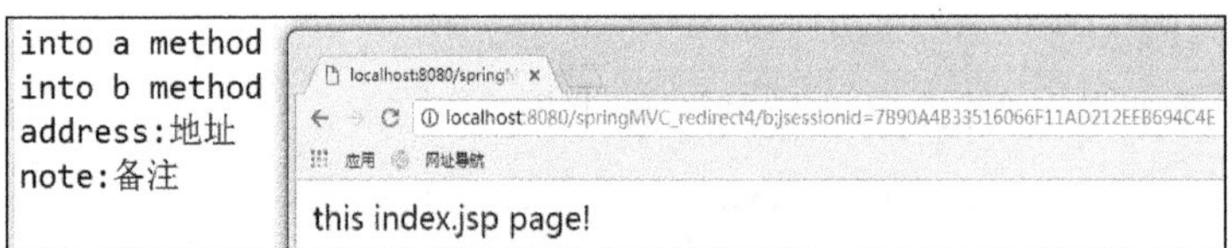

图 5-12 控制层重定向到控制层且传参（3）

5.3.16 解决转发到*.html 文件的 404 异常

在 Spring MVC 中转发到*.jsp 文件不出现异常，而转发到*.html 文件会出现异常。

创建测试用的项目 springMVC_forwardHTML。

（1）控制层代码如下：

```
@Controller
public class UserinfoController {
    @RequestMapping(value = "jsp_test")
    public String jsp_test() {
        System.out.println("into jsp_test method");
        return "index.jsp";
    }

    @RequestMapping(value = "html_test")
    public String html_test() {
        System.out.println("into html_test method");
        return "index.html";
    }
}
```

（2）部署项目，在浏览器输入网址：

```
http://localhost:8080/springMVC_forwardHTML/jsp_test
```

效果如图 5-13 所示。

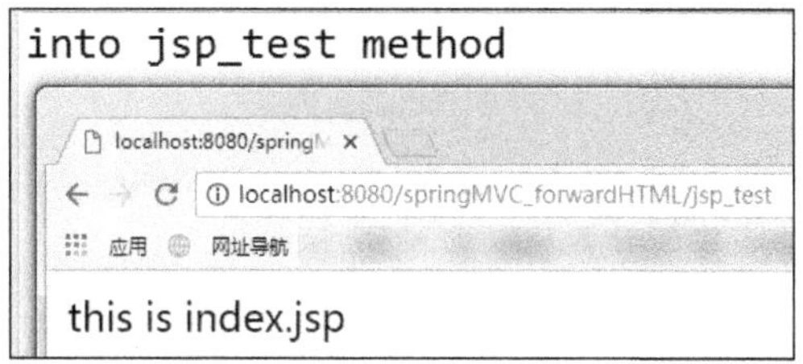

图 5-13　运行结果（1）

在浏览器输入网址：

```
http://localhost:8080/springMVC_forwardHTML/html_test
```

效果如图 5-14 所示。

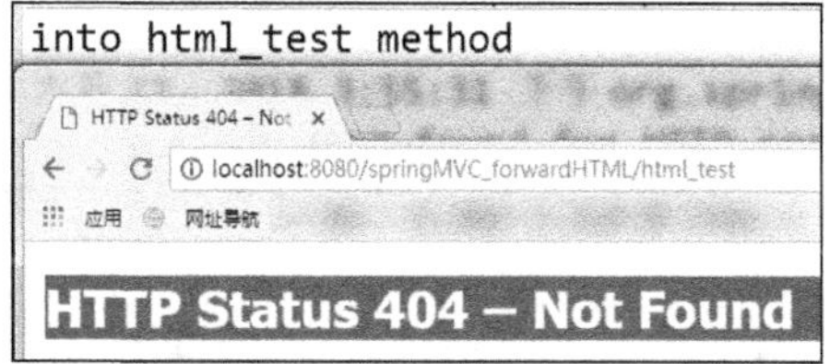

图 5-14　运行结果（2）

出现这样的问题是因为 Spring MVC 把 index.html 当成映射路径，并不是一个普通的 index.html 文件，所以要告诉 Spring MVC，一些静态的文件直接交给容器处理就可以了，不要当成 controller 的路径。

（3）添加如下配置代码：

```
<mvc:annotation-driven></mvc:annotation-driven>
<mvc:default-servlet-handler />
```

（4）在浏览器输入如下网址：

```
http://localhost:8080/springMVC_forwardHTML/
html_test
```

效果如图 5-15 所示。

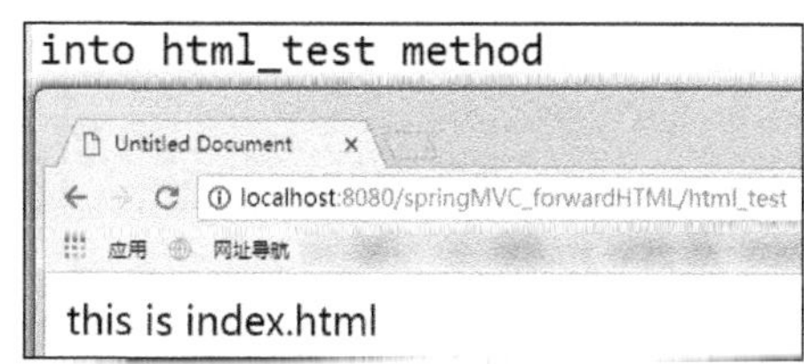

图 5-15　运行结果（3）

5.3.17　使用 fastjson 在服务端解析 JSON 字符串

下面将在服务器端取得客户端传递过来的 JSON 字符串，把 JSON 字符串转换成 JSON 对象并取得其中的属性值。

创建测试用的项目 springMVC_JSONString1。

（1）添加解析 JSON 字符串的 fastjson-1.2.47.jar。

（2）在 web.xml 中添加 Spring 的编码过滤器，代码如下：

```
<filter>
    <filter-name>charSetFilter</filter-name>
    <filter-class>org.springframework.web.filter.CharacterEncodingFilter</filter-class>
    <init-param>
        <param-name>encoding</param-name>
        <param-value>utf-8</param-value>
    </init-param>
</filter>
<filter-mapping>
    <filter-name>charSetFilter</filter-name>
    <url-pattern>/*</url-pattern>
</filter-mapping>
```

（3）在配置文件中添加如下配置代码：

```
<mvc:annotation-driven></mvc:annotation-driven>
<mvc:default-servlet-handler />
```

（4）新建 JSP 文件，代码如下：

```
<html>
    <head>
        <script src="jquery.js">
        </script>
        <script>
            function Userinfo(username, password){
                this.username = username;
                this.password = password;
            }

            function sendAjax(){
                var userinfo = new Userinfo("中国", "中国人");
                var jsonString = JSON.stringify(userinfo);
                $.post("test?t=" + new Date().getTime(), {
                    "jsonString": jsonString
                });
            }
        </script>
    </head>
    <body>
        <input type="button" value="sendAjax" onclick="javascript:sendAjax()">
    </body>
</html>
```

（5）控制层代码如下：

```
@Controller
public class UserinfoController {
    @RequestMapping(value = "test")
    public String test(String jsonString) {
        System.out.println("test run!");
        JSONObject jsonObject = JSONObject.parseObject(jsonString);
        System.out.println(jsonObject.get("username"));
        System.out.println(jsonObject.get("password"));
        return "index.jsp";
    }
}
```

（6）程序运行后单击界面中的“Button”按钮，在控制台输出 JSON 对象的属性值，结果如下：

```
test run!
中国
中国人
```

5.3.18 使用 jackson 在服务端将 JSON 字符串转换成各种 Java 数据类型

本节实现将 JSON 字符串转换成各种 Java 数据类型，但此功能需要依赖 jackson 库，而不是 fastjson 库，因为 Spring 5 MVC 在内部默认使用的 JSON 解析类库就是 jackson 库。

新建测试用的项目 getJSONStringToObject。

（1）添加 jackson 库中的 JAR 包。

（2）创建实体类，代码如下：

```
package entity;

public class Userinfo {
    private String username;
    private String password;
    //省略 set 和 get 方法
}
```

（3）控制层代码如下：

```
@Controller
public class UserinfoController {
    @RequestMapping(value = "test1")
    public String test1(@RequestBody Userinfo userinfo) {
        System.out.println(userinfo.getUsername());
        System.out.println(userinfo.getPassword());
        return "index.jsp";
    }

    @RequestMapping(value = "test2")
    public String test2(@RequestBody List<String> listData) {
        for (int i = 0; i < listData.size(); i++) {
            System.out.println(listData.get(i));
        }
        return "index.jsp";
    }

    @RequestMapping(value = "test3")
    public String test3(@RequestBody List<LinkedHashMap> listData) {
        for (int i = 0; i < listData.size(); i++) {
            Map map = listData.get(i);
            System.out.println(map.get("username") + " " + map.get("password"));
        }
        return "index.jsp";
    }

    @RequestMapping(value = "test4")
    public String test4(@RequestBody Map map) {
        System.out.println(map.get("username"));
        List<Map> workList = (List) map.get("work");
        for (int i = 0; i < workList.size(); i++) {
            Map eachWorkMap = workList.get(i);
            System.out.println(eachWorkMap.get("address"));

        }
        Map schoolMap = (Map) map.get("school");
        System.out.println(schoolMap.get("name"));
        System.out.println(schoolMap.get("address"));
        return "index.jsp";
    }

    @RequestMapping(value = "test5")
    public String test5(@RequestBody Map map) {
```

```
            List list1 = (List) map.get("myArray");
            System.out.println(((Map) list1.get(0)).get("username1"));
            System.out.println(((Map) list1.get(1)).get("username2"));
            List list2 = (List) list1.get(2);
            System.out.println(list2.get(0));
            System.out.println(list2.get(1));
            System.out.println(list2.get(2));

            List list3 = (List) list2.get(3);
            for (int i = 0; i < list3.size(); i++) {
                System.out.println(list3.get(i));
            }

            System.out.println(((Map) map.get("myObject")).get("username"));

            List<Map> list4 = (List) ((Map) map.get("myObject1")).get("address");
            for (int i = 0; i < list4.size(); i++) {
                Map eachMap = list4.get(i);
                System.out.println(eachMap.get("name"));

            }
            return "index.jsp";
        }

}
```

使用 @RequestBody 注解后，前台只需要向 Controller 提交一段符合 JSON 格式的 request body 体，Spring 5 MVC 就会自动将其转换成 Java 各种数据类型。

（4）创建 JSP 文件，代码如下：

```
<html>
    <head>
        <script src="jquery-1.4.2.js">
        </script>
        <script>
            function Userinfo(username, password){
                this.username = username;
                this.password = password;
            }

            function test1(){
                var userinfo = new Userinfo("中国", "中国人");
                var jsonString = JSON.stringify(userinfo);
                $.ajax({
                    "type": "post",
                    "url": "test1?t=" + new Date().getTime(),
                    "data": jsonString,
                    "contentType": "application/json"
                });
            }

            function test2(){
                var myArray = new Array();
                myArray[0] = "中国 1";
                myArray[1] = "中国 2";
                myArray[2] = "中国 3";
                myArray[3] = "中国 4";
```

```
    var jsonString = JSON.stringify(myArray);
    $.ajax({
        "type": "post",
        "url": "test2?t=" + new Date().getTime(),
        "data": jsonString,
        "contentType": "application/json"
    });
}

function test3(){
    var myArray = new Array();
    myArray[0] = new Userinfo("中国 1", "中国人 1");
    myArray[1] = new Userinfo("中国 2", "中国人 2");
    myArray[2] = new Userinfo("中国 3", "中国人 3");
    myArray[3] = new Userinfo("中国 4", "中国人 4");


    var jsonString = JSON.stringify(myArray);
    $.ajax({
        "type": "post",
        "url": "test3?t=" + new Date().getTime(),
        "data": jsonString,
        "contentType": "application/json"
    });
}

function test4(){
    var jsonObject = {
        "username": "accp",
        "work": [{
            "address": "address1"
        }, {
            "address": "address2"
        }],
        "school": {
            "name": "tc",
            "address": "pjy"
        }
    }

    var jsonString = JSON.stringify(jsonObject);
    $.ajax({
        "type": "post",
        "url": "test4?t=" + new Date().getTime(),
        "data": jsonString,
        "contentType": "application/json"
    });
}

function test5(){
    var userinfo = {
        "myArray". [{
            "username1": "usernameValue11"
        }, {
            "username2": "usernameValue22"
        }, ["abc", 123, true, [123, 456]]],
        "myObject": {
```

```
                    "username": "中国"
                },
                "myObject1": {
                    "address": [{
                        "name": "name1"
                    }, {
                        "name": "name2"
                    }]
                },
            };

            var jsonString = JSON.stringify(userinfo);
            $.ajax({
                "type": "post",
                "url": "test5?t=" + new Date().getTime(),
                "data": jsonString,
                "contentType": "application/json"
            });
        }
    </script>
</head>
<body>
    <input type="button" value="sendAjax1" onclick="javascript:test1()">
    <br/>
    <input type="button" value="sendAjax2" onclick="javascript:test2()">
    <br/>
    <input type="button" value="sendAjax3" onclick="javascript:test3()">
    <br/>
    <input type="button" value="sendAjax4" onclick="javascript:test4()">
    <br/>
    <input type="button" value="sendAjax5" onclick="javascript:test5()">
</body>
</html>
```

（5）运行项目，单击 5 个按钮后，分别在控制台输出相应的信息，说明 JSON 字符串成功转换成 Java 的不同数据类型。

5.3.19 在控制层返回 JSON 对象示例

有时需要在控制层以 response（响应）的方式返回 JSON 对象，比如返回学生信息列表等信息。

创建测试用的项目 springMVC_JSONString3。

（1）添加 jackson 库中的 JAR 包。

（2）控制层代码如下：

```
@Controller
public class UserinfoController {
    @RequestMapping(value = "test1", produces = "application/json")
    @ResponseBody
    public Userinfo test1(@RequestBody Userinfo userinfo) {
        System.out.println(userinfo.getUsername());
        System.out.println(userinfo.getPassword());
        Userinfo returnUserinfo = new Userinfo();
        returnUserinfo.setUsername("返回的账号");
```

```
            returnUserinfo.setPassword("返回的密码");
            return returnUserinfo;
        }
    }
```

控制层通过属性 produces = "application/json"将 Userinfo 对象转成 JSON 对象并回传给客户端，注解 @ResponseBody 指的是将 JSON 字符串作为响应处理。

（3）创建 JSP 文件，代码如下：

```
<html>
    <head>
        <script src="jquery-1.4.2.js">
        </script>
        <script>
            function Userinfo(username, password){
                this.username = username;
                this.password = password;
            }

            function test1(){
                var userinfo = new Userinfo("中国", "中国人");
                var jsonString = JSON.stringify(userinfo);
                $.ajax({
                    "type": "post",
                    "url": "test1?t=" + new Date().getTime(),
                    "data": jsonString,
                    "contentType": "application/json",
                    "success": function(data){
                        alert(data.username + " " + data.password);
                    }
                });
            }
        </script>
    </head>
    <body>
        <input type="button" value="sendAjax1" onclick="javascript:test1()">
    </body>
</html>
```

（4）运行项目，单击按钮后，分别在控制台和前台输出相应的信息，效果如图 5-16 所示。

图 5-16 运行结果

5.3.20 在控制层返回 JSON 字符串示例

前面实现的是在 response 对象中返回 JSON 对象，本示例要在 response 对象中返回 JSON 字符串，其实返回 JSON 对象和 JSON 字符串都可以在前端进行处理，只是每个程序员的习惯不一样，但结果都是相同的。

创建测试用的项目 springMVC_JSONString4。

（1）添加解析 JSON 字符串的 fastjson-1.2.47.jar。

（2）创建实体类，代码如下：

```
package entity;

import java.util.ArrayList;
import java.util.List;

public class Userinfo {
    private String username;
    private String password;
    private List xxxx = new ArrayList();
    //省略 set 和 get 方法
}
```

（3）新建 JSP 文件，代码如下：

```
<html>
    <head>
        <script src="jquery-1.4.2.js">
        </script>
        <script>
            function test1(){
                $.ajax({
                    "type": "post",
                    "url": "test1?t=" + new Date().getTime(),
                    "success": function(data){
                        var jsonObject = JSON.parse(data);
                        alert(jsonObject.username + " " + jsonObject.password);

                        var listString = jsonObject.xxxx;
                        for (var i = 0; i < listString.length; i++) {
                            alert(listString[i]);
                        }

                    }
                });
            }
        </script>
    </head>
    <body>
        <input type="button" value="sendAjax1" onclick="javascript:test1()">
    </body>
</html>
```

（4）控制层代码如下：

```
@Controller
public class UserinfoController {
    @RequestMapping(value = "test1", produces = "text/html;charset=utf-8")
    @ResponseBody
    public String test1() {
        Userinfo returnUserinfo = new Userinfo();
        returnUserinfo.setUsername("返回的账号");
        returnUserinfo.setPassword("返回的密码");

        returnUserinfo.getXxxx().add("中国 1");
```

```
            returnUserinfo.getXxxx().add("中国 2");
            returnUserinfo.getXxxx().add("中国 3");

            String jsonString = JSONObject.toJSONString(returnUserinfo);
            return jsonString;
        }
    }
```

（5）程序运行后，单击界面中的“Button”按钮，可在前端正确显示后台传递过来的数据。

5.3.21 使用 HttpServletResopnse 对象输出响应字符

可以使用 HttpServletResopnse 对象来输出字符串，把字符串数据作为 AJAX 请求的响应字符，从而在客户端进一步处理。

创建测试用的项目 springMVC_JSONString5。

（1）添加解析 JSON 字符串的 fastjson-1.2.47.jar。

（2）新建 JSP 文件，代码如下：

```
<html>
    <head>
        <script src="jquery-1.4.2.js">
        </script>
        <script>
            function test1(){
                $.ajax({
                    "type": "get",
                    "url": "test?t=" + new Date().getTime(),
                    "dataType": "json",
                    "success": function(data){
                        alert(data.username + " " + data.password);
                    }
                });
            }
        </script>
    </head>
    <body>
        <input type="button" value="sendAjax1" onclick="javascript:test1()">
    </body>
</html>
```

（3）控制层代码如下：

```
@Controller
public class UserinfoController {
    @RequestMapping(value = "test")
    public void test1(HttpServletRequest request, HttpServletResponse response) throws
IOException {
        Userinfo userinfo = new Userinfo();
        userinfo.setUsername("返回的账号");
        userinfo.setPassword("返回的密码");

        String returnJSONString = JSONObject.toJSONString(userinfo);

        response.setCharacterEncoding("utf-8");
        response.setContentType("text/html");
```

```
        PrintWriter out = response.getWriter();
        out.print(returnJSONString);
        out.flush();
        out.close();
    }
}
```

（4）程序运行后，单击界面中的“Button”按钮，可在前端正确显示后台传递过来的数据。

5.3.22 单文件上传 1：使用 MultipartHttpServletRequest

Spring 5 MVC 可以实现文件上传。

创建测试用的项目 springMVC_upload1。

（1）添加 commons-fileupload-1.3.1.jar 和 commons-io-2.5.jar 文件。

（2）JSP 文件代码如下：

```
<html>
    <head>
    </head>
    <body>
        <form action="upload.spring" method="post" enctype="multipart/form-data">
            username:<input type="text" name="username">
            <br/>
            username:<input type="file" name="uploadFile">
            <br/>
            <input type="submit" value="submit">
            <br/>
        </form>
    </body>
</html>
```

（3）控制层代码如下：

```
@Controller
public class UserinfoController {
    @RequestMapping(value = "upload")
    public String loginMethod(MultipartHttpServletRequest request) throws IOException {
        String username = request.getParameter("username");
        System.out.println("username=" + username);

        MultipartFile file = request.getFile("uploadFile");
        String uploadFileName = file.getOriginalFilename();
        System.out.println("原始文件名：" + uploadFileName);

        InputStream fileStream = file.getInputStream();

        String uploadPath = request.getSession().getServletContext().getRealPath("/upload");

        System.out.println(uploadPath);

        File destination = new File(uploadPath, uploadFileName);
        FileUtils.copyInputStreamToFile(fileStream, destination);

        fileStream.close();

        return "index.jsp";
```

```
    }
}
```

（4）添加如下配置代码：

```
<bean id="multipartResolver"
    class="org.springframework.web.multipart.commons.CommonsMultipartResolver">
    <property name="maxUploadSize" value="2048000000" />
</bean>
```

设置所有上传文件的总大小为 2GB，如果不在配置文件中添加上面的配置代码，那么运行时会出现异常：

```
    java.lang.IllegalStateException: Current request is not of type [org.springframework.
web.multipart.MultipartHttpServletRequest]: ServletWebRequest: uri=/springMVC_upload1/
upload.spring;client=0:0:0:0:0:0:0:1;session=E369781B4641C2DC7F541D1260F978E7
```

注意： id 属性值“multipartResolver”不能写错，这是固定的写法。

（5）程序运行后成功实现文件上传。

5.3.23 单文件上传 2：使用 MultipartFile

创建测试用的项目 springMVC_upload2。

（1）控制层代码如下：

```
@Controller
public class UserinfoController {
    @RequestMapping(value = "upload")
    public String loginMethod(String username, MultipartFile uploadFile, HttpServletRequest
request,
            HttpServletResponse response) throws IOException {
        System.out.println("username=" + username);

        String uploadFileName = uploadFile.getOriginalFilename();
        System.out.println("原始文件名: " + uploadFileName);

        InputStream fileStream = uploadFile.getInputStream();

        String uploadPath = request.getSession().getServletContext().getRealPath("/upload");
        System.out.println(uploadPath);

        File destination = new File(uploadPath, uploadFileName);
        FileUtils.copyInputStreamToFile(fileStream, destination);

        fileStream.close();
        return "index.jsp";
    }
}
```

MultipartFile uploadFile 参数名 uploadFile 一定要和前台<input type="file" name="uploadFile">文件域的 name 名称一样。

（2）程序运行后成功实现文件上传。

5.3.24 单文件上传 3：使用 MultipartFile 结合实体类

创建测试用的项目 springMVC_upload3。

（1）创建实体类，代码如下：

```
public class Userinfo {
    private String username;
    private MultipartFile file;
    //省略 set 和 get 方法
}
```

（2）控制层代码如下：

```
@Controller
public class UserinfoController {
    @RequestMapping(value = "upload")
    public String loginMethod(Userinfo userinfo, HttpServletRequest request) throws
IOException {
        System.out.println("username=" + userinfo.getUsername());

        MultipartFile uploadFile = userinfo.getFile();

        String uploadFileName = uploadFile.getOriginalFilename();
        System.out.println("原始文件名: " + uploadFileName);

        InputStream fileStream = uploadFile.getInputStream();

        String uploadPath = request.getSession().getServletContext().getRealPath("/upload");

        System.out.println(uploadPath);

        File destination = new File(uploadPath, uploadFileName);
        FileUtils.copyInputStreamToFile(fileStream, destination);

        fileStream.close();

        return "index.jsp";
    }
}
```

（3）程序运行后成功实现文件上传。

5.3.25 多文件上传 1：使用 MultipartHttpServletRequest

创建测试用的项目 springMVC_upload4。

（1）JSP 文件代码如下：

```
<html>
    <head>
    </head>
    <body>
        <form action="upload.spring" method="post" enctype="multipart/form-data">
            username:<input type="text" name="username">
            <br/>
```

```
            file1:<input type="file" name="uploadFile1">
            <br/>
            file2:<input type="file" name="uploadFile2">
            <br/>
            file3:<input type="file" name="uploadFile3">
            <br/>
            file4:<input type="file" name="uploadFile4">
            <br/>
            file5:<input type="file" name="uploadFile5">
            <br/>
            <input type="submit" value="submit">
            <br/>
        </form>
    </body>
</html>
```

（2）控制层代码如下：

```
@Controller
public class UserinfoController {

    @RequestMapping(value = "upload")
    public String loginMethod(MultipartHttpServletRequest request) throws IOException {

        String username = request.getParameter("username");
        System.out.println("username=" + username);

        SimpleDateFormat format = new SimpleDateFormat("yyyy-MM-dd");
        String uploadPath = request.getSession().getServletContext().getRealPath("/upload");

        Map<String, MultipartFile> fileMap = request.getFileMap();
        Iterator<String> iterator = fileMap.keySet().iterator();
        while (iterator.hasNext()) {
            String eachInputName = iterator.next();
            MultipartFile eachFile = fileMap.get(eachInputName);

            String eachFileName = eachFile.getOriginalFilename();
            InputStream eachFileStream = eachFile.getInputStream();

            String dateString = format.format(new Date());
            dateString = dateString + "_" + System.currentTimeMillis() + "_" + eachFileName;

            File destination = new File(uploadPath, dateString);
            FileUtils.copyInputStreamToFile(eachFileStream, destination);
            eachFileStream.close();
        }

        return "index.jsp";
    }
}
```

（3）程序运行后成功实现文件上传。

5.3.26 多文件上传 2：使用 MultipartFile[]

创建测试用的项目 springMVC_upload5。

（1）JSP 文件代码如下：

```
<html>
    <head>
    </head>
    <body>
        <form action="upload.spring" method="post" enctype="multipart/form-data">
            username:<input type="text" name="username">
            <br/>
            file1:<input type="file" name="uploadFile">
            <br/>
            file2:<input type="file" name="uploadFile">
            <br/>
            file3:<input type="file" name="uploadFile">
            <br/>
            file4:<input type="file" name="uploadFile">
            <br/>
            file5:<input type="file" name="uploadFile">
            <br/>
            <input type="submit" value="submit">
            <br/>
        </form>
    </body>
</html>
```

（2）控制层代码如下：

```
@Controller
public class UserinfoController {

    @RequestMapping(value = "upload")
    public String loginMethod(String username, MultipartFile uploadFile[],
HttpServletRequest request,
            HttpServletResponse response) throws IOException {

        System.out.println("username=" + username);

        SimpleDateFormat format = new SimpleDateFormat("yyyy-MM-dd");
        String uploadPath = request.getSession().getServletContext().getRealPath("/upload");

        System.out.println(uploadFile.length);

        for (int i = 0; i < uploadFile.length; i++) {
            MultipartFile eachFile = uploadFile[i];

            String eachFileName = eachFile.getOriginalFilename();
            InputStream eachFileStream = eachFile.getInputStream();

            String dateString = format.format(new Date());
            dateString = dateString + "_" + System.currentTimeMillis() + "_" + eachFileName;

            File destination = new File(uploadPath, dateString);
            FileUtils.copyInputStreamToFile(eachFileStream, destination);
            eachFileStream.close();
        }

        return "index.jsp";
    }
}
```

（3）程序运行后成功实现文件上传。

5.3.27 多文件上传 3：使用 MultipartFile[]结合实体类

创建测试用的项目 springMVC_upload6。

（1）创建实体类，代码如下：

```
public class Userinfo {
    private String username;
    private MultipartFile uploadFile[];
    //省略 set 和 get 方法
}
```

（2）JSP 文件代码如下：

```
<html>
    <head>
    </head>
    <body>
        <form action="upload.spring" method="post" enctype="multipart/form-data">
            username:<input type="text" name="username">
            <br/>
            file1:<input type="file" name="uploadFile">
            <br/>
            file2:<input type="file" name="uploadFile">
            <br/>
            file3:<input type="file" name="uploadFile">
            <br/>
            file4:<input type="file" name="uploadFile">
            <br/>
            file5:<input type="file" name="uploadFile">
            <br/>
            <input type="submit" value="submit">
            <br/>
        </form>
    </body>
</html>
```

（3）控制层代码如下：

```
@Controller
public class UserinfoController {
    @RequestMapping(value = "upload")
    public String loginMethod(Userinfo userinfo, HttpServletRequest request) throws
IOException {
        System.out.println(userinfo.getUsername());
        SimpleDateFormat sdf = new SimpleDateFormat("yyyy_MM_dd_hh_mm_ss");
        MultipartFile[] files = userinfo.getUploadFile();
        for (int i = 0; i < files.length; i++) {
            MultipartFile file = files[i];
            String uploadFileName = file.getOriginalFilename();
            InputStream isRef = file.getInputStream();
            String targetDir = request.getSession().getServletContext().getRealPath("/upload");
            System.out.println(targetDir);
            String getDateString = sdf.format(new Date());
            File targetFile = new File(targetDir, "" + getDateString + "_" + System.
nanoTime() + "_" + uploadFileName);
```

```
                FileOutputStream fosRef = new FileOutputStream(targetFile);
                IOUtils.copy(isRef, fosRef);
                isRef.close();
                fosRef.close();
            }
            return "index.jsp";
        }
    }
```

（4）程序运行后成功实现文件上传。

5.3.28 支持文件名为中文的文件的下载

创建测试用的项目 springMVC_download1。

（1）创建 JSP 文件，代码如下：

```
<body>
    <a href="downloadFile.spring?fileName=<%=java.net.URLEncoder.encode("中国.rar",
"utf-8").toString().replace("%", "_")%>">中国.rar</a>
    <br/>
    <a href="downloadFile.spring?fileName=postTest.rar">postTest.rar</a>
    <br/>
</body>
```

（2）控制层代码如下：

```
@Controller
public class UserinfoController {
    // 下载文件的文件名支持存在空格
    @RequestMapping(value = "downloadFile")
    public void testA(String fileName, HttpServletRequest request, HttpServletResponse response)
            throws UnsupportedEncodingException {
        try {
            String downPath = request.getSession().getServletContext().getRealPath("/");
            fileName = fileName.replace("_", "%");
            fileName = java.net.URLDecoder.decode(fileName, "utf-8");
            System.out.println(fileName);
            String downfileName = "";
            if (request.getHeader("USER-AGENT").indexOf("Trident") > 0) {// IE
                System.out.println("IE");
                downfileName = new String(fileName.getBytes("GBK"), "ISO-8859-1");
            } else {
                System.out.println("NOT IE");
                downfileName = new String(fileName.getBytes("UTF-8"), "ISO-8859-1");
            }
            System.out.println(downPath + fileName);
            File downloadFile = new File(downPath + fileName);
            response.setContentType("application/octet-stream;");
            response.setHeader("Content-disposition", String.format("attachment;
filename=\"%s\"", downfileName)); // 文件名外的双引号处理 Firefox 的空格截断问题
            response.setHeader("Content-Length", String.valueOf(downloadFile.length()));
            FileInputStream fis = new FileInputStream(downloadFile);
            ServletOutputStream out = response.getOutputStream();
            IOUtils.copy(fis, out);
        } catch (FileNotFoundException e) {
            e.printStackTrace();
        } catch (IOException e) {
            e.printStackTrace();
```

```
        }
    }
}
```

（3）程序运行后成功下载文件名为中文和文件名为英文的文件。

5.4 扩展技术

本节丰富了 Spring MVC 的案例，对开发软件项目起到辅助的作用。

5.4.1 使用 InternalResourceViewResolver 简化返回的视图名称

如果所有的 JSP 文件都存储在 jsp 文件夹中，则转发到 JSP 文件需要写上完整路径，代码如下：

```
@RequestMapping(value = "login")
public String loginMethod(String username, String password) {
    if (username.equals("a") && password.equals("aa")) {
        return "jsp/ok.jsp";
    } else {
        return "jsp/no.jsp";
    }
}
```

返回值中都有“jsp/”字符，写法比较烦琐，可以使用 InternalResourceViewResolver 简化返回的视图名称。

创建测试用的项目 springMVC_shortViewName。

（1）在 WebContent 路径下的 myview 文件夹中有 showThisView.jsp 文件。

（2）添加如下配置代码：

```
<bean
    class="org.springframework.web.servlet.view.InternalResourceViewResolver">
    <property name="prefix" value="myview/"></property>
    <property name="suffix" value=".jsp"></property>
</bean>
```

属性 prefix 代表 JSP 文件存储在哪个路径，suffix 代表视图文件的扩展名。

（3）控制层代码如下：

```
@Controller
public class UserinfoController {
    @RequestMapping(value = "test")
    public String testController() {
        return "showThisView";
    }
}
```

（4）运行程序，成功转发到 showThisView.jsp 文件。

5.4.2 控制层返回 List 对象及实体的效果

在 Spring 5 MVC 中的控制层还可以返回 Java 的数据类型，比如 List 或 Userinfo.java 自定义实体等，而且还可以自动转发到 JSP 页面，本测试在项目 springMVC_returnJavaType1 中进行。

（1）新建控制层，代码如下：

```
@Controller
public class UserinfoController {
    @RequestMapping(value = "listMethod")
    public List<String> listMethodXXXXXXXX() {
        List list = new ArrayList();
        list.add("中国 1");
        list.add("中国 2");
        list.add("中国 3");
        list.add("中国 4");
        return list;
    }

    @RequestMapping(value = "getUserinfo")
    public Userinfo getUserinfoXXXXXXXXX() {
        Userinfo userinfo = new Userinfo("100", "中国");
        return userinfo;
    }
}
```

（2）创建 JSP 文件 listMethod.jsp，核心代码如下：

```
<body>
    <%
    Enumeration enum1 = request.getAttributeNames();
    while (enum1.hasMoreElements()) {
    String key = (String) enum1.nextElement();
    out.println("key=" + key + "<br/>");
    } %>
    <br/>
    <br/>
    <c:forEach var="eachString" items="${stringList}">
        ${eachString}
        <br/>
    </c:forEach>
</body>
```

（3）创建 JSP 文件 getUserinfo.jsp，核心代码如下：

```
<body>
    <%
    Enumeration enum1 = request.getAttributeNames();
    while (enum1.hasMoreElements()) {
    String key = (String) enum1.nextElement();
    out.println("key=" + key + "<br/>");
    } %>
    <br/>
    <br/>
    ${userinfo.id}__${userinfo.username}
</body>
```

（4）配置文件 Spring5MVC-servlet.xml 的配置更改如下：

```
<bean
    class="org.springframework.web.servlet.view.InternalResourceViewResolver"
    p:prefix="/" p:suffix=".jsp" />
<context:component-scan
    base-package="controller"></context:component-scan>
```

（5）部署项目运行程序，输入 URL：

```
http://localhost:8080/springMVC_returnJavaType1/listMethod
```

程序运行的效果如图 5-17 所示。

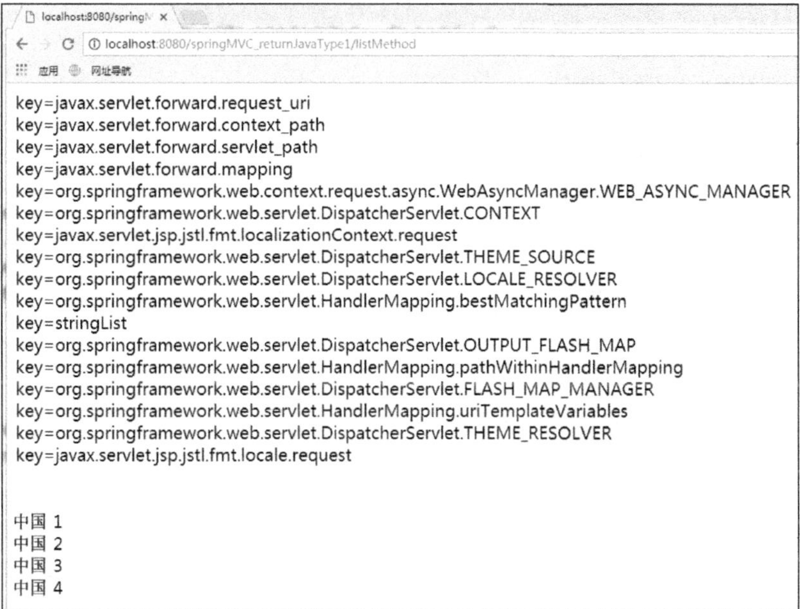

图 5-17　输出列表数据

（6）继续输入 URL：

```
http://localhost:8080/springMVC_returnJavaType1/getUserinfo
```

程序运行的效果如图 5-18 所示。

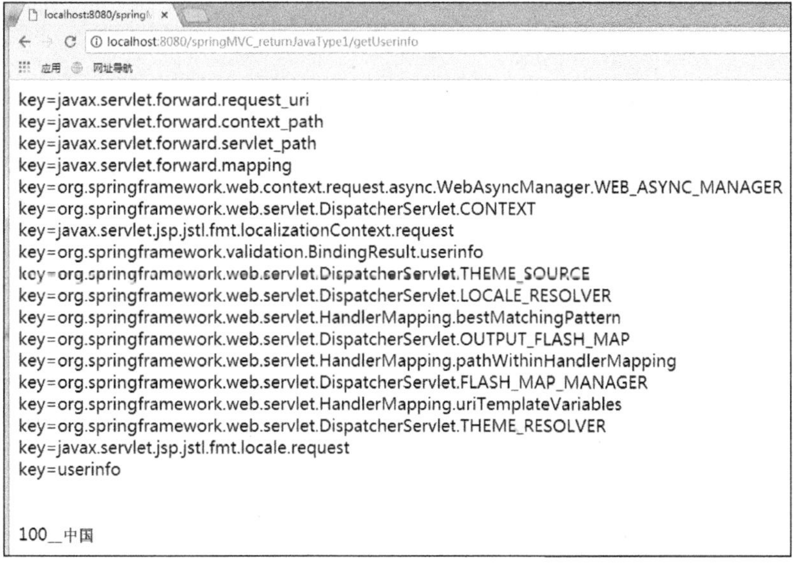

图 5-18　输出实体数据

自动放入 request 的 key 并不是自定义的，但是可以变成自定义的。

创建测试用的项目 springMVC_returnJavaType2。

（7）控制层代码更改如下：

```
@Controller
public class UserinfoController {
    @ModelAttribute(name = "AAA")
    @RequestMapping(value = "listMethod")
    public List<String> listMethodXXXXXXXX() {
        List list = new ArrayList();
        list.add("中国 1");
        list.add("中国 2");
        list.add("中国 3");
        list.add("中国 4");
        return list;
    }

    @ModelAttribute(name = "BBB")
    @RequestMapping(value = "getUserinfo")
    public Userinfo getUserinfoXXXXXXXXX() {
        Userinfo userinfo = new Userinfo("100", "中国");
        return userinfo;
    }
}
```

（8）加入注解：

```
@ModelAttribute(name = "AAA")
```

相当于如下代码：

```
request.setAttribute("AAA",list);
```

将对象 list 以 key 为 AAA 放入 request 作用域中，此时前台 JSP 也要改成如下代码来取得对应的值：

```
${AAA}
```

（9）更改 JSP 文件 listMethod.jsp，核心代码如下：

```
<body>
    <%
        Enumeration enum1 = request.getAttributeNames();
        while (enum1.hasMoreElements()) {
            String key = (String) enum1.nextElement();
            out.println("key=" + key + "<br/>");
        }
    %>
    <br />
    <br />
    <c:forEach var="eachString" items="${AAA}">
            ${eachString}
            <br />
    </c:forEach>
</body>
```

（10）更改 JSP 文件 getUserinfo.jsp，核心代码如下：

```
<body>
    <%
    Enumeration enum1 = request.getAttributeNames();
    while (enum1.hasMoreElements()) {
    String key = (String) enum1.nextElement();
    out.println("key=" + key + "<br/>");
    } %>
    <br/>
    <br/>
    ${BBB.id}__${BBB.username}
</body>
```

（11）程序运行后可以从 request 中获得自定义 attributeKey 对应的值。

5.4.3 实现国际化

国际化的优势是能根据浏览器语言的不同而显示对应语言的信息，它是开发多语种软件项目的必备技术。

1. 在 JSP 文件中向国际化文本传入参数

创建测试用的项目 springMVC_i18n_1。

（1）创建 MyI18N_en_US.properties 资源文件，内容如下：

```
hasParamStatic=I am {0} ,age {1}
name=jack
age=100
```

（2）创建 MyI18N_zh_CN.properties 资源文件，内容如下：

```
hasParamStatic=\u6211\u662F {0}\uFF0C\u5E74\u9F84 {1}
name=\u4E2D\u56FD
age=100
```

属性文件名中包含 zh 和 CN，以及 en 和 US，其中 zh、en 等代表语言，CN、US 等代表国家或地区。语言和国家（或地区）代码的获取来自于如下程序：

```
public class Test {
    public static void main(String[] args) {
        Locale[] localeArray = Locale.getAvailableLocales();
        for (int i = 0; i < localeArray.length; i++) {
            Locale locale = localeArray[i];
            System.out.println(locale.getCountry() + " " + locale.getLanguage());
        }
    }
}
```

（3）配置文件的内容如下：

```
<context:component-scan
    base-package="controller"></context:component-scan>
<mvc:annotation-driven></mvc:annotation-driven>
<mvc:default-servlet-handler />
```

```
<!-- id="messageSource"必须有，固定值 -->
<bean id="messageSource"
    class="org.springframework.context.support.ReloadableResourceBundleMessageSource">
    <property name="basename" value="classpath:MyI18N" />
</bean>
```

（4）控制层代码如下：

```
@Controller
public class UserinfoController {
    @RequestMapping(value = "test")
    public String test() {
        System.out.println("test run !");
        return "test.jsp";
    }
}
```

（5）test.jsp 文件代码如下：

```
<body>
    参数值非国际化的示例：
    <br/>
    <spring:message code="hasParamStatic">
        <spring:argument>
            <spring:message text="姓名非国际化">
            </spring:message>
        </spring:argument>
        <spring:argument>
            <spring:message text="年龄非国际化">
            </spring:message>
        </spring:argument>
    </spring:message>
    <br/>
    <br/>
    参数值国际化的示例：
    <br/>
    <spring:message code="hasParamStatic">
        <spring:argument>
            <spring:message code="name">
            </spring:message>
        </spring:argument>
        <spring:argument>
            <spring:message code="age">
            </spring:message>
        </spring:argument>
    </spring:message>
</body>
```

（6）运行项目，执行网址：

```
http://localhost:8080/springMVC_i18n_1/test
```

程序运行后根据浏览器语言的不同会显示对应语言的消息，效果如图 5-19 所示。

（7）在 Chrome 浏览器切换语言，如图 5-20 所示。

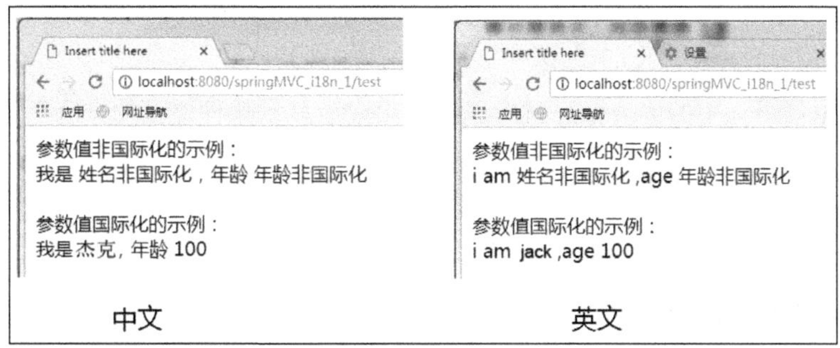

图 5-19 运行结果

图 5-20 切换语言

2. 识别不同语言的原理

服务端如何识别出客户端使用不同的语言呢？

创建测试用的项目 springMVC_i18n_lang。

创建 Servlet 类，代码如下：

```
public class test extends HttpServlet {
    protected void doGet(HttpServletRequest request, HttpServletResponse response)
            throws ServletException, IOException {
        Locale locale = request.getLocale();
        System.out.println("getLanguage=" + locale.getLanguage());
    }
}
```

当执行网址并设置浏览器使用不同的语言时，在服务端的控制台也会输出指定的语言信息。

```
http://localhost:8080/springMVC_i18n_lang/test
```

输出的结果如下：

```
getLanguage=zh
getLanguage=en
```

语言信息 zh 和 en 是包含在 request 请求头中被发送给服务端的。

在设置浏览器语言为英文时，请求头内容如下：

```
Accept-Language: en-US,en;q=0.9,zh;q=0.8,zh-CN;q=0.7
```

在设置浏览器语言为中文时，请求头内容如下：

```
Accept-Language: zh,en-US;q=0.9,en;q=0.8,zh-CN;q=0.7
```

服务端根据请求头中的语言信息就可以识别出应该显示哪种语言的文字。

3. 使用超链接实现语言的切换：使用 HttpSession

创建测试用的项目 springMVC_i18n_2。

（1）创建 MyI18N_en_US.properties 资源文件，内容如下：

```
hasParamStatic=i am {0} ,age {1}
name=jack
age=100

language.en=en
language.cn=\u4E2D\u6587
```

（2）创建 MyI18N_zh_CN.properties 资源文件，内容如下：

```
hasParamStatic=\u6211\u662F {0}\uFF0C\u5E74\u9F84 {1}
name=\u4E2D\u56FD
age=100

language.en=en
language.cn=\u4E2D\u6587
```

（3）配置文件的内容如下：

```
<context:component-scan
    base-package="controller"></context:component-scan>
<mvc:annotation-driven></mvc:annotation-driven>
<mvc:default-servlet-handler />

<bean id="messageSource"
    class="org.springframework.context.support.ReloadableResourceBundleMessageSource">
    <property name="basename" value="classpath:MyI18N" />
</bean>
<!-- id="localeResolver"必须有，固定值 -->
<bean id="localeResolver"
    class="org.springframework.web.servlet.i18n.SessionLocaleResolver"></bean>

<mvc:interceptors>
    <bean
        class="org.springframework.web.servlet.i18n.LocaleChangeInterceptor">
        <property name="paramName" value="lang"></property>
    </bean>
</mvc:interceptors>
```

（4）控制层代码如下：

```
@Controller
public class UserinfoController {
    @RequestMapping(value = "test")
    public String test() {
        System.out.println("test run !");
        return "test.jsp";
```

```
        }
    }
```

（5）JSP 文件代码如下：

```
<body>
    <a href="?lang=zh_CN"><spring:message code="language.cn" /></a>____-<a href="?lang=
en_US"><spring:message code="language.en" /></a>
    <br/>
    <br/>
    参数值非国际化的示例：
    <br/>
    <spring:message code="hasParamStatic">
        <spring:argument>
            <spring:message text="姓名非国际化">
            </spring:message>
        </spring:argument>
        <spring:argument>
            <spring:message text="年龄非国际化">
            </spring:message>
        </spring:argument>
    </spring:message>
    <br/>
    <br/>
    参数值国际化的示例：
    <br/>
    <spring:message code="hasParamStatic">
        <spring:argument>
            <spring:message code="name">
            </spring:message>
        </spring:argument>
        <spring:argument>
            <spring:message code="age">
            </spring:message>
        </spring:argument>
    </spring:message>
</body>
```

（6）运行项目，执行网址：

```
http://localhost:8080/springMVC_i18n_2/test
```

程序运行后，单击中文或英文的超链接会显示对应语言的消息，效果如图 5-21 所示。

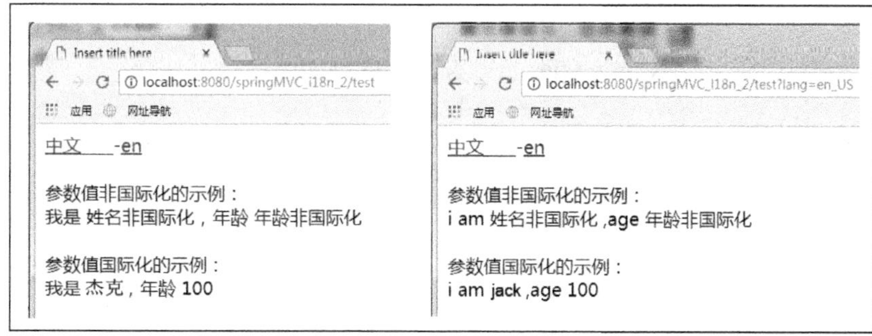

图 5-21 运行结果

（7）从以下配置文件代码中可以分析出，语言信息是保存在 HttpSession 中的，但在打开新的浏览器进程时所产生的新的 sessionId 会导致语言的状态没有被保留，而是使用浏览器默认的语言，这时可以将语言状态保存在 Cookie 中，这就避免了保存在 HttpSession 中的语言状态丢失的问题。

```
<bean id="localeResolver"
    class="org.springframework.web.servlet.i18n.SessionLocaleResolver"></bean>
```

4. 使用超链接后实现语言的切换：使用 Cookie

创建测试用的项目 springMVC_i18n_3。

（1）配置文件的内容如下：

```
<bean id="localeResolver"
    class="org.springframework.web.servlet.i18n.CookieLocaleResolver">
    <property name="cookieMaxAge" value="36000"></property>
</bean>
```

（2）运行项目后，单击切换中文或英文语言的超链接可以改变显示的语言种类，而且在打开新的浏览器进程时还是使用原有语言的消息，因为语言信息已经存储在 Cookie 中了。

5. 在控制层中处理国际化消息

本部分将使用登录功能作为国际化测试的场景，登录失败会显示不同语言的信息。

创建测试用的项目 springMVC_i18n_4。

（1）创建 MyI18N_en_US.properties 资源文件，内容如下：

```
username=username
password=password
submit=submit

hasParamStatic=i am {0} ,age {1}

name=jack
age=100

usernameisnull=username is null
passwordisnull=password is null

language.en=en
language.cn=\u4E2D\u6587
```

（2）创建 MyI18N_zh_CN.properties 资源文件，内容如下：

```
username=\u8D26\u53F7
password=\u5BC6\u7801
submit=\u63D0\u4EA4

hasParamStatic=\u6211\u662F {0}\uFF0C\u5E74\u9F84 {1}

name=\u4E2D\u56FD
age=100

usernameisnull=\u8D26\u53F7\u4E3A\u7A7A
```

```
passwordisnull=\u5BC6\u7801\u4E3A\u7A7A

language.en=en
language.cn=\u4E2D\u6587
```

（3）配置文件的内容如下：

```
<context:component-scan
    base-package="controller"></context:component-scan>
<mvc:annotation-driven></mvc:annotation-driven>
<mvc:default-servlet-handler />

<bean id="messageSource"
    class="org.springframework.context.support.ReloadableResourceBundleMessageSource">
    <property name="basename" value="classpath:MyI18N" />
</bean>

<!-- 配置 SessionLocaleResolver 用于将 Locale 对象存储于 Session 中供后续使用 -->
<bean id="localeResolver"
    class="org.springframework.web.servlet.i18n.SessionLocaleResolver"></bean>
<!-- 配置 LocaleChangeInterceptor 主要用于获取请求中的 locale 信息，将期转为 Locale 对象，获取
LocaleResolver 对象 -->
<mvc:interceptors>
    <bean
        class="org.springframework.web.servlet.i18n.LocaleChangeInterceptor">
        <property name="paramName" value="lang"></property>
    </bean>
</mvc:interceptors>
```

（4）控制层代码如下：

```
@Controller
public class UserinfoController {
    @RequestMapping(value = "login")
    public String login(String username, String password, HttpServletRequest request,
RedirectAttributes attr) {
        System.out.println("login run !");

        RequestContext context = new RequestContext(request);

        if (username == null || "".equals(username)) {
            attr.addFlashAttribute("usernameisnull", context.getMessage("usernameisnull"));
        }
        if (password == null || "".equals(password)) {
            attr.addFlashAttribute("passwordisnull", context.getMessage("passwordisnull"));
        }
        return "redirect:/showLogin";
    }

    @RequestMapping(value = "showLogin")
    public String showLogin(HttpServletRequest request) {
        System.out.println("test run !");
        return "showLogin.jsp";
    }
}
```

（5）JSP 文件代码如下：

```
<body>
    <a href="?lang=zh_CN"><spring:message code="language.cn" /></a>_____-<a
href="?lang=en_US"><spring:message code="language.en" /></a>
```

```
    <br/>
    <br/>
    <%
    Enumeration enum1 = session.getAttributeNames();
    while (enum1.hasMoreElements()) {
    String key = "" + enum1.nextElement();
    out.println(key + "<br/>");
    } %>
    <br/>
    <br/>
    <form action="login" method="post">
        <spring:message code="username">
        </spring:message>
        : <input type="text" name="username" />${usernameisnull}
        <br/>
        <spring:message code="password">
        </spring:message>
        : <input type="text" name="password" />${passwordisnull}
        <br/>
        <input type="submit" value='<spring:message code="submit"></spring:message>' />
    </form>
</body>
```

（6）运行项目并执行以下网址：

```
http://localhost:8080/springMVC_i18n_4/showLogin
```

在登录失败时，会根据语言显示不同信息，效果如图 5-22 所示。

图 5-22 运行结果

5.4.4 处理异常

Spring 5 MVC 框架针对异常的处理有多种方式。

1. 使用 @ResponseStatus 自定义响应码和异常信息

创建测试用的项目 springMVC_exception2。

（1）控制层代码如下：

```
@Controller
public class UserinfoController {

    @RequestMapping(value = "test1")
    public String test1() {
        System.out.println("public String test1()");
```

```
        int i = 10;
        int j = 0;
        int result = i / j;
        return "index.jsp";
    }

    @RequestMapping(value = "test2")
    public String test2() throws Exception {
        System.out.println("public String test2()");
        if (1 == 1) {
            throw new IOException("找不到文件");
        }
        return "index.jsp";
    }

    @RequestMapping(value = "test3")
    public String test3() throws Exception {
        System.out.println("public String test3()");
        if (1 == 1) {
            throw new SQLException("SQL 语句错误");
        }
        return "index.jsp";
    }

    @RequestMapping(value = "test4")
    public String test4() throws Exception {
        System.out.println("public String test4()");
        if (1 == 1) {
            throw new LoginException("SQL 语句错误");
        }
        return "index.jsp";
    }
}
```

（2）创建自定义异常类，代码如下：

```
@ResponseStatus(value = HttpStatus.UNAUTHORIZED, reason = "登录失败，请重新登录！")
public class LoginException extends RuntimeException {
    public LoginException() {
        super();
    }

    public LoginException(String message) {
        super(message);
    }
}
```

（3）配置文件的代码如下：

```
<context:component-scan
    base-package="controller"></context:component-scan>
```

（4）分别输入如下网址：

```
http://localhost:8080/springMVC_exception2/test1
http://localhost:8080/springMVC_exception2/test2
http://localhost:8080/springMVC_exception2/test3
```

显示效果如图 5-23 所示。

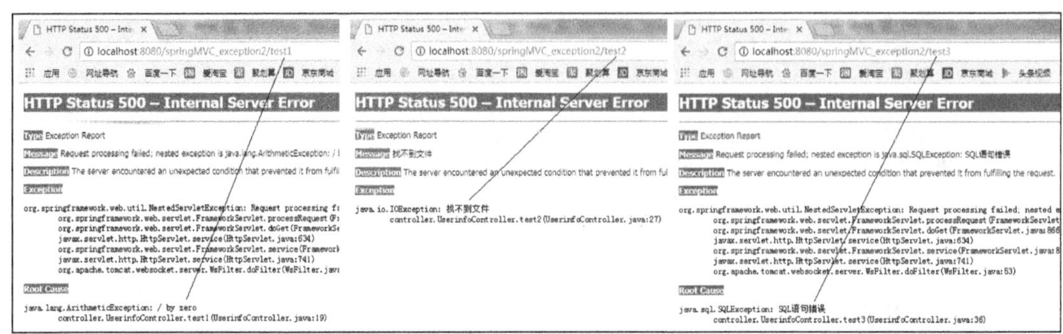

图 5-23 出现异常

在控制台中显示了异常信息，但 response code（响应码）都是 500，如果想自定义响应码和异常信息，可以输入如下网址：

```
http://localhost:8080/springMVC_exception2/test4
```

显示效果如图 5-24 所示。

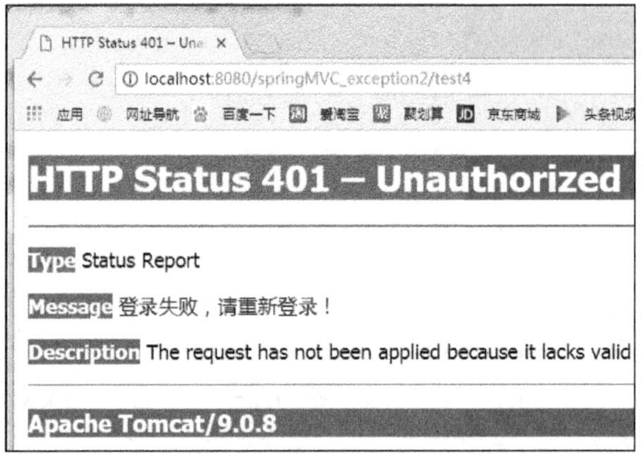

图 5-24 出现异常

本实验实现了响应码和异常信息的可定制化，但出错的界面对用户来讲还是非常粗糙的。如果能使用自定义的视图来显示出错信息就好了。

2．使用 SimpleMappingExceptionResolver 自定义显示异常的视图

创建测试用的项目 springMVC_exception1。

（1）控制层代码如下：

```java
@Controller
public class UserinfoController {

    @RequestMapping(value = "test1")
    public String test1() {
        System.out.println("public String test1()");
        int i = 10;
```

```
        int j = 0;
        int result = i / j;
        return "index.jsp";
    }

    @RequestMapping(value = "test2")
    public String test2() throws Exception {
        System.out.println("public String test2()");
        if (1 == 1) {
            throw new IOException("找不到文件");
        }
        return "index.jsp";
    }

    @RequestMapping(value = "test3")
    public String test3() throws Exception {
        System.out.println("public String test3()");
        if (1 == 1) {
            throw new SQLException("SQL 语句错误");
        }
        return "index.jsp";
    }

    @RequestMapping(value = "test4")
    public String test4() throws Exception {
        System.out.println("public String test3()");
        if (1 == 1) {
            throw new LoginException("SQL 语句错误");
        }
        return "index.jsp";
    }
}
```

（2）创建自定义异常类，代码如下：

```
public class LoginException extends RuntimeException {
    public LoginException() {
        super();
    }

    public LoginException(String message) {
        super(message);
    }
}
```

（3）配置文件的代码如下：

```
<context:component-scan
    base-package="controller"></context:component-scan>
<bean
    class="org.springframework.web.servlet.handler.SimpleMappingExceptionResolver">
    <property name="defaultErrorView" value="errorPage.jsp"></property>
    <property name="exceptionAttribute" value="ex"></property>
    <property name="exceptionMappings">
        <props>
            <prop key="SQLException">sqlerror.jsp</prop>
            <prop key="IOException">ioerror.jsp</prop>
            <prop key="myexception.LoginException">loginerror.jsp</prop>
        </props>
```

```
    </property>
</bean>
```

（4）JSP 文件 errorPage.jsp 的代码如下：

```
<body>
    到达了 errorPage.jsp 页面，异常信息为：${ex.message}
</body>
```

（5）JSP 文件 ioerror.jsp 的代码如下：

```
<body>
    出现 IO 异常，异常信息为：${ex.message}
</body>
```

（6）JSP 文件 sqlerror.jsp 的代码如下：

```
<body>
    出现 SQL 异常，异常信息为：${ex.message}
</body>
```

（7）JSP 文件 loginerror1.jsp 的代码如下：

```
    <body>
    登录异常，异常信息为：${ex.message}
</body>
```

（8）输入如下网址后都可以在自定义的异常视图中显示出错信息，代码如下：

```
http://localhost:8080/springMVC_exception1/test1
http://localhost:8080/springMVC_exception1/test2
http://localhost:8080/springMVC_exception1/test3
http://localhost:8080/springMVC_exception1/test4
```

3. 在控制层中使用 @ExceptionHandler 注解实现局部异常处理

创建测试用的项目 springMVC_exception3。

（1）控制层代码如下：

```
@Controller
public class UserinfoControllerA {

    @RequestMapping(value = "test1")
    public String test1() {
        System.out.println("public String test1()");
        int i = 10;
        int j = 0;
        int result = i / j;
        return "index.jsp";
    }

    @RequestMapping(value = "test2")
    public String test2() throws Exception {
        System.out.println("public String test2()");
        if (1 == 1) {
            throw new SQLException("SQL 语句错误");
        }
        return "index.jsp";
```

```
    }

    @RequestMapping(value = "test3")
    public String test3() throws LoginException1 {
        System.out.println("public String test3()");
        if (1 == 1) {
            throw new LoginException1("账号 XX 登录失败，请重新登录！");
        }
        return "index.jsp";
    }

    @RequestMapping(value = "test4")
    public String test4() throws LoginException2 {
        System.out.println("public String test4()");
        if (1 == 1) {
            throw new LoginException2("账号 XX 登录失败，请重新登录！");
        }
        return "index.jsp";
    }

    @ExceptionHandler(value = Exception.class)
    public ModelAndView processException(Exception ex) {
        ModelAndView mav = new ModelAndView();
        mav.addObject("ex", ex);
        mav.setViewName("errorPage.jsp");
        return mav;
    }

    @ExceptionHandler(value = SQLException.class)
    public ModelAndView processSQLException(SQLException ex) {
        ModelAndView mav = new ModelAndView();
        mav.addObject("ex", ex);
        mav.setViewName("sqlerror.jsp");
        return mav;
    }

    @ExceptionHandler(value = LoginException1.class)
    public ModelAndView processLoginException1(LoginException1 ex) {
        ModelAndView mav = new ModelAndView();
        mav.addObject("ex", ex);
        mav.setViewName("loginException1.jsp");
        return mav;
    }

    @ExceptionHandler(value = LoginException2.class)
    @ResponseBody
    public Map processLoginException2(LoginException2 ex) {
        Map map = new HashMap();
        map.put("controllerName", "UserinfoController");
        map.put("message", ex.getMessage());
        return map;
    }
}
```

（2）控制层代码如下：

```
@Controller
public class UserinfoControllerB {
```

```
    @RequestMapping(value = "test5")
    public String test5() throws LoginException1 {
        System.out.println("public String test5()");
        if (1 == 1) {
            throw new LoginException1("账号 XX 登录失败，请重新登录!");
        }
        return "index.jsp";
    }
}
```

注意：在 UserinfoControllerB 类的方法中抛出的异常类型为自定义异常，执行时会在浏览器中显示原始的异常界面。这是因为在 UserinfoControllerB 类中并没有针对这个自定义异常使用 @ExceptionHandler 注解进行处理。但是，如果在 UserinfoControllerB 类的方法中，抛出了在其他控制层中已经使用 @ExceptionHandler 注解关联的非自定义异常，那么会显示与之关联的自定义异常视图。此结论说明非自定义异常关联的 @ExceptionHandler 处理器是共享的。

（3）创建自定义异常类，代码如下：

```
public class LoginException1 extends RuntimeException {
    public LoginException1(String message) {
        super(message);
    }
}
```

（4）创建自定义异常类，代码如下：

```
public class LoginException2 extends RuntimeException {
    public LoginException2(String message) {
        super(message);
    }
}
```

（5）配置文件的代码如下：

```
<context:component-scan
    base-package="controller"></context:component-scan>
<mvc:annotation-driven></mvc:annotation-driven>
```

（6）JSP 文件 errorPage.jsp 的代码如下：

```
<body>
    到达了 errorPage.jsp 页面，异常信息为：${ex.message}
</body>
```

（7）JSP 文件 sqlerror.jsp 的代码如下：

```
<body>
    出现 SQL 异常，异常信息为：${ex.message}
</body>
```

（8）JSP 文件 loginerror1.jsp 的代码如下：

```
    <body>
    登录异常，异常信息为：${ex.message}
</body>
```

（9）输入如下网址后，可以在自定义的异常视图中显示出错信息。

```
http://localhost:8080/springMVC_exception3/test1
http://localhost:8080/springMVC_exception3/test2
http://localhost:8080/springMVC_exception3/test3
```

输入如下网址后，出错信息被转换成 JSON 格式并交给客户端。

```
http://localhost:8080/springMVC_exception3/test4
```

输入如下网址后在浏览器中显示原始错误界面，而并没有在自定义的视图中显示异常信息。

```
http://localhost:8080/springMVC_exception3/test5
```

如果想在多个控制层之间共享使用 @ExceptionHandler 关联的自定义异常处理器，那么要结合@ControllerAdvice 注解。

4. 使用 @ControllerAdvice 注解实现全局异常处理器

创建测试用的项目 springMVC_exception4。

（1）异常处理类的代码如下：

```
@org.springframework.web.bind.annotation.ControllerAdvice
public class ControllerAdvice {
    @ExceptionHandler(value = Exception.class)
    public ModelAndView processException(Exception ex) {
        ModelAndView mav = new ModelAndView();
        mav.addObject("ex", ex);
        mav.setViewName("errorPage.jsp");
        return mav;
    }

    @ExceptionHandler(value = SQLException.class)
    public ModelAndView processSQLException(SQLException ex) {
        ModelAndView mav = new ModelAndView();
        mav.addObject("ex", ex);
        mav.setViewName("sqlerror.jsp");
        return mav;
    }

    @ExceptionHandler(value = LoginException1.class)
    public ModelAndView processLoginException1(LoginException1 ex) {
        ModelAndView mav = new ModelAndView();
        mav.addObject("ex", ex);
        mav.setViewName("loginException1.jsp");
        return mav;
    }

    @ExceptionHandler(value = LoginException2.class)
    @ResponseBody
    public Map processLoginException2(LoginException2 ex) {
        Map map = new HashMap();
        map.put("controllerName", "UserinfoController");
        map.put("message", ex.getMessage());
        return map;
```

```
    }
}
```

（2）配置代码如下：

```
<context:component-scan
    base-package="controller"></context:component-scan>
<context:component-scan
    base-package="controlleradvice"></context:component-scan>
<mvc:annotation-driven></mvc:annotation-driven>
```

（3）输入如下网址后在自定义的视图中显示了异常信息，成功实现了全局的异常处理器。

```
http://localhost:8080/springMVC_exception3/test5
```

5. 使用 @RestControllerAdvice 注解实现全局异常处理器

注解 @RestController 的作用是隐式地使用了 @ResponseBody。
创建测试用的项目 springMVC_exception5。
（1）控制层代码如下：

```
@RestController
public class UserinfoController {

    @RequestMapping(value = "test1")
    public String test1() {
        System.out.println("public String test1()");
        int i = 10;
        int j = 0;
        int result = i / j;
        return "responseBody1";
    }

    @RequestMapping(value = "test2")
    public String test2() throws Exception {
        System.out.println("public String test2()");
        if (1 == 1) {
            throw new SQLException("SQL 语句错误");
        }
        return "responseBody2";
    }

    @RequestMapping(value = "test3")
    public String test3() throws LoginException1 {
        System.out.println("public String test3()");
        if (1 == 1) {
            throw new LoginException1("账号 XX 登录失败，请重新登录！");
        }
        return "responseBody3";
    }

    @RequestMapping(value = "test4")
    public String test4() throws LoginException2 {
        System.out.println("public String test4()");
        if (1 == 1) {
            throw new LoginException2("账号 XX 登录失败，请重新登录！");
```

```
        }
        return "responseBody4";
    }

}
```

（2）异常处理器的代码如下：

```
@RestControllerAdvice
public class ControllerAdvice {
    @ExceptionHandler(value = Exception.class)
    public ModelAndView processException(Exception ex) {
        ModelAndView mav = new ModelAndView();
        mav.addObject("ex", ex);
        mav.setViewName("errorPage.jsp");
        return mav;
    }

    @ExceptionHandler(value = SQLException.class)
    public ModelAndView processSQLException(SQLException ex) {
        ModelAndView mav = new ModelAndView();
        mav.addObject("ex", ex);
        mav.setViewName("sqlerror.jsp");
        return mav;
    }

    @ExceptionHandler(value = LoginException1.class)
    public ModelAndView processLoginException1(LoginException1 ex) {
        ModelAndView mav = new ModelAndView();
        mav.addObject("ex", ex);
        mav.setViewName("loginException1.jsp");
        return mav;
    }

    @ExceptionHandler(value = LoginException2.class)
    public Map processLoginException2(LoginException2 ex) {
        Map map = new HashMap();
        map.put("controllerName", "UserinfoController");
        map.put("message", ex.getMessage());
        return map;
    }
}
```

（3）配置文件的代码如下：

```
<context:component-scan
    base-package="controller"></context:component-scan>
<context:component-scan
    base-package="controlleradvice"></context:component-scan>
<mvc:annotation-driven></mvc:annotation-driven>
```

（4）输入如下网址后在控制台成功使用 JSON 格式显示出异常信息，并且没有显式地使用 @ResponseBody 注解。

```
http://localhost:8080/springMVC_exception5/test4
```

5.4.5 配置文件的不同使用方式

前面都是在 web.xml 中添加如下的 Spring 5 MVC 配置映射。

```
<servlet>
    <servlet-name>springMVC</servlet-name>
    <servlet-class>org.springframework.web.servlet.DispatcherServlet</servlet-class>
    <load-on-startup>1</load-on-startup>
</servlet>
<servlet-mapping>
    <servlet-name>springMVC</servlet-name>
    <url-pattern>/</url-pattern>
</servlet-mapping>
```

然后，Spring 框架自动在 WEB-INF 文件夹中寻找 springMVC-servlet.xml 配置文件，这是默认的使用方式，其实在 Spring 5 MVC 中还有其他几种 web.xml 配置映射的方法。

1. 存放于 src 资源路径中

创建测试用的项目 springMVC_xmlPosition1。

将配置文件存放于 src 资源路径中，代码如下：

```
<servlet>
    <servlet-name>springMVC</servlet-name>
    <servlet-class>org.springframework.web.servlet.DispatcherServlet</servlet-class>
    <init-param>
        <param-name>contextConfigLocation</param-name>
        <param-value>classpath*:/abc.xml</param-value>
    </init-param>
    <load-on-startup>1</load-on-startup>
</servlet>
<servlet-mapping>
    <servlet-name>springMVC</servlet-name>
    <url-pattern>/</url-pattern>
</servlet-mapping>
```

2. 多个配置文件存放于 src 资源路径中

创建测试用的项目 springMVC_xmlPosition2。

将多个配置文件存放于 src 资源路径中，代码如下：

```
<servlet>
    <servlet-name>springMVC</servlet-name>
    <servlet-class>org.springframework.web.servlet.DispatcherServlet</servlet-class>
    <init-param>
        <param-name>contextConfigLocation</param-name>
        <param-value>classpath*:/a.xml,classpath*:/b.xml</param-value>
    </init-param>
    <load-on-startup>1</load-on-startup>
</servlet>
<servlet-mapping>
    <servlet-name>springMVC</servlet-name>
    <url-pattern>/</url-pattern>
</servlet-mapping>
```

3. 将多个配置文件存放于指定的路径中

创建测试用的项目 springMVC_xmlPosition3。

两个配置文件存放的位置如下。

（1）WebContent/a/a.xml。

（2）WebContent/b/b.xml。

配置代码如下：

```
<servlet>
    <servlet-name>springMVC</servlet-name>
    <servlet-class>org.springframework.web.servlet.DispatcherServlet</servlet-class>
    <init-param>
        <param-name>contextConfigLocation</param-name>
        <param-value>a/a.xml,b/b.xml</param-value>
    </init-param>
    <load-on-startup>1</load-on-startup>
</servlet>
<servlet-mapping>
    <servlet-name>springMVC</servlet-name>
    <url-pattern>/</url-pattern>
</servlet-mapping>
```

5.4.6 方法参数是 Model 数据类型

向 Model 对象中存储数据相当于向 request 作用域中存储数据。

创建测试用的项目 springMVC_Model。

（1）创建控制层，代码如下：

```
@Controller
public class UserinfoController {
    @RequestMapping(value = "test")
    public String testMethod(Model model) {
        model.addAttribute("myKey1", "myValue1");
        return "index.jsp";
    }
}
```

（2）JSP 文件的代码如下：

```
<body>
    <%
    Enumeration enum1 = request.getAttributeNames();
    while (enum1.hasMoreElements()) {
    String key = "" + enum1.nextElement();
    out.print(key + "<br/>");
    } %>
    <br/>
    ${myKey1}
</body>
```

（3）程序运行的效果如图 5-25 所示。

```
javax.servlet.forward.request_uri
javax.servlet.forward.context_path
javax.servlet.forward.servlet_path
javax.servlet.forward.mapping
org.springframework.web.context.request.async.WebAsyncManager.WEB_ASYNC_MANAGER
org.springframework.web.servlet.DispatcherServlet.CONTEXT
javax.servlet.jsp.jstl.fmt.localizationContext.request
org.springframework.web.servlet.DispatcherServlet.THEME_SOURCE
myKey1
org.springframework.web.servlet.DispatcherServlet.LOCALE_RESOLVER
org.springframework.web.servlet.HandlerMapping.bestMatchingPattern
org.springframework.web.servlet.DispatcherServlet.OUTPUT_FLASH_MAP
org.springframework.web.servlet.HandlerMapping.pathWithinHandlerMapping
org.springframework.web.servlet.DispatcherServlet.FLASH_MAP_MANAGER
org.springframework.web.servlet.HandlerMapping.uriTemplateVariables
org.springframework.web.servlet.DispatcherServlet.THEME_RESOLVER
javax.servlet.jsp.jstl.fmt.locale.request

myValue1
```

图 5-25 运行结果

5.4.7 方法参数是 ModelMap 数据类型

向 ModelMap 对象中存储数据相当于向 request 作用域中存储数据。ModelMap 类比 Model 类的 API 更加丰富，两者作用一样。

创建测试用的项目 springMVC_paramModelMap。

（1）创建控制层，代码如下：

```
@Controller
public class UserinfoController {
    // ModelMap 比 Model 对象在 API 上更加丰富
    @RequestMapping(value = "test")
    public String testMethod(ModelMap modelMap) {
        List list = new ArrayList();
        list.add("中国 1");
        list.add("中国 2");
        list.add("中国 3");
        list.add("中国 4");
        modelMap.addAttribute("listStringKey", list);
        return "listString.jsp";
    }
}
```

（2）JSP 文件代码如下：

```
<body>
    <%
    Enumeration enum1 = request.getAttributeNames();
    while (enum1.hasMoreElements()) {
    String key = (String) enum1.nextElement();
    out.print("key=" + key + "<br/>");
    } %>
    <br/>
    <c:forEach var="eachString" items="${listStringKey}">
```

```
        ${eachString}
        <br/>
    </c:forEach>
</body>
```

（3）程序运行的效果如图 5-26 所示。

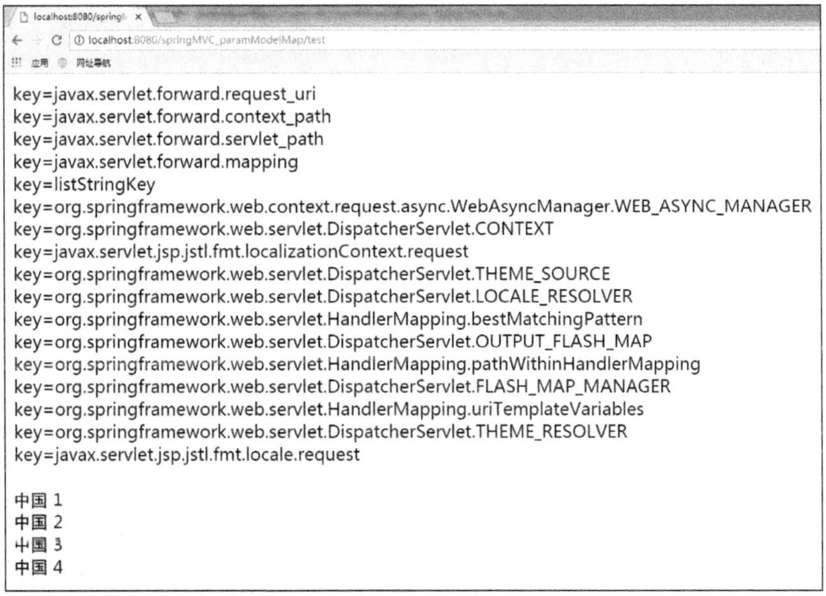

```
localhost:8080/springMVC_paramModelMap/test

key=javax.servlet.forward.request_uri
key=javax.servlet.forward.context_path
key=javax.servlet.forward.servlet_path
key=javax.servlet.forward.mapping
key=listStringKey
key=org.springframework.web.context.request.async.WebAsyncManager.WEB_ASYNC_MANAGER
key=org.springframework.web.servlet.DispatcherServlet.CONTEXT
key=javax.servlet.jsp.jstl.fmt.localizationContext.request
key=org.springframework.web.servlet.DispatcherServlet.THEME_SOURCE
key=org.springframework.web.servlet.DispatcherServlet.LOCALE_RESOLVER
key=org.springframework.web.servlet.HandlerMapping.bestMatchingPattern
key=org.springframework.web.servlet.DispatcherServlet.OUTPUT_FLASH_MAP
key=org.springframework.web.servlet.HandlerMapping.pathWithinHandlerMapping
key=org.springframework.web.servlet.DispatcherServlet.FLASH_MAP_MANAGER
key=org.springframework.web.servlet.HandlerMapping.uriTemplateVariables
key=org.springframework.web.servlet.DispatcherServlet.THEME_RESOLVER
key=javax.servlet.jsp.jstl.fmt.locale.request

中国 1
中国 2
中国 3
中国 4
```

图 5-26 运行结果

5.4.8 方法返回值是 ModelMap 数据类型

创建测试用的项目 springMVC_returnModelMap。

（1）创建控制层，代码如下：

```
@Controller
public class UserinfoController {
    @RequestMapping(value = "test")
    public ModelMap testMethod() {
        List list = new ArrayList();
        list.add("中国 1");
        list.add("中国 2");
        list.add("中国 3");
        list.add("中国 4");

        ModelMap map = new ModelMap();
        map.addAttribute("listString", list);

        return map;
    }
}
```

（2）配置文件代码如下：

```
<bean
    class="org.springframework.web.servlet.view.InternalResourceViewResolver"
```

```
    p:prefix="/" p:suffix=".jsp" />
<context:component-scan
    base-package="controller"></context:component-scan>
```

（3）JSP 文件代码如下：

```
<body>
    <%
    Enumeration enum1 = request.getAttributeNames();
    while (enum1.hasMoreElements()) {
    String key = (String) enum1.nextElement();
    out.println("key=" + key + "<br/>");
    } %>
    <br/>
    <c:forEach var="eachString" items="${listString}">
        ${eachString}
        <br/>
    </c:forEach>
</body>
```

（4）程序运行的效果如图 5-27 所示。

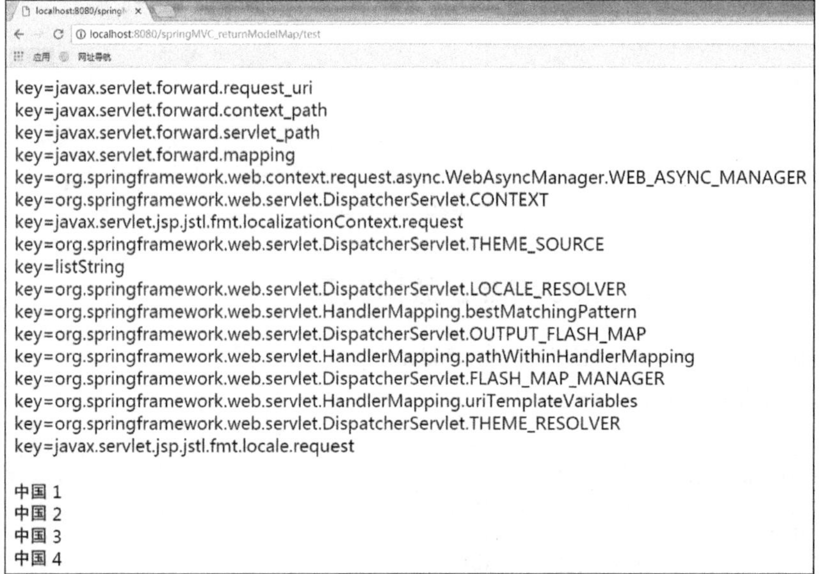

图 5-27　运行结果

5.4.9　方法返回值是 ModelAndView 数据类型

返回 ModelMap 类的缺点是不能指定视图，而 ModelAndView 对象可以实现。

创建测试用的项目 springMVC_returnModelAndView。

（1）创建控制层，代码如下：

```
@Controller
public class UserinfoController {
    @RequestMapping(value = "test")
    public ModelAndView test() {
        ModelAndView view = new ModelAndView();
```

```
        view.setViewName("index.jsp");
        view.addObject("myKey", "中国");
        return view;
    }
}
```

（2）JSP 文件代码如下：

```
    <body>
        <%
        Enumeration enum1 = request.getAttributeNames();
        while (enum1.hasMoreElements()) {
        String key = (String) enum1.nextElement();
        out.print("key=" + key + "<br/>");
        } %>
        <br/>
        ${myKey}
</body>
```

（3）程序运行的效果如图 5-28 所示。

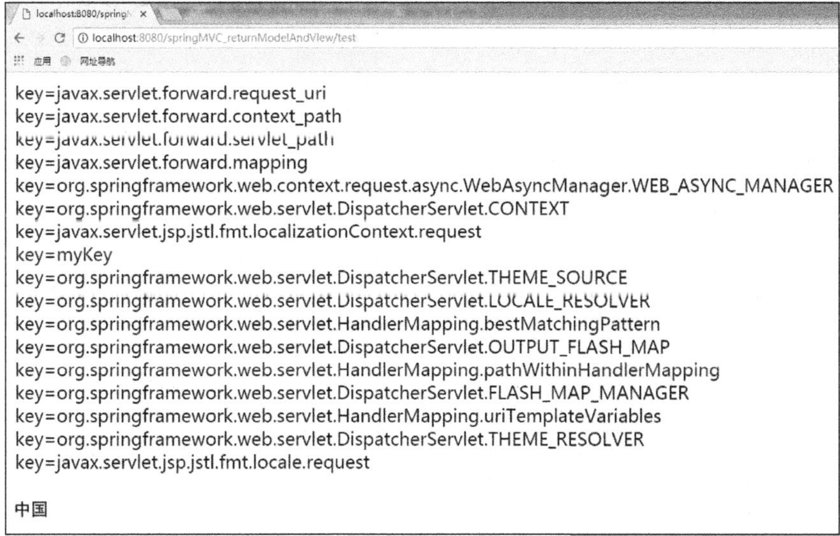

图 5-28 运行结果

5.4.10 方法返回值是 ModelAndView 实现重定向

创建测试用的项目 springMVC_ModelAndView_forward_redirect。

（1）创建控制层，代码如下：

```
@Controller
public class UserinfoController {
    @RequestMapping(value = "test1")
    public ModelAndView test1() {
        System.out.println("test1");
        ModelAndView mav = new ModelAndView();
        mav.setViewName("index.jsp");
        return mav;
    }
```

```
    @RequestMapping(value = "test2")
    public ModelAndView test2() {
        System.out.println("test2");
        ModelAndView mav = new ModelAndView();
        mav.setViewName("redirect:/index.jsp");
        return mav;
    }
}
```

（2）输入如下网址，实现转发操作：

```
http://localhost:8080/springMVC_ModelAndView_forward_redirect/test1
```

输入如下网址实现重定向操作：

```
http://localhost:8080/springMVC_ModelAndView_forward_redirect/test2
```

5.4.11 使用 @RequestAttribute 和 @SessionAttribute 注解

使用 @RequestAttribute 和 @SessionAttribute 注解可以访问 request 和 session 作用域中的数据。

创建测试用的项目 springMVC_request_session_attr。

（1）创建控制层，代码如下：

```
@Controller
public class UserinfoController {
    @RequestMapping(value = "test")
    public String test(HttpServletRequest request, HttpSession session) {
        request.setAttribute("requestKey", "requestValue 值 1");
        request.getSession().setAttribute("sessionKey", "sessionValue 值 2");
        return "printInfo";
    }

    @RequestMapping(value = "printInfo")
    public String printInfo(@RequestAttribute("requestKey") String value1,
            @SessionAttribute("sessionKey") String value2) {
        System.out.println("value1=" + value1);
        System.out.println("value2=" + value2);
        return "index.jsp";
    }
}
```

（2）输入如下网址：

```
http://localhost:8080/springMVC_request_session_attr/test
```

控制台输出的结果如下：

```
value1=requestValue 值 1
value2=sessionValue 值 2
```

5.4.12 使用 @CookieValue 和 @RequestHeader 注解

注解 @CookieValue 用于获得 Cookie 值。

注解 @RequestHeader 用于获得 request header 值。

创建测试用的项目 springMVC_requestHeader_Cookie。

（1）创建控制层，代码如下：

```
@Controller
public class UserinfoController {
    @RequestMapping(value = "test")
    public String test(HttpServletRequest request, HttpServletResponse response,
HttpSession session) {
        Cookie cookie = new Cookie("myCookieName", "我是 Cookie 值");
        cookie.setMaxAge(36000);
        response.addCookie(cookie);
        return "redirect:/printInfo";
    }

    @RequestMapping(value = "printInfo")
    public String printInfo(@CookieValue(name = "myCookieName") String cookieValue,
            @RequestHeader(name = "Accept-Language") String acceptLanguage, @RequestHeader
(name = "Host") String host) {
        System.out.println("cookieValue=" + cookieValue);
        System.out.println("Accept-Language=" + acceptLanguage);
        System.out.println("Host=" + host);
        return "index.jsp";
    }
}
```

（2）输入如下网址：

```
http://localhost:8080/springMVC_requestHeader_Cookie/test
```

控制台输出的结果如下：

```
cookieValue=我是 Cookie 值
Accept-Language=zh,en-US;q=0.9,en;q=0.8,zh-CN;q=0.7
Host=localhost:8080
```

5.4.13 使用 @SessionAttributes 注解

注解 @SessionAttribute 常用于从 HttpSession 中获取数据，而注解 @SessionAttributes 用于向 HttpSession 中存放数据。

创建测试用的项目 springMVC_sessionAttributes。

（1）创建控制层，代码如下：

```
@Controller
@SessionAttributes(names = "myKey2")
public class UserinfoController {
    @RequestMapping(value = "test1")
    public String test1(Model model) {
        model.addAttribute("myKey1", "我在 request 存在");
        return "index1.jsp";
    }

    @RequestMapping(value = "test2")
    public String test2(Model model) {
        model.addAttribute("myKey2", "我在 request 和 session 中都存在");
        return "index2.jsp";
    }

    @RequestMapping(value = "test3")
    public String test3(@ModelAttribute("myKey2") String mySessionValue) {
```

```
        System.out.println("mySessionValue=" + mySessionValue);
        return "index1.jsp";
    }
}
```

（2）JSP 文件 index1.jsp 代码如下：

```
<body>
    request_keys:
    <br />
    <%
        Enumeration enum1 = request.getAttributeNames();
        while (enum1.hasMoreElements()) {
            String key = "" + enum1.nextElement();
            out.print(key + "<br/>");
        }
    %>
    <br />
    <br /> session_keys:
    <br />
    <%
        Enumeration enum2 = request.getSession().getAttributeNames();
        while (enum2.hasMoreElements()) {
            String key = "" + enum2.nextElement();
            out.print(key + "<br/>");
        }
    %>
    <br />
    <br /> 从 request 获取值：${requestScope.myKey1}
</body>
```

（3）JSP 文件 index2jsp 代码如下：

```
<body>
    request_keys:
    <br />
    <%
        Enumeration enum1 = request.getAttributeNames();
        while (enum1.hasMoreElements()) {
            String key = "" + enum1.nextElement();
            out.print(key + "<br/>");
        }
    %>
    <br />
    <br /> session_keys:
    <br />
    <%
        Enumeration enum2 = request.getSession().getAttributeNames();
        while (enum2.hasMoreElements()) {
            String key = "" + enum2.nextElement();
            out.print(key + "<br/>");
        }
    %>
    <br />
    <br /> 从 request 获取值：${requestScope.myKey2}
    <br /> 从 session 获取值：${sessionScope.myKey2}
</body>
```

（4）输入如下网址：

```
http://localhost:8080/springMVC_sessionAttributes/test1
```

浏览器输出内容如图 5-29 所示。

```
request_keys:
javax.servlet.forward.request_uri
javax.servlet.forward.context_path
javax.servlet.forward.servlet_path
javax.servlet.forward.mapping
org.springframework.web.context.request.async.WebAsyncManager.WEB_ASYNC_MANAGER
org.springframework.web.servlet.DispatcherServlet.CONTEXT
javax.servlet.jsp.jstl.fmt.localizationContext.request
org.springframework.web.servlet.DispatcherServlet.THEME_SOURCE
myKey1
org.springframework.web.servlet.DispatcherServlet.LOCALE_RESOLVER
org.springframework.web.servlet.HandlerMapping.bestMatchingPattern
org.springframework.web.servlet.DispatcherServlet.OUTPUT_FLASH_MAP
org.springframework.web.servlet.HandlerMapping.pathWithinHandlerMapping
org.springframework.web.servlet.DispatcherServlet.FLASH_MAP_MANAGER
org.springframework.web.servlet.HandlerMapping.uriTemplateVariables
org.springframework.web.servlet.DispatcherServlet.THEME_RESOLVER
javax.servlet.jsp.jstl.fmt.locale.request


session_keys:


从request获取值：我在request存在
```

图 5-29　运行结果

（5）输入如下网址：

```
http://localhost:8080/springMVC_sessionAttributes/test2
```

浏览器输出的内容如图 5-30 所示。

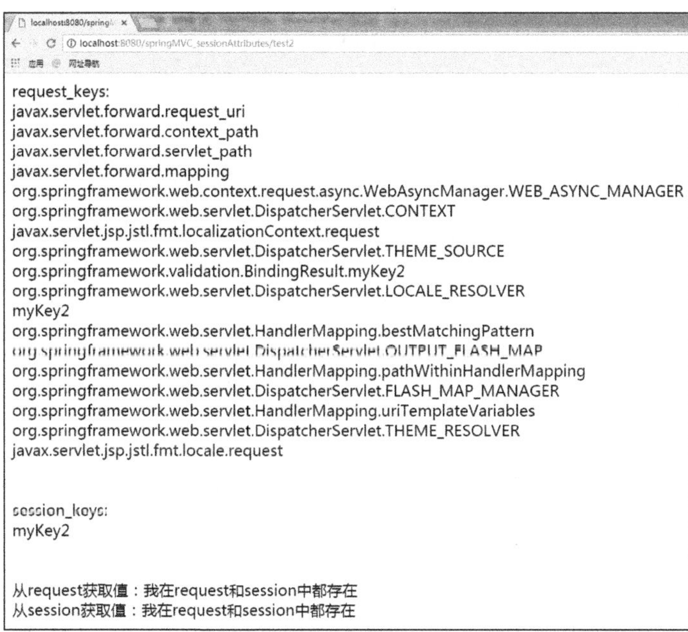

图 5-30　运行结果

（6）输入如下网址：

```
http://localhost:8080/springMVC_sessionAttributes/test3
```

浏览器输出的内容如图 5-31 所示。

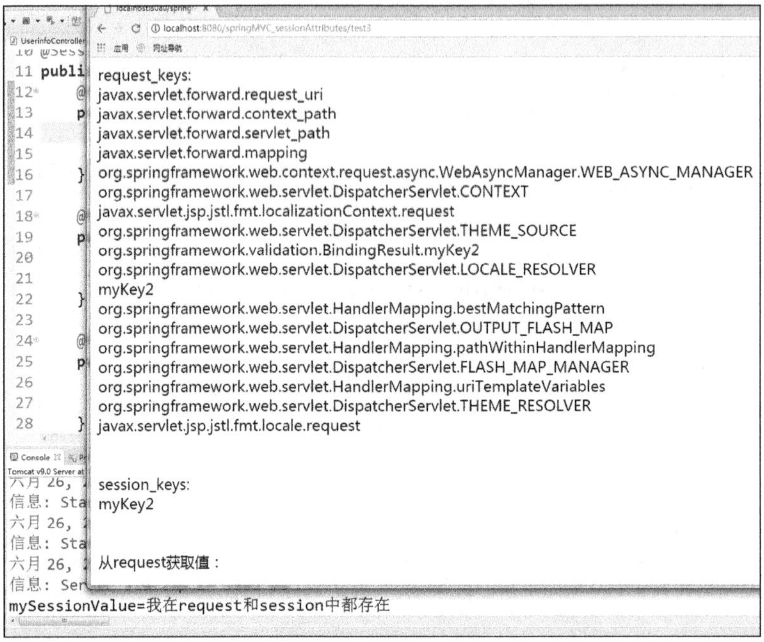

图 5-31　运行结果

5.4.14　使用 @ModelAttribute 注解

创建测试用的项目 springMVC_ModelAttribute。

1. 自动运行特性与存储 1 个值

（1）创建控制层，代码如下：

```
@Controller
public class UserinfoController1 {
    @ModelAttribute(name = "usernameKey")
    public String firstRun(String username) {
        System.out.println("public String firstRun(String username) username=" + username);
        return username;
    }

    @RequestMapping(value = "test1")
    public String test1(ModelMap map) {
        System.out.println("public String test1(ModelMap map) usernameKey=" + map.get
("usernameKey"));
        return "index1.jsp";
    }
}
```

使用 usernameKey 作为 key，将 username 参数值放入模型（model）中。

（2）JSP 文件 index1.jsp 代码如下：

```
<body>
    <%
        Enumeration enum1 = request.getAttributeNames();
        while (enum1.hasMoreElements()) {
            String key = "" + enum1.nextElement();
            out.print(key + "<br/>");
        }
    %>
</body>
```

（3）输入如下网址：

```
http://localhost:8080/springMVC_ModelAttribute/test1?username=abc
```

控制台和浏览器输出的内容如图 5-32 所示。

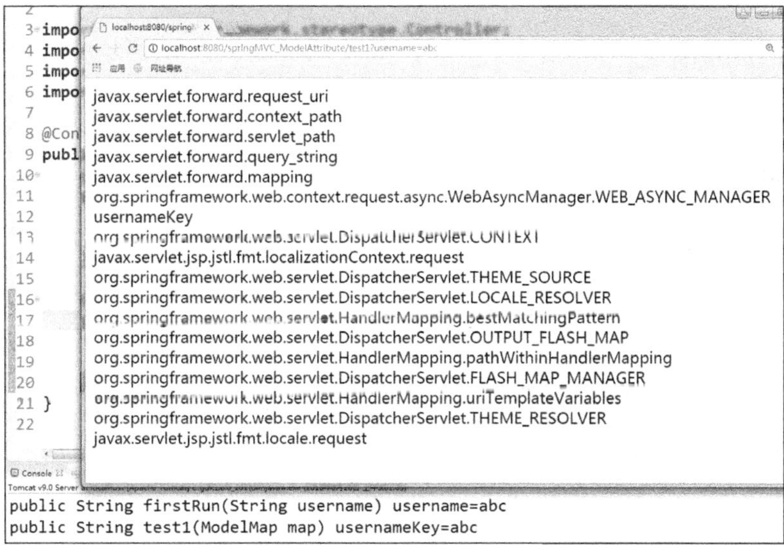

图 5-32 运行结果

2. 向模型中存储多个值

上一个实验的 ModelMap 中只能存储 1 个值，要存储多个值可查看下面的实验。

（1）创建控制层，代码如下：

```
@Controller
public class UserinfoController2 {
    @ModelAttribute
    public void firstRun(String username, String password, ModelMap map) {
        System.out.println("public String firstRun(String username, String password)");
        System.out.println("username=" + username);
        System.out.println("password=" + password);
        map.put("usernameKey", username);
        map.put("passwordKey", password);
    }

    @RequestMapping(value = "test2")
```

```
    public String test2(ModelMap map) {
        System.out.println("public String test2(ModelMap map)");
        System.out.println("usernameKey=" + map.get("usernameKey"));
        System.out.println("passwordKey=" + map.get("passwordKey"));
        return "index2.jsp";
    }
}
```

（2）JSP 文件 index2.jsp 代码如下：

```
<body>
    <%
        Enumeration enum1 = request.getAttributeNames();
        while (enum1.hasMoreElements()) {
            String key = "" + enum1.nextElement();
            out.print(key + "<br/>");
        }
    %>
</body>
```

（3）输入如下网址：

```
http://localhost:8080/springMVC_ModelAttribute/test2?username=abc&password=123
```

控制台和浏览器输出的内容如图 5-33 所示。

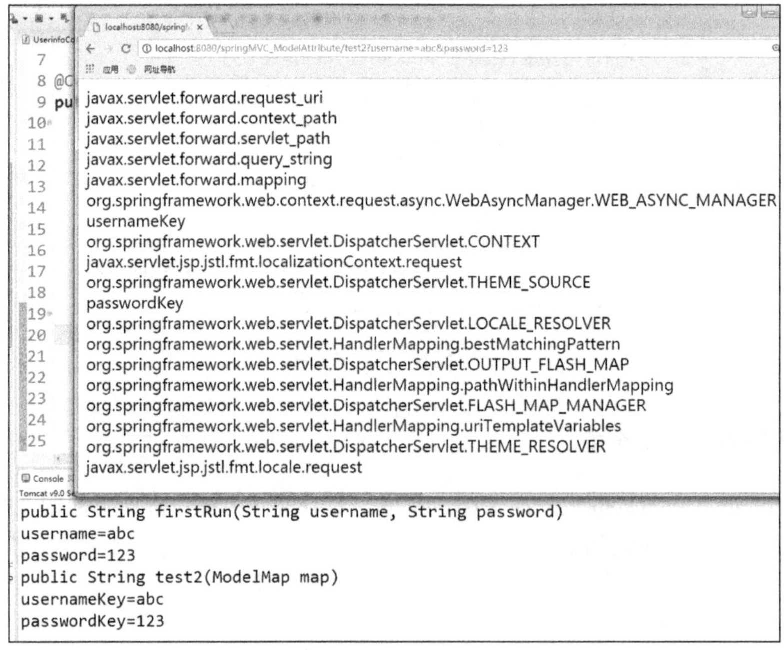

图 5-33　运行结果

3. 向模型中存储任意数据类型

（1）创建控制层，代码如下：

```
@Controller
public class UserinfoController3 {
```

```
    @ModelAttribute(name = "showList")
    public List getList(String username, String password) {
        System.out.println("模拟以 username 和 password 为查询条件取得 List");
        System.out.println("getList username=" + username + " password=" + password);
        List list = new ArrayList();
        list.add("中国 1");
        list.add("中国 2");
        list.add("中国 3");
        list.add("中国 4");
        return list;
    }

    @RequestMapping(value = "test3")
    public String test3(ModelMap map) {
        System.out.println("public String test3(ModelMap map)");
        System.out.println("showList size=" + ((List) map.get("showList")).size());
        return "index3.jsp";
    }
}
```

（2）JSP 文件 index2.jsp 代码如下：

```
<body>
    <%
        Enumeration enum1 = request.getAttributeNames();
        while (enum1.hasMoreElements()) {
            String key = "" + enum1.nextElement();
            out.print(key + "<br/>");
        }
    %>
</body>
```

（3）输入如下网址：

```
http://localhost:8080/springMVC_ModelAttribute/test3?username=abc&password=123
```

控制台和浏览器输出的内容如图 5-34 所示。

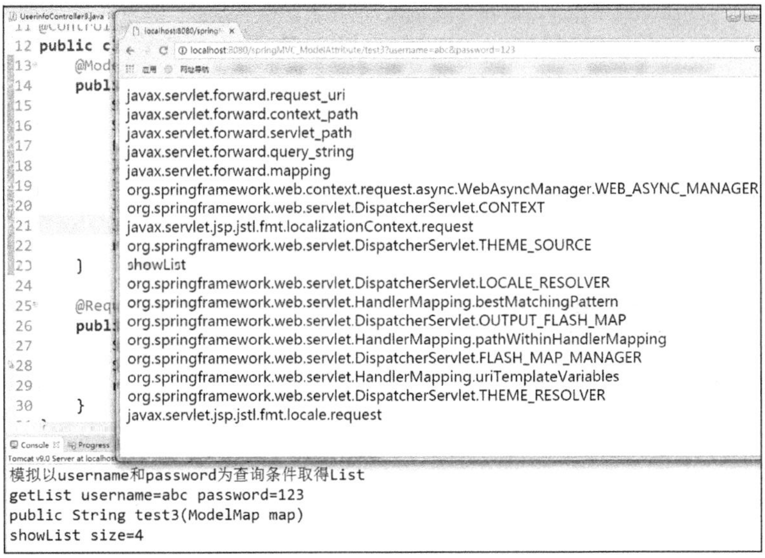

图 5-34　运行结果

4. 作用于控制层方法上

（1）创建控制层，代码如下：

```
@Controller
public class UserinfoController4 {
    @RequestMapping(value = "test4")
    @ModelAttribute(name = "showList")
    public List test4() {
        System.out.println("public String test4(ModelMap map)");
        List list = new ArrayList();
        list.add("中国 1");
        list.add("中国 2");
        list.add("中国 3");
        list.add("中国 4");
        return list;
    }
}
```

（2）配置文件代码如下：

```
<context:component-scan
    base-package="controller"></context:component-scan>
<bean
    class="org.springframework.web.servlet.view.InternalResourceViewResolver">
    <property name="prefix" value="/"></property>
    <property name="suffix" value=".jsp"></property>
</bean>
```

（3）JSP 文件 test4.jsp 代码如下：

```
<body>
    <%
        Enumeration enum1 = request.getAttributeNames();
        while (enum1.hasMoreElements()) {
            String key = "" + enum1.nextElement();
            out.print(key + "<br/>");
        }
    %>
</body>
```

（4）输入如下网址：

```
http://localhost:8080/springMVC_ModelAttribute/test4
```

浏览器输出的内容如图 5-35 所示。

5. 作用于控制层方法参数上

（1）创建控制层，代码如下：

```
@Controller
public class UserinfoController5 {
    @ModelAttribute
    public void firstRun(String username, String password, ModelMap map) {
        System.out.println("public String firstRun(String username, String password)");
        System.out.println("username=" + username);
        System.out.println("password=" + password);
```

```
            map.put("usernameKey", username);
            map.put("passwordKey", password);
        }

        @RequestMapping(value = "test5")
        public String test5(@ModelAttribute("usernameKey") String u, @ModelAttribute
("passwordKey") String p) {
            System.out.println("public String test5");
            System.out.println("u=" + u);
            System.out.println("p=" + p);
            return "index2.jsp";
        }
    }
```

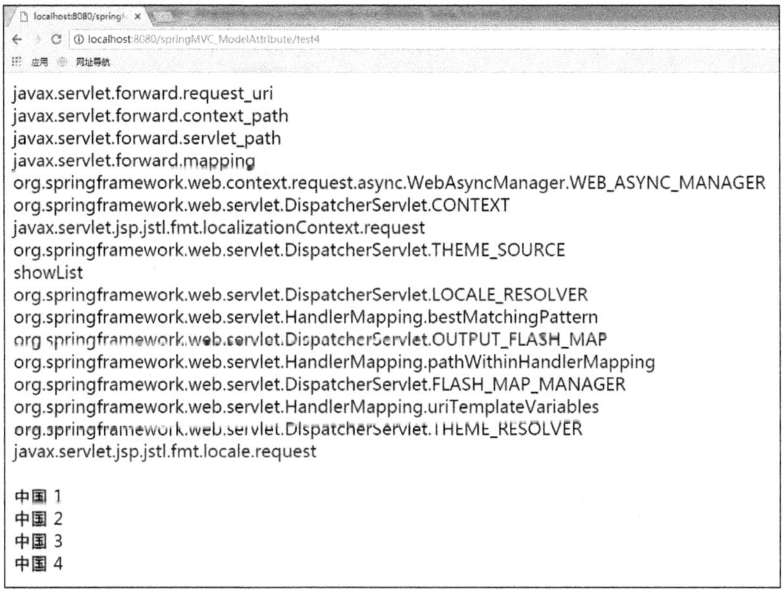

```
localhost:8080/springMVC_ModelAttribute/test4

javax.servlet.forward.request_uri
javax.servlet.forward.context_path
javax.servlet.forward.servlet_path
javax.servlet.forward.mapping
org.springframework.web.context.request.async.WebAsyncManager.WEB_ASYNC_MANAGER
org.springframework.web.servlet.DispatcherServlet.CONTEXT
javax.servlet.jsp.jstl.fmt.localizationContext.request
org.springframework.web.servlet.DispatcherServlet.THEME_SOURCE
showList
org.springframework.web.servlet.DispatcherServlet.LOCALE_RESOLVER
org.springframework.web.servlet.HandlerMapping.bestMatchingPattern
org.springframework.web.servlet.DispatcherServlet.OUTPUT_FLASH_MAP
org.springframework.web.servlet.HandlerMapping.pathWithinHandlerMapping
org.springframework.web.servlet.DispatcherServlet.FLASH_MAP_MANAGER
org.springframework.web.servlet.HandlerMapping.uriTemplateVariables
org.springframework.web.servlet.DispatcherServlet.THEME_RESOLVER
javax.servlet.jsp.jstl.fmt.locale.request

中国 1
中国 2
中国 3
中国 4
```

图 5-35 运行结果

（2）删除配置文件中的代码：

```
<context:component-scan
    base-package="controller"></context:component-scan>
<bean
    class="org.springframework.web.servlet.view.InternalResourceViewResolver">
    <property name="prefix" value="/"></property>
    <property name="suffix" value=".jsp"></property>
</bean>
```

（3）JSP 文件 test5.jsp 代码如下：

```
<body>
    <%
        Enumeration enum1 = request.getAttributeNames();
        while (enum1.hasMoreElements()) {
            String key = "" + enum1.nextElement();
            out.print(key + "<br/>");
        }
    %>
</body>
```

（4）输入如下网址：

```
http://localhost:8080/springMVC_ModelAttribute/test5?username=123&password=abc
```

控制台和浏览器输出的内容如图 5-36 所示。

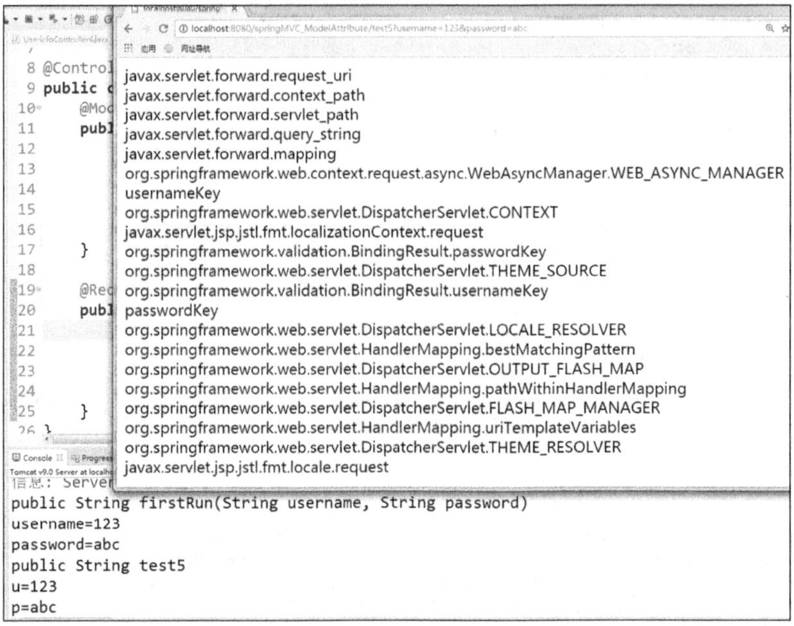

图 5-36 运行结果

5.4.15 在路径中添加通配符的功能

还可以在访问映射路径中添加通配符。

创建测试用的项目 springMVC_url_xing。

（1）控制层核心，代码如下：

```java
@Controller
public class UserinfoController {
    @RequestMapping(value = "findById_*")
    public String test(HttpServletRequest request, HttpServletResponse response) {
        String servletPath = request.getServletPath();
        servletPath = servletPath.substring(1);
        int beginIndex = servletPath.indexOf("_");
        servletPath = servletPath.substring(beginIndex + 1);
        System.out.println(servletPath);
        return "index.jsp";
    }
}
```

（2）可以通过如下的 URL 进行访问：

```
http://localhost:8080/springMVC_url_xing/findById_1
http://localhost:8080/springMVC_url_xing/findById_100
```

控制台输出的内容如下：

```
1
100
```

5.4.16 控制层返回 void 数据的情况

前面示例的控制层大多数返回的是 String 数据类型，代表转发到指定名称的 JSP 文件，控制层还可以返回 void 数据类型，存在两种情况。

（1）使用默认的 JSP。

（2）通过 HttpServletResponse 打印输出。

创建测试用的项目 springMVC_returnNull。

（1）创建控制层文件，核心代码如下：

```
@Controller
public class UserinfoController {
    @RequestMapping(value = "test")
    public void test() {
        System.out.println("test run !");
        // test.jsp
    }

    // 如果方法存在 request 和 response 参数
    // 不进行转发操作
    @RequestMapping(value = "testHasParam")
    public void testHasParam(HttpServletRequest request, HttpServletResponse response) {
        System.out.println("testHasParam run !");
    }

    @RequestMapping(value = "getUsername")
    public void getUsername(HttpServletRequest request, HttpServletResponse response) {
        try {
            response.setCharacterEncoding("utf-8");
            response.setContentType("text/html;charset=utf-8");
            PrintWriter out = response.getWriter();
            out.print("中国");
            out.flush();
            out.close();
        } catch (IOException e) {
            e.printStackTrace();
        }
    }
}
```

（2）文件 test.jsp 的核心代码如下：

```
<body>
    test.jsp page!
</body>
```

（3）更改配置文件，核心代码如下：

```
<context:component-scan
    base-package="controller"></context:component-scan>
<bean
    class="org.springframework.web.servlet.view.InternalResourceViewResolver">
    <property name="prefix" value="/"></property>
```

```
    <property name="suffix" value=".jsp"></property>
</bean>
```

（4）部署项目并运行程序，输入 URL：

```
http://localhost:8081/springMVC_returnNull/test
```

程序运行的结果如图 5-37 所示。

图 5-37 程序运行结果

（5）输入 URL：

```
http://localhost:8080/springMVC_returnNull/testHasParam
```

程序运行结果如图 5-38 所示。

（6）输入 URL：

```
http://localhost:8080/springMVC_returnNull/getUsername
```

程序运行的结果如图 5-39 所示。

图 5-38 程序运行结果

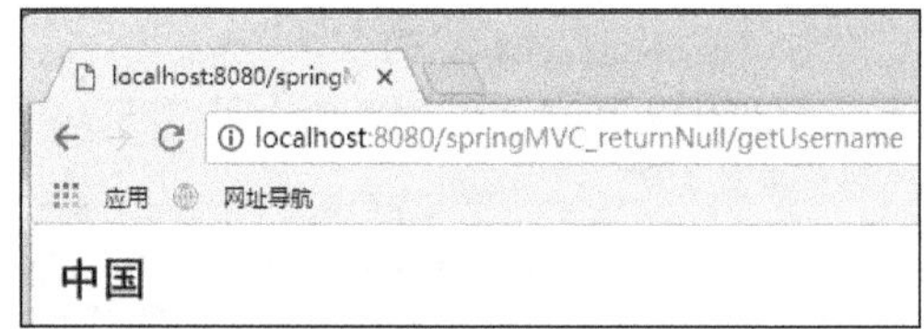

图 5-39 默认转发到 index.jsp 文件中

5.4.17 解决多人开发路径可能重复的问题

在开发 Java EE 项目时，分组开发、分工协作是软件公司常用的工作方式，这时就会出现一些问题，比如 A 开发前台登录功能，路径为 login；B 开发后台登录功能，路径也是 login。这种情况就会出现错误。

创建测试用的项目 springMVC_pathSame。

（1）A 开发的前台登录功能代码如下：

```
@Controller
public class UserinfoControllerA {
    @RequestMapping(value = "login")
    public String listStringMethod() {
        System.out.println("a login new");
        return "index.jsp";
    }
}
```

（2）B 开发的后台登录功能代码如下：

```
@Controller
public class UserinfoControllerB {
    @RequestMapping(value = "login")
    public String listStringMethod() {
        System.out.println("b login new");
        return "index.jsp";
    }
}
```

（3）在这个示例中需要扫描两个包，代码如下：

```
<context:component-scan base-package="a.controller"></context:component-scan>
<context:component-scan base-package="b.controller"></context:component-scan>
```

（4）把这个项目部署到 Tomcat 中，启动时已经出现异常，信息如下：

```
    Caused by: java.lang.IllegalStateException: Ambiguous mapping. Cannot map
'userinfoControllerB' method
    public java.lang.String b.controller.UserinfoControllerB.listStringMethod()
    to {[/login]}: There is already 'userinfoControllerA' bean method
    public java.lang.String a.controller.UserinfoControllerA.listStringMethod() mapped.
```

从出错提示可以看到，路径/login 已经被注册，不能重复注册。如果遇到这种情况，该怎么办呢？解决办法就是限定各模块的访问路径。

（5）更改 A 模块的控制层代码，代码如下：

```
@Controller
@RequestMapping("/a")
public class UserinfoControllerA {
    @RequestMapping(value = "login")
    public String listStringMethod() {
        System.out.println("a login new");
        return "../index.jsp";
    }
}
```

更改 B 模块的控制层代码，代码如下：

```
@Controller
@RequestMapping("/b")
public class UserinfoControllerB {
    @RequestMapping(value = "login")
    public String listStringMethod() {
        System.out.println("b login new");
        return "../index.jsp";
    }
}
```

如果在类的上方使用 @RequestMapping 注解，表示首先定义了相对的父路径，然后在方法上定义的路径是相对于类级别上的。

（6）重新启动 Tomcat 时并没有出现异常，并且在控制台（console）启动日志中不同的 login 被划分到不同模块的工作路径中：

```
    信息: Mapped "{[/a/login]}" onto public java.lang.String a.controller.UserinfoControllerA.
listStringMethod()
    org.springframework.web.servlet.handler.AbstractHandlerMethodMapping$MappingRegistry
register
    信息: Mapped "{[/b/login]}" onto public java.lang.String b.controller.UserinfoControllerB.
listStringMethod()
```

（7）在浏览器输入以下两个网址后，成功执行不同模块在相同控制层路径对应的方法：

```
http://localhost:8080/springMVC_pathSame/a/login
http://localhost:8080/springMVC_pathSame/b/login
```

（8）通过在类的上方加入 @RequestMapping（"/a"）注解，可以在 Spring MVC 中进行模块化开发：

```
@RequestMapping("/a")
public class UserinfoControllerA {
```

虽然现在的状态正确地实现了多人开发，但是控制层中的代码 “return "../index.jsp";” 感觉处理得不太美观，因为如果路径过多，那么 “../” 符号也会增多，如何解决呢？在 springMVC-servlet.xml 配置文件中加入如下代码即可：

```
<bean
    class="org.springframework.web.servlet.view.InternalResourceViewResolver"
    p:prefix="/" />
```

它的功能就是限定默认的访问资源路径是根路径（“/”），也就是相对于 WebContent 路径。将控制层代码改成如下形式：

```
return "index.jsp";
```

（9）再次运行这两个网址，也能在控制台中正确输出想要的字符串：

```
http://localhost:8080/springMVC_pathSame/a/login
http://localhost:8080/springMVC_pathSame/b/login
```

5.4.18　@PathVariable 注解的使用

Spring 5 MVC 框架还提供一种功能，非常类似于 Android 中的 Intent 技术，也就是将指定 URL 模式地址与指定访问 Controller 的路径进行关联与匹配。如果某一个访问 Controller（控制层）的 URL 与该 URL 模式相匹配，那么 Controller 即被调用。

创建测试用的项目 springMVC_PathVariable。

（1）控制层代码如下：

```
@Controller
public class UserinfoControllerA {
    @RequestMapping(value = "findUserinfo1/{userId}")
    public String findUserinfo1(@PathVariable("userId") String xxxxxx) {
        System.out.println(xxxxxx);
        return "index.jsp";
    }

    @RequestMapping(value = "findUserinfo2/{userId}")
    public String findUserinfo2(@PathVariable String userId) {
```

```
        System.out.println(userId);
        return "index.jsp";
    }

    @RequestMapping(value = "findUserinfo3/username/{username}/age/{age}")
    public String findUserinfo2(@PathVariable String username, @PathVariable String age) {
        System.out.println(username + " " + age);
        return "index.jsp";
    }
}
```

（2）控制层代码如下：

```
@Controller
@RequestMapping(value = "findUserinfo4/username/{username}")
public class UserinfoControllerB {
    @RequestMapping(value = "address/{address}")
    public String findUserinfo1(@PathVariable String username, @PathVariable String
address) {
        System.out.println(username + " " + address);
        return "index.jsp";
    }
}
```

（3）部署项目，输入如下网址：

```
http://localhost:8080/springMVC_PathVariable/findUserinfo1/123
```

控制台输出的结果如下：

```
123
```

（4）输入如下网址：

```
http://localhost:8080/springMVC_PathVariable/findUserinfo2/456
```

控制台输出的结果如下：

```
456
```

（5）输入如下网址：

```
http://localhost:8080/springMVC_PathVariable/findUserinfo3/username/abc/age/123
```

控制台输出的结果如下：

```
abc 123
```

（6）输入如下网址：

```
http://localhost:8080/springMVC_PathVariable/findUserinfo4/username/abc/address/bj
```

控制台输出的结果如下：

```
abc bj
```

5.4.19 通过 URL 参数访问指定的业务方法

可以实现在访问同一个 URL 地址的同时，以传递参数的方式来调用指定控制层中的指定方法，使用 @RequestMapping 注解很容易实现这样的需求。

创建测试用的项目 springMVC_urlParam_Method。

（1）控制层代码如下：

```
@Controller
public class UserinfoController {
    @RequestMapping(value = "listInfo", params = "type=A")
    public String listInfoAAA() {
        System.out.println("AAA");
        return "index.jsp";
    }

    @RequestMapping(value = "listInfo", params = "type=B")
    public String listInfoBBB() {
        System.out.println("BBB");
        return "index.jsp";
    }
}
```

（2）控制层中有两个业务方法，如何调用这两个业务方法呢？

可以使用如下 URL：

```
http://localhost:8080/springMVC_urlParam_Method/listInfo?type=A
http://localhost:8080/springMVC_urlParam_Method/listInfo?type=B
```

5.4.20 @RestController 注解的使用

使用 @RestController 注解相当于同时使用 @Controller 和 @ResponseBody 注解。

创建测试用的项目 springMVC_restController。

（1）控制层代码如下：

```
@RestController
public class UserinfoController {
    class Userinfo {
        private String username;
        private String password;

        public String getUsername() {
            return username;
        }

        public void setUsername(String username) {
            this.username = username;
        }

        public String getPassword() {
            return password;
        }

        public void setPassword(String password) {
            this.password = password;
        }
    }

    @RequestMapping(value = "test")
```

```
    public Userinfo printInfo() {
        Userinfo userinfo = new Userinfo();
        userinfo.setUsername("中国");
        userinfo.setPassword("中国人");
        return userinfo;
    }
}
```

（2）配置文件代码如下：

```
<context:component-scan
    base-package="controller"></context:component-scan>
<mvc:annotation-driven></mvc:annotation-driven>
```

（3）输入如下网址：

```
http://localhost:8080/springMVC_restController/test
```

浏览器输出的结果如图 5-40 所示。

图 5-40 运行结果

5.4.21 @GetMapping、@PostMapping、@PutMapping 和 @DeleteMapping 注解的使用

在 HTTP 中，除了 get 和 post 提交类型外，还可以使用常见的 put 和 delete，在 Spring 5 MVC 中分别使用 @GetMapping、@PostMapping、@PutMapping 和 @DeleteMapping 注解进行对应。

（1）注解 @GetMapping 的作用是查询。

（2）注解 @PostMapping 的作用是添加。

（3）注解 @PutMapping 的作用是更新。

（4）注解 @DeleteMapping 的作用是删除。

下面会使用两种方式来测试这 4 个注解的使用，即<form>（表单）和 AJAX。

1. 使用<form>

创建测试用的项目 springMVC_form_methodType。

（1）控制层代码如下：

```
@Controller
public class UserinfoController {
    @GetMapping(value = "get")
    public String get(String username) {
        System.out.println("get username=" + username);
        return "index.jsp";
```

```
    }

    @PostMapping(value = "post")
    public String post(String username) {
        System.out.println("post username=" + username);
        return "redirect:/get?username=getValue";
    }

    @PutMapping(value = "put")
    public String put(String username) {
        System.out.println("put username=" + username);
        return "redirect:/get?username=getValue";
    }

    @DeleteMapping(value = "delete")
    public String delete(String username) {
        System.out.println("delete username=" + username);
        return "redirect:/get?username=getValue";
    }
}
```

（2）JSP 文件代码如下：

```
<body>
    <form action="get" method="get">
        username:<input type="text" name="username" value="getValue">
        <br/>
        <input type="submit" value="get">
    </form>
    <form action="post" method="post">
        username:<input type="text" name="username" value="postValue">
        <br/>
        <input type="submit" value="post">
    </form>
    <form action="put" method="post">
        <input type="hidden" name="_method" value="put" />
        <br/>
        username:<input type="text" name="username" value="putValue">
        <br/>
        <input type="submit" value="put">
    </form>
    <form action="delete" method="post">
        <input type="hidden" name="_method" value="delete" />
        <br/>
        username:<input type="text" name="username" value="deleteValue">
        <br/>
        <input type="submit" value="delete">
    </form>
</body>
```

（3）在 web.xml 中配置 Filter（过滤器），代码如下：

```
<filter>
    <filter-name>HiddenHttpMethodFilter</filter-name>
    <filter-class>org.springframework.web.filter.HiddenHttpMethodFilter</filter-class>
</filter>
<filter-mapping>
    <filter-name>HiddenHttpMethodFilter</filter-name>
```

```
    <url-pattern>/*</url-pattern>
</filter-mapping>
```

（4）运行项目，执行 index.jsp 文件，按顺序单击 4 个按钮，控制台输出的结果如下：

```
get username=getValue
post username=postValue
get username=getValue
put username=putValue
get username=getValue
delete username=deleteValue
get username=getValue
```

2. 使用 AJAX

创建测试用的项目 springMVC_ajax_methodType。

（1）控制层代码如下：

```
@Controller
public class UserinfoController {
    @GetMapping(value = "get")
    @ResponseBody
    public String get(String username) {
        System.out.println("get username=" + username);
        return "getReturn";
    }

    @PostMapping(value = "post")
    @ResponseBody
    public String post(String username) {
        System.out.println("post username=" + username);
        return "postReturn";
    }

    @PutMapping(value = "put")
    @ResponseBody
    public String put(String username) {
        System.out.println("put username=" + username);
        return "putReturn";
    }

    @DeleteMapping(value = "delete")
    @ResponseBody
    public String delete(String username) {
        System.out.println("delete username=" + username);
        return "deleteReturn";
    }
}
```

（2）HTML 文件代码如下：

```
<!DOCTYPE HTML PUBLIC "-//W3C//DTD HTML 4.01 Transitional//EN">
<html>
    <head>
        <script src="jquery-1.4.2.js">
        </script>
```

```
    <script>
        function getMethod(){
            $.ajax({
                "type": "get",
                "url": "get?t=" + new Date().getTime(),
                "data": {
                    "username": "getValue"
                },
                success: function(data){
                    alert(data);
                }
            });
        }

        function postMethod(){
            $.ajax({
                "type": "post",
                "url": "post?t=" + new Date().getTime(),
                "data": {
                    "username": "postValue"
                },
                success: function(data){
                    alert(data);
                }
            });
        }

        function putMethod(){
            $.ajax({
                "type": "post",
                "url": "put?t=" + new Date().getTime(),
                "data": {
                    "_method": "put",
                    "username": "putValue"
                },
                success: function(data){
                    alert(data);
                }
            });
        }

        function deleteMethod(){
            $.ajax({
                "type": "post",
                "url": "delete?t=" + new Date().getTime(),
                "data": {
                    "_method": "delete",
                    "username": "deleteValue"
                },
                success: function(data){
                    alert(data);
                }
            });
        }
    </script>
</head>
<body>
```

```
        <input type="button" value="get" onclick="getMethod()"/>
        <br/>
        <input type="button" value="post" onclick="postMethod()"/>
        <br/>
        <input type="button" value="put" onclick="putMethod()"/>
        <br/>
        <input type="button" value="delete" onclick="deleteMethod()"/>
    </body>
</html>
```

（3）在 web.xml 中配置 Filter，代码如下：

```
<filter>
    <filter-name>HiddenHttpMethodFilter</filter-name>
    <filter-class>org.springframework.web.filter.HiddenHttpMethodFilter</filter-class>
</filter>
<filter-mapping>
    <filter-name>HiddenHttpMethodFilter</filter-name>
    <url-pattern>/*</url-pattern>
</filter-mapping>
```

（4）运行项目，执行 index.jsp 文件，按顺序单击 4 个按钮，控制台输出如下结果，而且在浏览器中成功接收从服务端返回的数据：

```
get username=getValue
post username=postValue
put username=putValue
delete username=deleteValue
```

5.4.22 Spring 5 MVC 与 Spring 5 的整合及应用 AOP 切面

前面已经将 MyBatis 3 与 Spring 5 进行了整合，本节将要对 Spring 5 MVC 与 Spring 5 进行整合，为第 6 章的学习做好准备。

整合 Spring 5 MVC 与 Spring 5 的目的是将 Spring 5 MVC 的控制层交由 Spring 5 框架处理，以对控制层进行有效管理，也方便对控制层应用切面。

创建 Java 项目 Spring5MVC_Spring5。

（1）添加 Spring 5 MVC 和 Spring 5 的基础 JAR 文件。

（2）添加 aspectj 相关的 JAR 文件。

（3）在 web.xml 文件中添加如下代码：

```
<servlet>
    <servlet-name>springMVC</servlet-name>
    <servlet-class>org.springframework.web.servlet.DispatcherServlet</servlet-class>
    <load-on-startup>1</load-on-startup>
</servlet>
<servlet-mapping>
    <servlet-name>springMVC</servlet-name>
    <url-pattern>/</url-pattern>
</servlet-mapping>

<listener>
    <listener-class>org.springframework.web.context.ContextLoaderListener</listener-class>
```

```
</listener>

<context-param>
    <param-name>contextConfigLocation</param-name>
    <param-value>\WEB-INF\classes\applicationContext.xml</param-value>
</context-param>
```

（4）创建 applicationContext.xml 配置文件，核心代码如下：

```
<context:component-scan base-package="service"></context:component-scan>
<context:component-scan base-package="myaspect"></context:component-scan>
```

（5）配置文件 springMVC-servlet.xml 的核心代码如下：

```
<context:component-scan
    base-package="controller"></context:component-scan>
<aop:aspectj-autoproxy></aop:aspectj-autoproxy>
```

（6）设计服务层，代码如下：

```
@Service
public class UserinfoService {
    public void printInfoService() {
        System.out.println("printInfoService run !");
    }
}
```

（7）设计切面类，代码如下：

```
@Component
@Aspect
public class MyAspect {

    @Around(value = "execution(* controller..*.*(..))")
    public Object myaround(ProceedingJoinPoint point) throws Throwable {
        System.out.println("begin " + System.currentTimeMillis());
        Object returnValue = point.proceed();
        System.out.println("  end " + System.currentTimeMillis());
        return returnValue;
    }
}
```

（8）设计控制层，代码如下：

```
@Controller
public class UserinfoController {

    @Autowired
    private UserinfoService userinfoService;

    public UserinfoService getUserinfoService() {
        return userinfoService;
    }

    public void setUserinfoService(UserinfoService userinfoService) {
        this.userinfoService = userinfoService;
    }
```

```
    @RequestMapping(value = "test")
    public String test() {
        userinfoService.printInfoService();
        return "test.jsp";
    }
}
```

（9）执行程序，在控制台输出的结果说明切面成功应用到控制层上，效果如下：

```
begin 1530780307941
printInfoService run !
  end 1530780307955
begin 1530780308278
printInfoService run !
  end 1530780308279
```

第 6 章　MyBatis 3、Spring 5 和 Spring 5 MVC 的整合

了解了 MyBatis 3、Spring 5 MVC 及 Spring 5 框架后，可以将它们进行整合。整合的目的是令事务自动地提交或回滚，从而生成 MyBatis 3 操作数据库接口的实现类，支持切面编程，使程序员在开发项目时写法更加统一，便于维护。

6.1　准备 MyBatis 3、Spring 5 和 Spring 5 MVC 框架的 JAR 包文件

下载 MyBatis 3、Spring 5 和 Spring 5 MVC 框架的 JAR 包文件。

6.2　准备 MyBatis 3 与 Spring 5 整合的插件

在 Spring 5 框架中，默认不支持与 MyBatis 3 进行直接整合，因此 MyBatis 官方发布了整合 MyBatis 与 Spring 的插件。此插件就是 1 个 JAR 文件 mybatis-spring-1.3.2.jar，通过此 JAR 文件即可将 MyBatis 3 与 Spring 5 进行整合。

6.3　创建 Web 项目

创建名称为 SpringMVC5_MyBatis3_Spring5 的 Web 项目，将所有涉及的 JAR 文件放入 Web 项目的 lib 文件夹中。注意，这里并不局限于前面介绍的 JAR 文件。

6.4　配置 web.xml 文件

web.xml 配置的核心代码如下：

```
<servlet>
    <servlet-name>springMVC</servlet-name>
    <servlet-class>org.springframework.web.servlet.DispatcherServlet</servlet-class>
    <load-on-startup>1</load-on-startup>
</servlet>
<servlet-mapping>
```

```
    <servlet-name>springMVC</servlet-name>
    <url-pattern>/</url-pattern>
</servlet-mapping>

<listener>
    <listener-class>org.springframework.web.context.ContextLoaderListener</listener-class>
</listener>

<context-param>
    <param-name>contextConfigLocation</param-name>
    <param-value>\WEB-INF\classes\applicationContext.xml</param-value>
</context-param>
```

6.5 配置 springMVC-servlet.xml 文件

文件 springMVC-servlet.xml 的核心代码如下：

```
<context:component-scan
    base-package="controller" />
<aop:aspectj-autoproxy proxy-target-class="true"></aop:aspectj-autoproxy>
```

使用<context:component-scan base-package="controller" />配置代码扫描 controller 包中带有@Controller 注解的控制层类。

使用<aop:aspectj-autoproxy proxy-target-class="true" />配置代码为控制层提供事务代理 AOP 支持。

6.6 MyBatis 配置文件

创建 MyBatis 框架的配置文件 mybatis-3-config.xml，代码如下：

```
<?xml version="1.0" encoding="UTF-8" ?>
<!DOCTYPE configuration PUBLIC "-//mybatis.org//DTD Config 3.0//EN" "mybatis-3-config.dtd">
<configuration>
</configuration>
```

因为连接数据库等信息是在 Spring 5 配置文件中进行定义的，所以此文件中的配置代码极少，但要保持 MyBatis 的一些默认行为，还是可以在此文件中进行声明定义的。

6.7 创建 MyBatis 映射的相关文件

SQL 映射文件 IUserinfoMapping.xml 的代码如下：

```
<mapper namespace="mapping.IUserinfoMapping">
    <insert id="save" parameterType="entity.Userinfo">
        <selectKey resultType="java.lang.Long" order="BEFORE"
            keyProperty="id">
            select idauto.nextval from dual
        </selectKey>
        insert into userinfo(id,username)
        values(#{id},#{username})
    </insert>
</mapper>
```

实体类 Userinfo 的代码如下：

```
public class Userinfo implements java.io.Serializable {

    private Long id;
    private String username;
    private String password;
    private Long age;
    private Timestamp insertdate;

    public Userinfo() {
    }

    public Userinfo(String username, String password, Long age, Timestamp insertdate) {
        this.username = username;
        this.password = password;
        this.age = age;
        this.insertdate = insertdate;
    }

    //get 和 set 方法省略

}
```

再创建 mapping 的映射接口 IUserinfoMapping，代码如下：

```
package mapping;

import entity.Userinfo;

public interface IUserinfoMapping {
    public void save(Userinfo userinfo);
}
```

接口仅仅是定义，并不能实现任何功能，此接口的实现类是由 Spring 5 框架生成的代理类。创建 AllMapping，代码如下：

```
@Component
public class AllMapping {
    @Autowired
    private IUserinfoMapping userinfoMapping;

    public IUserinfoMapping getUserinfoMapping() {
        return userinfoMapping;
    }

    public void setUserinfoMapping(IUserinfoMapping userinfoMapping) {
        this.userinfoMapping = userinfoMapping;
    }
}
```

6.8 配置 applicationContext.xml 文件

现在 Spring 5 MVC、Spring 5 及 MyBatis 3 的基础环境已经准备就绪，使用 applicationContext.xml 文件可以对这 3 个模块进行整合。applicationContext.xml 文件的核心配置代码如下：

```
<context:component-scan base-package="mapping"></context:component-scan>
<context:component-scan base-package="service"></context:component-scan>

<bean id="dataSource"
    class="org.springframework.jdbc.datasource.DriverManagerDataSource">
    <property name="url"
        value="jdbc:oracle:thin:@localhost:1521:orcl"></property>
    <property name="driverClassName"
        value="oracle.jdbc.OracleDriver"></property>
    <property name="username" value="y2"></property>
    <property name="password" value="123"></property>
</bean>

<bean id="sqlSessionFactory"
    class="org.mybatis.spring.SqlSessionFactoryBean">
    <property name="dataSource" ref="dataSource"></property>
</bean>

<bean class="org.mybatis.spring.mapper.MapperScannerConfigurer">
    <property name="basePackage" value="mapping"></property>
    <property name="sqlSessionFactoryBeanName"
        value="sqlSessionFactory"></property>
</bean>

<bean id="transactionManager"
    class="org.springframework.jdbc.datasource.DataSourceTransactionManager">
    <property name="dataSource" ref="dataSource"></property>
</bean>
<tx:annotation-driven
    transaction-manager="transactionManager" />
```

在上面的配置中使用以下代码对 service 服务层中的包进行扫描：

```
<context:component-scan base-package="service"></context:component-scan>
```

使用以下代码创建数据源 dataSource 对象：

```
<bean id="dataSource" class="org.springframework.jdbc.datasource.DriverManagerDataSource">
```

再将数据源 dataSource 对象注入 SqlSessionFactoryBean 对象中：

```
<bean id="sqlSessionFactory"
    class="org.mybatis.spring.SqlSessionFactoryBean">
    <property name="dataSource" ref="dataSource"></property>
</bean>
```

还要添加 dataSource 对象关联事务的功能，配置代码如下：

```
<bean id="transactionManager"
    class="org.springframework.jdbc.datasource.DataSourceTransactionManager">
    <property name="dataSource" ref="dataSource"></property>
</bean>
<tx:annotation-driven
    transaction-manager="transactionManager" />
```

将 sqlSessionFactory 对象注入 UserinfoMapper.java 接口的代理实现类中，这就需要告诉 Spring 这些映射接口的位置，使用如下代码定义：

```
<bean class="org.mybatis.spring.mapper.MapperScannerConfigurer">
    <property name="basePackage" value="mapping"></property>
```

```
    <property name="sqlSessionFactoryBeanName"
        value="sqlSessionFactory"></property>
</bean>
```

6.9 创建 Service 对象

创建 service 包并创建 UserinfoService 类，代码如下：

```
@Service
public class UserinfoService {

    @Autowired
    private AllMapping allMapping;

    public AllMapping getAllMapping() {
        return allMapping;
    }

    public void setAllMapping(AllMapping allMapping) {
        this.allMapping = allMapping;
    }

    public void saveServiceMethod() {
        Userinfo userinfo1 = new Userinfo();
        userinfo1.setUsername("中国");

        Userinfo userinfo2 = new Userinfo();
        userinfo2.setUsername("公司");

        allMapping.getUserinfoMapping().save(userinfo1);
        allMapping.getUserinfoMapping().save(userinfo2);
    }

}
```

创建 AllService.java 文件，代码如下：

```
@Service
public class AllService {

    @Autowired
    private UserinfoService userinfoService;

    public UserinfoService getUserinfoService() {
        return userinfoService;
    }

    public void setUserinfoService(UserinfoService userinfoService) {
        this.userinfoService = userinfoService;
    }

}
```

6.10 创建 Controller 对象

创建 Controller 对象以及控制层的 UserinfoController 类，代码如下：

```
@Transactional
```

```
@Controller
public class UserinfoController {
    @Autowired
    private AllService allService;

    public AllService getAllService() {
        return allService;
    }

    public void setAllService(AllService allService) {
        this.allService = allService;
    }

    @RequestMapping(value = "test")
    public String test() {
        this.getAllService().getUserinfoService().saveServiceMethod();
        return "index.jsp";
    }
}
```

至此，整合的步骤结束。

6.11 测试正常的效果

将项目部署到 Tomcat 中，执行控制层后，在控制台并没有输出异常，而在数据表 userinfo 中增加了两条记录。

6.12 测试回滚的效果

在 MyBatis 3、Spring 5 和 Spring 5 MVC 的整合过程中，因为考虑到操作数据库如果出现异常事务应该回滚，所以需要更改 UserinfoService 类中的代码：

```
public void saveServiceMethod() {
    Userinfo userinfo1 = new Userinfo();
    userinfo1.setUsername("中国");

    Userinfo userinfo2 = new Userinfo();
    userinfo2.setUsername("公司公司公司公司公司公司公司公司公司公司公司公司公司公司公司公司公司公司公司公司公司公司公司公司公司公司公司公司公司公司公司公司公司公司公司公司公司公司公司公司公司公司公司公司公司公司公司公司公司公司公司公司公司公司公司公司");

    allMapping.getUserinfoMapping().save(userinfo1);
    allMapping.getUserinfoMapping().save(userinfo2);
}
```

程序运行后，在控制台输出的异常信息如下：

```
java.sql.SQLException: ORA-12899: 列 "Y2"."USERINFO"."USERNAME" 的值太大 (实际值: 168, 最大值: 50)
```

数据表 userinfo 中并没有添加新的记录，说明回滚成功，也就是 MyBatis 3、Spring 5 和 Spring 5 MVC 的整合是成功的。

第 7 章　前沿技术 Spring Boot

本章主要介绍如何使用 Spring Boot 开发 Web 软件项目。

本章目的：

- 安装与配置 Maven
- 测试 Maven 环境是否正确
- 在 Eclipse 中关联 Maven 工具
- 模板引擎 Thymeleaf 的使用
- 创建并配置 Maven Web Project 或 Maven Project
- 使用 Maven 工具来搭建 Spring Boot 开发环境
- 整合 Thymeleaf 模板引擎
- 在 Spring Boot 中应用 IoC 和 AOP
- 整合 MyBatis 并处理事务

Spring Boot 框架是现在主流的 Java EE 开发框架，它简化了开发 Spring 项目的配置，还可以方便地与各种主流框架进行整合，Spring Boot 框架其实就是一个整合器。Spring Boot 可以集成很多的框架，本书篇幅有限，只将前面介绍的 Spring MVC 和 MyBatis 框架与 Spring Boot 进行整合。有关 Spring Boot 整合其他框架的内容可参考相关图书或官方的帮助文档。

7.1　搭建 Maven 开发环境

在开发 Spring Boot 项目之前需要创建 Maven 环境，可以借助 Maven 工具管理相关的资源，以方便项目资源的维护。

7.1.1　Maven 介绍

使用 Maven 时可以通过一小段配置信息来管理项目的构建和依赖等。

Maven 是优秀的构建工具，它支持自动化构建，其中包括从清理、编译、测试到生成报告，再到打包和部署的整个过程。

Maven 是跨平台的，无论是 Windows，还是 Linux 或 Mac 操作系统，都使用相同的命令。

Maven 可以帮助我们管理构建过程，但它又不仅仅是构建工具，还提供依赖管理工具，包括中央仓库和自动下载，以及下载 source 和 doc 的功能组件。

Maven 仓库就是放置所有 JAR 文件（WAR、ZIP 和 POM 等）的地方，所有 Maven 项目都可以从同一个 Maven 仓库中获取自己所需要的依赖 JAR 包文件。

Maven 的核心配置文件是 pom.xml 文件。pom 全称是（project object model），即项目对象模型。

在 Eclipse 中提供了支持 Maven 的插件 E2M，下载 E2M 插件就可以在 Eclipse 中使用 Maven 工具。最新版本的 Eclipse 均自带该插件，直接在 Eclipse 中使用即可。

7.1.2 搭建 Maven 环境

本节的主要目的就是搭建 Maven 环境。

1. 下载 Maven

进入 Maven 官网下载 Maven 工具。

单击图 7-1 中的“Download”链接下载 Maven 框架。

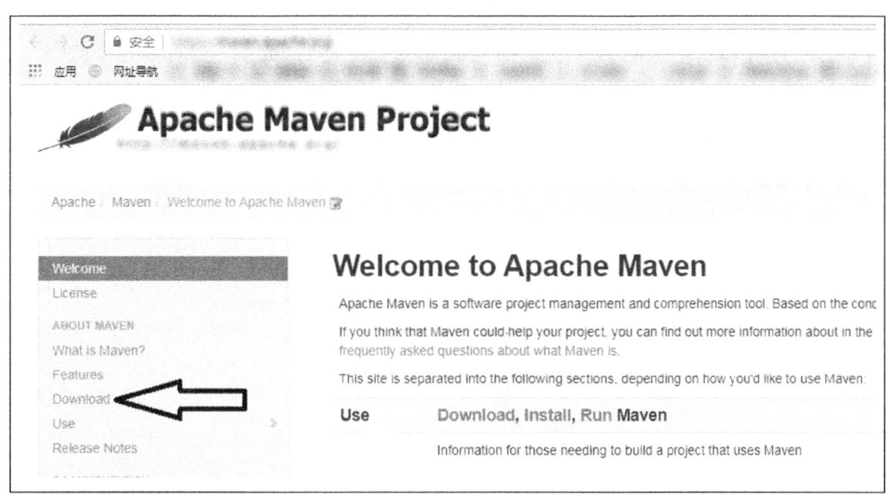

图 7-1 单击“Download”下载链接

下载成功后解压缩到硬盘。

2. 配置 Maven

在系统变量 Path 中添加路径：

```
C:\maven-3.5.4\bin;
```

效果如图 7-2 所示。

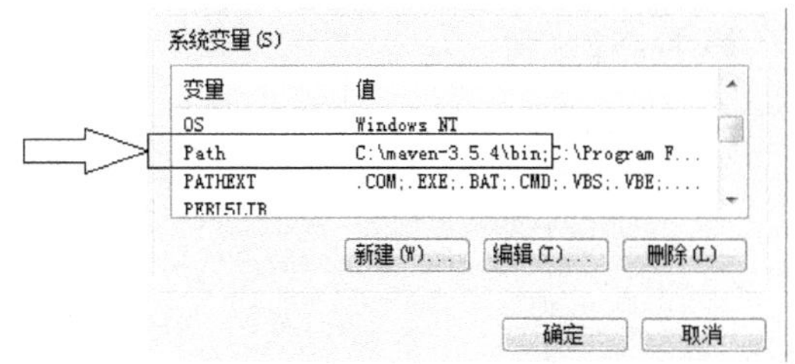

图 7-2 在系统变量 Path 中配置路径

3. 测试 Maven 环境

在 CMD 中输入命令：

```
mvn -v
```

控制台输出的结果如图 7-3 所示。

```
C:\Users\gaohongyan>mvn -v
Apache Maven 3.5.4 (1edded0938998edf8bf061f1ceb3cfdeccf443fe;
Maven home: C:\maven-3.5.4\bin\..
Java version: 1.8.0_161, vendor: Oracle Corporation, runtime: C:\jdk1.8.0_161\jre
Default locale: zh_CN, platform encoding: GBK
OS name: "windows 7", version: "6.1", arch: "amd64", family: "windows"

C:\Users\gaohongyan>
```

图 7-3 Maven 环境成功搭建

说明 Maven 环境搭建成功。

4. 配置 Maven 本地仓库路径

“本地仓库”是指存在于当前计算机的仓库，在加入依赖时，首先会去本地仓库里进行查找，如果找不到，就会到远程仓库中查找。“本地仓库”用来存储所有项目中具有依赖关系的文件，当建立一个 Maven 项目时，所有相关的文件将被存储在 Maven 本地仓库中。

“远程仓库”是指其他服务器上的仓库，包括全球中央仓库、公司内部的私服，或者其他公司提供的公共库。

在默认情况下，Maven 本地仓库的路径在哪里呢？打开 C:\maven-3.5.4\conf\settings.xml 文件可以看到，效果如图 7-4 所示。

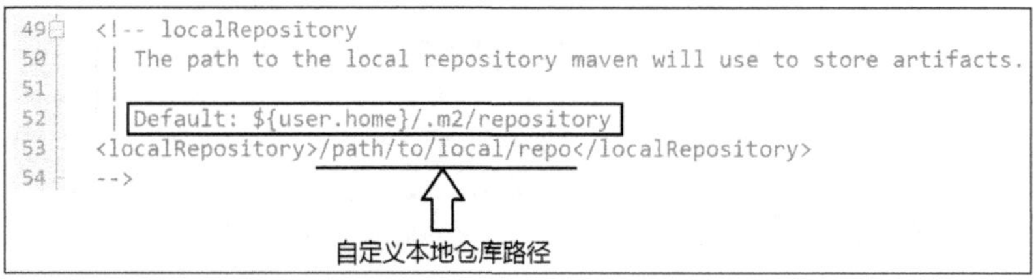

```
<!-- localRepository
 | The path to the local repository maven will use to store artifacts.
 |
 | Default: ${user.home}/.m2/repository
<localRepository>/path/to/local/repo</localRepository>
-->
```

自定义本地仓库路径

图 7-4 本地仓库路径

在 settings.xml 配置文件中已经说明，默认路径在“${user.home}/.m2/repository”中，变量${user.home}可以使用代码来获得，创建运行类，代码如下：

```
package test;

public class Test {
    public static void main(String[] args) {
        String path = System.getProperty("user.home");
        System.out.println(path);
    }
}
```

控制台输出的结果如下：

```
C:\Users\gaohongyan
```

说明本地仓库在默认的情况下被存储在路径“C:\Users\gaohongyan\.m2\repository”中。此路径可以更改。更改 settings.xml 配置文件，代码如下：

```
<localRepository>C:\mvnrepository</localRepository>
```

本地仓库路径被设置在 C:\mvnrepository 文件夹中。

5. 配置 Maven 镜像

Maven 官方仓库默认在国外，由于网络环境等因素，快速下载依赖 JAR 包不太方便，因此可以使用国内阿里云提供的 Maven 仓库的镜像，在 settings.xml 配置文件中添加如下代码：

```
<mirrors>
    <mirror>
        <id>aliyun</id>
        <name>aliyun Maven</name>
        <mirrorOf>central</mirrorOf>
        <url>阿里云的Maven 仓库镜像的网址</url>
    </mirror>
</mirrors>
```

配置代码如下：

```
<id>aliyun</id>
```

代表镜像库在当前 settings.xml 配置文件中的 ID 值，它是镜像库的唯一标识。值可以是任意的，但要有意义。

配置代码如下：

```
<name>aliyun Maven</name>
```

代表镜像库的名称。值可以是任意的，但要有意义。

配置代码如下：

```
<mirrorOf>central</mirrorOf>
```

代表该配置为 Maven 中央仓库的镜像，任何对于 Maven 中央仓库的请求都会被转至该镜像。

配置代码如下：

```
<url>阿里云的Maven 仓库镜像的网址</url>
```

代表镜像库的 URL。

7.1.3 在 Eclipse 中关联 Maven

下载 Eclipse，解压缩后启动 Eclipse。

单击“Preferences”菜单项，如图 7-5 所示。

进入如图 7-6 所示的“Preferences”窗口，然后在“Maven”→“Installations”中，添加最新版的 Maven。

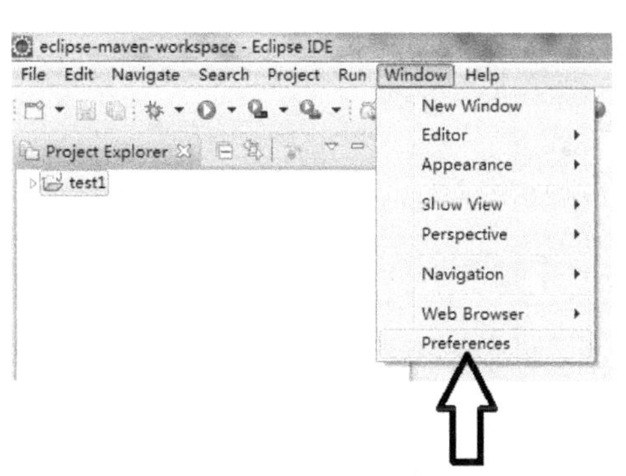

图 7-5 单击“Preferences”菜单项

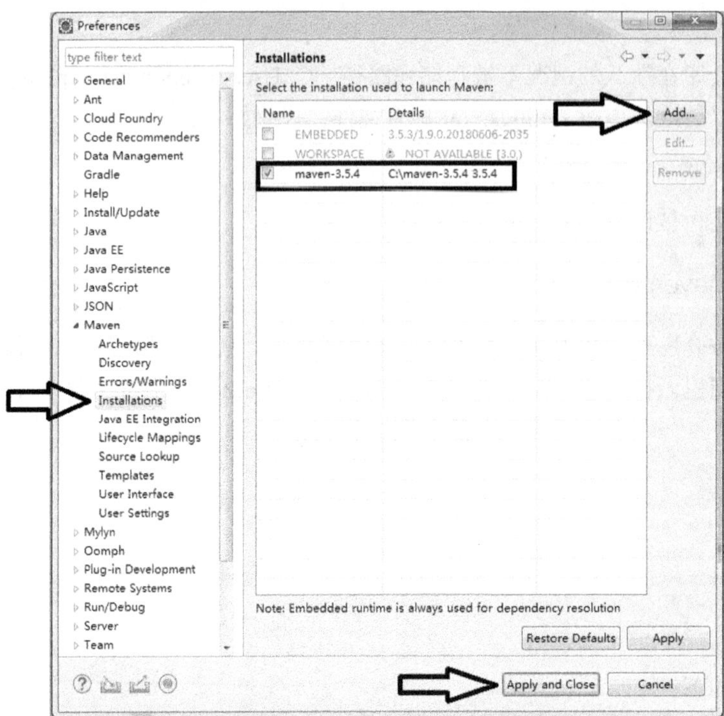

图 7-6　单击“Add”按钮添加最新版的 Maven 框架

进入“User Settings”界面，设置如图 7-7 所示。

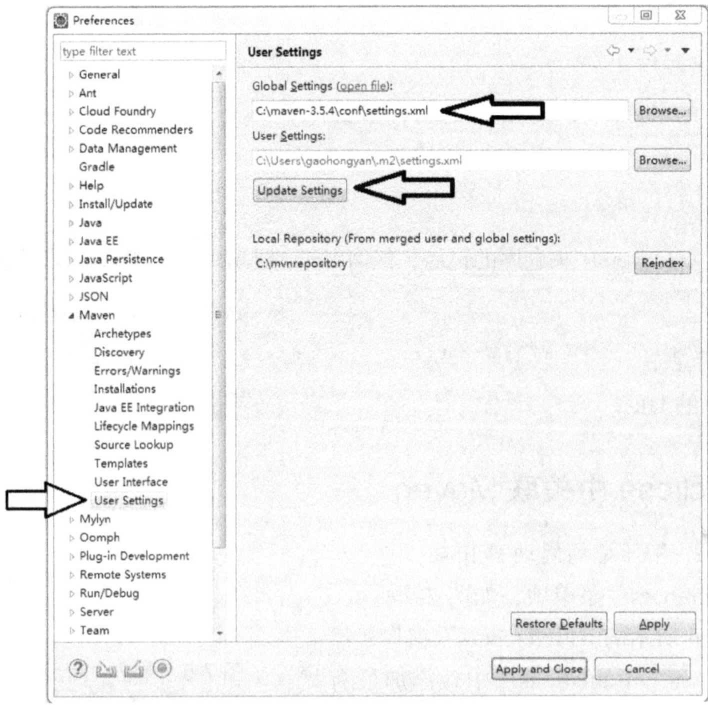

图 7-7　关联 settings.xml 配置文件

单击“Update Settings”按钮更新设置，然后单击“Apply and Close”按钮关闭设置界面。

单击“Other…”菜单项，如图 7-8 所示。

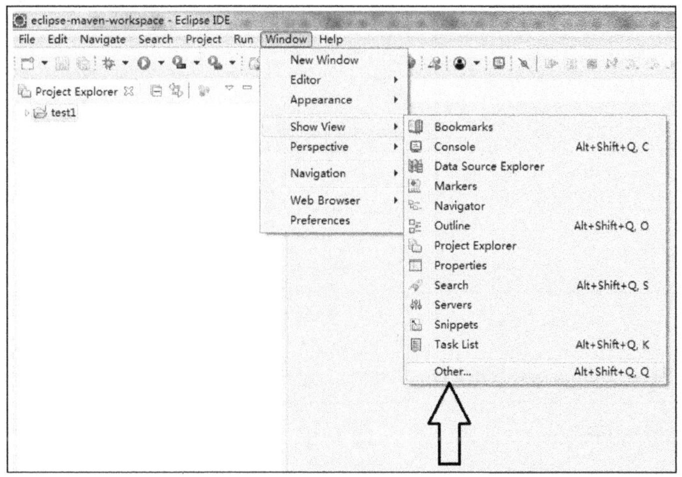

图 7-8 单击“Other…”菜单项

单击“Maven Repositories”选项，如图 7-9 所示。

弹出的“Maven Repositories”选项面板如图 7-10 所示。

成功创建“阿里云”远程仓库的镜像连接。

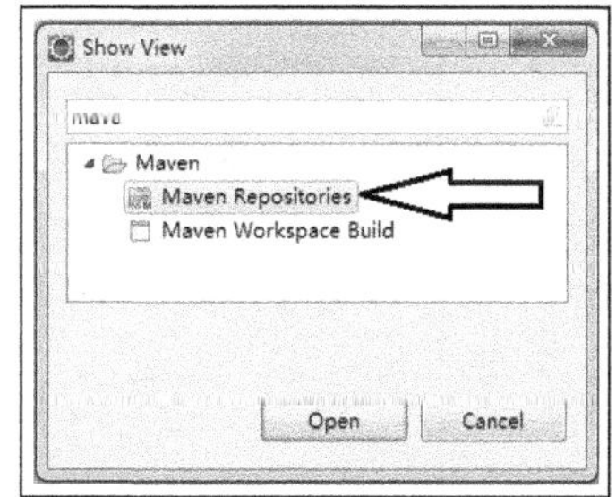

图 7-9 单击“Maven Repositories”选项

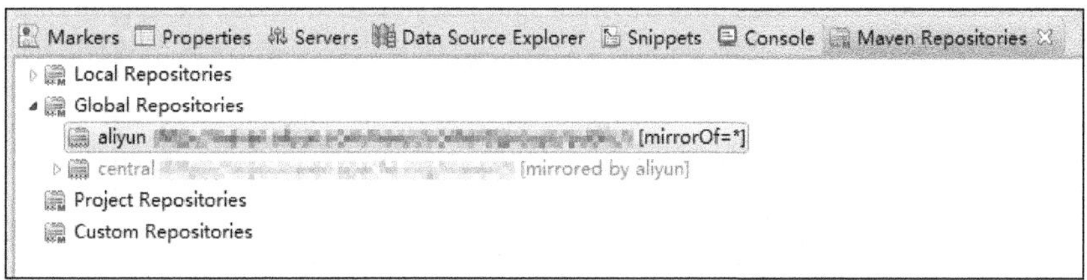

图 7-10 “Maven Repositories”选项面板

7.1.4 创建 Maven 项目

创建“Maven Project”项目，如图 7-11 所示。

“New Maven Project”的界面设置如图 7-12 所示。

单击“Next”按钮出现如图 7-13 所示的界面。

其中“Group Id”常常代表公司的名称或域名，“Artifact Id”常常代表项目名称。在一个“Group Id”中，常常有不同的“Artifact Id”，也就是 1 个公司或 1 个域名下有多个子项目。

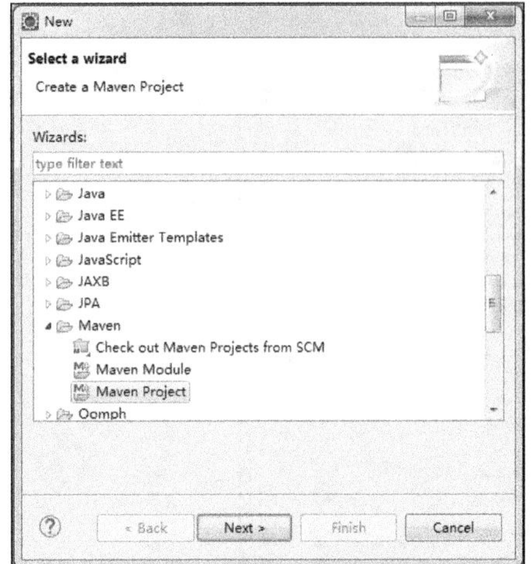

图 7-11　创建“Maven Project”项目

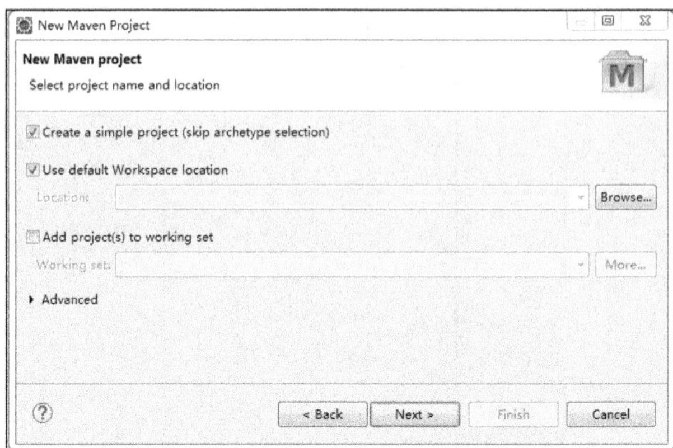

图 7-12　“Select project name and location”界面

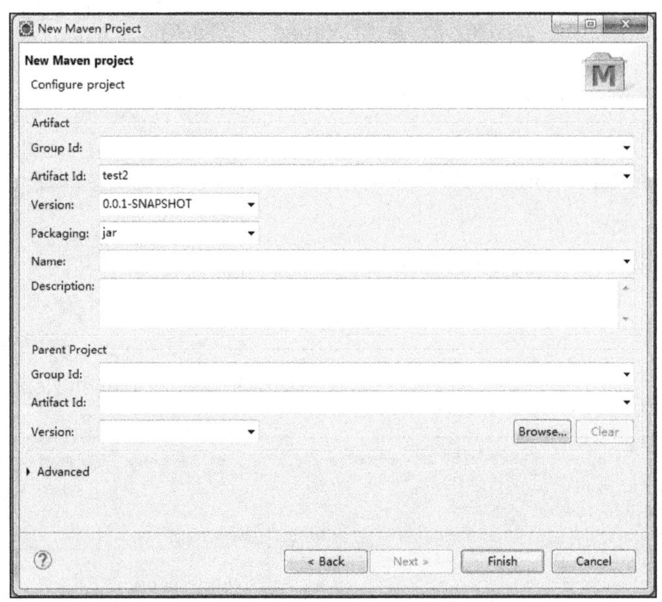

图 7-13　“Configure project”界面

单击“Finish”按钮完成项目的创建，并在 C:\mvnrepository 中下载相关的依赖 JAR 包资源，效果如图 7-14 所示。

图 7-14　下载相关的资源

下载成功后更改编译级别，效果如图 7-15 所示。

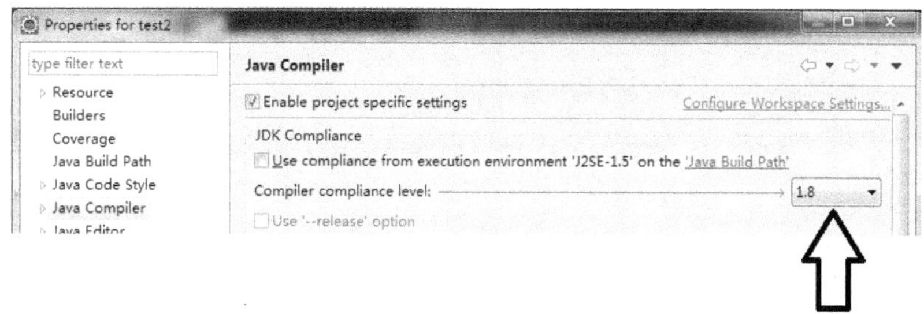

图 7-15　设置编译级别

7.1.5　使用 Maven 工具下载 Spring 框架（JAR 包、源代码和帮助文档）

Spring 框架的常用模块如图 7-16 所示。

```
spring-aop          spring-context-indexer  spring-instrument  spring-orm   spring-webflux
spring-aspects      spring-context-support  spring-jcl         spring-oxm   spring-webmvc
spring-beans        spring-core             spring-jdbc        spring-test  spring-websocket
spring-expression   spring-jms              spring-tx
spring-context      spring-messaging        spring-web
```

图 7-16　Spring 框架常用模块

根据模块列表中的条目信息，更改 test2 项目中的 pom.xml 文件，内容如下：

```
<project xmlns="http://maven.apache.org/POM/4.0.0"
    xmlns:xsi="http://www.w3.org/2001/XMLSchema-instance"
    xsi:schemaLocation="http://maven.apache.org/POM/4.0.0 http://maven.apache.org/xsd
/maven-4.0.0.xsd">
    <modelVersion>4.0.0</modelVersion>
    <groupId>公司名或域名</groupId>
    <artifactId>test2</artifactId>
    <version>0.0.1-SNAPSHOT</version>

    <properties>
        <org.springframework.version>5.0.7.RELEASE</org.springframework.version>
    </properties>

    <dependencies>
        <dependency>
            <groupId>org.springframework</groupId>
            <artifactId>spring-core</artifactId>
            <version>${org.springframework.version}</version>
        </dependency>
        <dependency>
            <groupId>org.springframework</groupId>
            <artifactId>spring-aop</artifactId>
            <version>${org.springframework.version}</version>
        </dependency>
        <dependency>
            <groupId>org.springframework</groupId>
            <artifactId>spring-context-indexer</artifactId>
            <version>${org.springframework.version}</version>
        </dependency>
        <dependency>
            <groupId>org.springframework</groupId>
            <artifactId>spring-instrument</artifactId>
            <version>${org.springframework.version}</version>
```

```
</dependency>
<dependency>
    <groupId>org.springframework</groupId>
    <artifactId>spring-orm</artifactId>
    <version>${org.springframework.version}</version>
</dependency>
<dependency>
    <groupId>org.springframework</groupId>
    <artifactId>spring-webflux</artifactId>
    <version>${org.springframework.version}</version>
</dependency>
<dependency>
    <groupId>org.springframework</groupId>
    <artifactId>spring-aspects</artifactId>
    <version>${org.springframework.version}</version>
</dependency>
<dependency>
    <groupId>org.springframework</groupId>
    <artifactId>spring-context-support</artifactId>
    <version>${org.springframework.version}</version>
</dependency>
<dependency>
    <groupId>org.springframework</groupId>
    <artifactId>spring-jcl</artifactId>
    <version>${org.springframework.version}</version>
</dependency>
<dependency>
    <groupId>org.springframework</groupId>
    <artifactId>spring-oxm</artifactId>
    <version>${org.springframework.version}</version>
</dependency>
<dependency>
    <groupId>org.springframework</groupId>
    <artifactId>spring-webmvc</artifactId>
    <version>${org.springframework.version}</version>
</dependency>
<dependency>
    <groupId>org.springframework</groupId>
    <artifactId>spring-jdbc</artifactId>
    <version>${org.springframework.version}</version>
</dependency>
<dependency>
    <groupId>org.springframework</groupId>
    <artifactId>spring-test</artifactId>
    <version>${org.springframework.version}</version>
</dependency>
<dependency>
    <groupId>org.springframework</groupId>
    <artifactId>spring-websocket</artifactId>
    <version>${org.springframework.version}</version>
</dependency>
<dependency>
    <groupId>org.springframework</groupId>
    <artifactId>spring-beans</artifactId>
    <version>${org.springframework.version}</version>
</dependency>
<dependency>
    <groupId>org.springframework</groupId>
    <artifactId>spring-expression</artifactId>
    <version>${org.springframework.version}</version>
```

```
        </dependency>
        <dependency>
            <groupId>org.springframework</groupId>
            <artifactId>spring-jms</artifactId>
            <version>${org.springframework.version}</version>
        </dependency>
        <dependency>
            <groupId>org.springframework</groupId>
            <artifactId>spring-tx</artifactId>
            <version>${org.springframework.version}</version>
        </dependency>
        <dependency>
            <groupId>org.springframework</groupId>
            <artifactId>spring-context</artifactId>
            <version>${org.springframework.version}</version>
        </dependency>
        <dependency>
            <groupId>org.springframework</groupId>
            <artifactId>spring-messaging</artifactId>
            <version>${org.springframework.version}</version>
        </dependency>
        <dependency>
            <groupId>org.springframework</groupId>
            <artifactId>spring-web</artifactId>
            <version>${org.springframework.version}</version>
        </dependency>
    </dependencies>
</project>
```

保存 pom.xml 文件时 Eclipse 自动下载依赖 JAR 包，下载 JAR 包成功后的 test2 项目结构如图 7-17 所示。

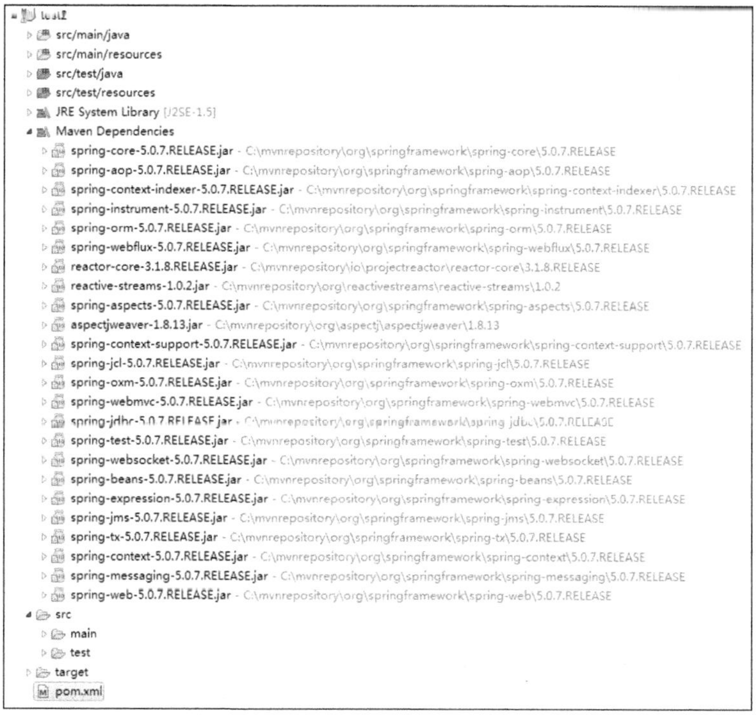

图 7-17　成功下载 JAR 包文件

Eclipse 会自动下载与 JAR 包有关的 source 和 javadoc。

至此，成功在 Eclipse 中创建了 Maven Project 项目。

7.1.6　向仓库中添加自定义的 JAR 包

在 CMD 中输入如下命令，向本地仓库中添加 JAR 文件：

```
mvn install:install-file -Dfile=UserinfoService.jar -DgroupId=com.ghy.www -DartifactId=UserinfoService -Dversion=1.0 -Dpackaging=jar
```

在 pom.xml 文件中通过如下配置代码引用即可：

```
<dependency>
    <groupId>com.ghy.www</groupId>
    <artifactId>UserinfoService</artifactId>
    <version>1.0</version>
</dependency>
```

7.1.7　查看依赖关系

在 Eclipse 中使用 Maven 插件可以通过图形的方式查看 pom.xml 文件的依赖关系，效果如图 7-18 所示。

图 7-18　查看依赖关系

左侧“Dependency Hierarchy”栏代表当前 pom.xml 文件依赖的资源，单击右侧“Resolved Dependencies”栏中的 JAR 包文件后，在左侧“Dependency Hierarchy”栏中显示哪些资源引用

了这个 JAR 包。

7.2 使用 Thymeleaf 模板引擎

Thymeleaf 是一款用于渲染 XML/XHTML/HTML5 内容的模板引擎，它类似于 JSP、Velocity 和 FreeMarker 等，可以轻易地与 Spring MVC 等 Web 框架进行集成以作为 Web 应用的模板引擎。与其他模板引擎相比，Thymeleaf 的特点是能够直接在浏览器中打开并正确显示模板页面，而不需要启动整个 Web 应用，方便原型设计。

在进行实验之前，首先要创建测试用的 Web 项目 thymeleafTest，并创建实体类，代码如下：

```
public class Userinfo {
    private String username;
    private long age;

    public Userinfo() {
    }

    public Userinfo(String username, long age) {
        super();
        this.username = username;
        this.age = age;
    }

    //忽略 set 和 get 方法

}
```

创建嵌套类型的实体类，如图 7-19 所示。

```
A.java
package entity;

public class A {
    private B b = new B();

    public B getB() {
        return b;
    }

    public void setB(B b) {
        this.b = b;
    }
}
```

```
B.java
package entity;

public class B {
    private C c = new C();

    public C getC() {
        return c;
    }

    public void setC(C c) {
        this.c = c;
    }
}
```

```
C.java
package entity;

public class C {
    private String username = "我是C类中的username属性值";

    public String getUsername() {
        return username;
    }

    public void setUsername(String username) {
        this.username = username;
    }

}
```

图 7-19 嵌套形式的实体类

7.2.1 常见的使用方式

示例代码如下：

```
public class test110 extends HttpServlet {
    protected void doGet(HttpServletRequest request, HttpServletResponse response)
            throws ServletException, IOException {
        ServletContextTemplateResolver templateResolver = new ServletContextTemplateResolver(
                request.getServletContext());
        templateResolver.setTemplateMode(TemplateMode.HTML);
```

```
            templateResolver.setPrefix("/");
            templateResolver.setCharacterEncoding("utf-8");

            TemplateEngine templateEngine = new TemplateEngine();
            templateEngine.setTemplateResolver(templateResolver);

            response.setContentType("text/html;charset=UTF-8");
            response.setHeader("Pragma", "no-cache");
            response.setHeader("Cache-Control", "no-cache");
            response.setDateHeader("Expires", 0);

            //Pragma:no-cache 与 Cache-Control:no-cache 作用相同
            //Pragma:no-cache 兼容 HTTP 1.0
            //Cache-Control:no-cache 是 HTTP 1.1 提供的
            //Pragma:no-cache 可以应用到 HTTP 1.0 和 HTTP 1.1 中
            //而 Cache-Control:no-cache 只能应用于 HTTP 1.1 中
            //Expires 设置过期时间

            WebContext webContext = new WebContext(request, response, request.getServletContext(),
request.getLocale());
            // （1）简单数据类型
            webContext.setVariable("firstShowMessage", "第\\\\\"一次显示消息，成功了！");
            （1）显示字符串
            <br/>
            [[${firstShowMessage}]]
            <br/>
            <span th:text="${firstShowMessage}"></span>
            <br/>
            <br/>
            <span th:text="我是普通的文本"></span>
            <br/>
            <span th:text="${'我是普通的文本'}"></span>
            <br/>
            <br/>
            <span th:text="'我\\是&quot;普通的文本'"></span><!--想显示特殊符号就要加单引号-->
            <br/>
            <span th:text="${'我\\是&quot;普通的文本'}"></span>

            // （2）复杂数据类型
            webContext.setVariable("nowDate", new Date());
            （2）显示日期
            <br/>
            [[${nowDate}]]
            <br/>
            <span th:text="${nowDate}"></span>

            // （3）自定义数据类型
            webContext.setVariable("userinfo", new Userinfo("中国", 123L));
            （3）显示实体类属性值
            <br/>
            [[${userinfo.username}]]
            <br/>
            [[${userinfo.age}]]
            <br/>
            <br/>
            [[${userinfo['username']}]]
            <br/>
            [[${userinfo['age']}]]
```

```
<br/>
<br/>
[[${userinfo.getUsername()}]]
<br/>
[[${userinfo.getAge()}]]
<br/>
<br/>
<th:block th:object="${userinfo}">
    [[*{username}]]___[[*{age}]]
</th:block>
<br/>
<th:block th:object="${userinfo}">
    [[*{#object.username}]]___[[*{#object.age}]]
</th:block>
<br/>
<th:block th:object="${userinfo}">
[[${#object.username}]]___[[${#object.age}]]
</th:block>
<br/>
<span th:text="${userinfo.username}"></span>
<br/>
<span th:text="${userinfo.age}"></span>
<br/>
<br/>
<span th:text="${userinfo['username']}"></span>
<br/>
<span th:text="${userinfo['age']}"></span>
<br/>
<br/>
<span th:text="${userinfo.getUsername()}"></span>
<br/>
<span th:text="${userinfo.getAge()}"></span>
<br/>
<br/>
<span th:object="${userinfo}" th:text="*{username}"></span>
<br/>
<span th:object="${userinfo}" th:text="*{age}"></span>
<br/>
<br/>
<span th:object="${userinfo}" th:text="*{#object.username}"></span>
<br/>
<span th:object="${userinfo}" th:text="*{#object.age}"></span>
<br/>
<br/>
<span th:object="${userinfo}" th:text="${#object.username}"></span>
<br/>
<span th:object="${userinfo}" th:text="${#object.age}"></span>
```

```
// （4）自定义嵌套数据类型
webContext.setVariable("a", new A());
```

```
（4）显示嵌套结构的属性值
<br/>
[[${a.b.c.username}]]
<br/>
[[${a['b']['c']['username']}]]
<br/>
<span th:text="${a.b.c.username}"></span>
<br/>
```

```
<span th:text="${a['b']['c']['username']}"></span>

// （5）数组
String[] stringArray = new String[] { "a", "b", "c" };
Userinfo[] userinfoArray = new Userinfo[] { new Userinfo("账号1", 1), new Userinfo
("账号2", 2),
        new Userinfo("账号3", 3) };
webContext.setVariable("stringArray", stringArray);
webContext.setVariable("userinfoArray", userinfoArray);
（5）显示String[]和Userinfo[]中的数据
<br/>
[[${stringArray[0]}]]
<br/>
[[${stringArray[1]}]]
<br/>
[[${stringArray[2]}]]
<br/>
<br/>
<span th:text="${stringArray[0]}"></span>
<br/>
<span th:text="${stringArray[1]}"></span>
<br/>
<span th:text="${stringArray[2]}"></span>
<br/>
<br/>
[[${userinfoArray[0].username}]]
<br/>
[[${userinfoArray[1].username}]]
<br/>
[[${userinfoArray[2].username}]]
<br/>
<br/>
<span th:text="${userinfoArray[0].username}"></span>
<br/>
<span th:text="${userinfoArray[1].username}"></span>
<br/>
<span th:text="${userinfoArray[2].username}"></span>

// （6）List
ArrayList listString = new ArrayList();
listString.add("listString1");
listString.add("listString2");
listString.add("listString3");
ArrayList listUserinfo = new ArrayList();
listUserinfo.add(new Userinfo("list Userinfo1", 1));
listUserinfo.add(new Userinfo("list Userinfo2", 2));
listUserinfo.add(new Userinfo("list Userinfo3", 3));
webContext.setVariable("listString", listString);
webContext.setVariable("listUserinfo", listUserinfo);
（6）显示List&lt;String&gt;和List&lt;Userinfo&gt;中的数据
<br/>
[[${listString[0]}]]
<br/>
[[${listString[1]}]]
<br/>
[[${listString[2]}]]
<br/>
<br/>
```

```
<span th:text="${listString[0]}"></span>
<br/>
<span th:text="${listString[1]}"></span>
<br/>
<span th:text="${listString[2]}"></span>
<br/>
<br/>
[[${listUserinfo[0].username}]]
<br/>
[[${listUserinfo[1].username}]]
<br/>
[[${listUserinfo[2].username}]]
<br/>
<br/>
<span th:text="${listUserinfo[0].username}"></span>
<br/>
<span th:text="${listUserinfo[1].username}"></span>
<br/>
<span th:text="${listUserinfo[2].username}"></span>

// (7) Map
Map mapString = new LinkedHashMap();
mapString.put("key1", "mapString1");
mapString.put("key2", "mapString2");
mapString.put("key3", "mapString3");
Map mapUserinfo = new LinkedHashMap();
mapUserinfo.put("key1", new Userinfo("map Userinfo1", 1));
mapUserinfo.put("key2", new Userinfo("map Userinfo2", 2));
mapUserinfo.put("key3", new Userinfo("map Userinfo3", 3));
webContext.setVariable("mapString", mapString);
webContext.setVariable("mapUserinfo", mapUserinfo);
(7) Map&lt;String,String&gt;和Map&lt;String,Userinfo&gt;中的数据
<br/>
[[${mapString['key1']}]]
<br/>
[[${mapString['key2']}]]
<br/>
[[${mapString['key3']}]]
<br/>
<br/>
<span th:text="${mapString['key1']}"></span>
<br/>
<span th:text="${mapString['key2']}"></span>
<br/>
<span th:text="${mapString['key3']}"></span>
<br/>
<br/>
[[${mapUserinfo['key1'].username}]]
<br/>
[[${mapUserinfo['key2'].username}]]
<br/>
[[${mapUserinfo['key3'].username}]]
<br/>
<br/>
<span th:text="${mapUserinfo['key1'].username}"></span>
<br/>
<span th:text="${mapUserinfo['key2'].username}"></span>
<br/>
```

```
<span th:text="${mapUserinfo['key3'].username}"></span>

// （8）访问 request-session-application 作用域中的值
webContext.getRequest().setAttribute("requestKey", "requestValue 值");
webContext.getRequest().getSession().setAttribute("sessionKey", "sessionValue 值");
webContext.getServletContext().setAttribute("applicationKey", "applicationValue 值");
（8）从 request-session-application 作用域中取值
<br/>
requestValue:
<br/>
[[${requestKey}]]
<br/>
<span th:text="${requestKey}"></span>
<br/>
<br/>
sessionValue:
<br/>
[[${session.sessionKey}]]
<br/>
<span th:text="${session.sessionKey}"></span>
<br/>
<br/>
applicationValue:
<br/>
[[${application.applicationKey}]]
<br/>
<span th:text="${application.applicationKey}"></span>

（9）获得 URL 参数 username 值:<span th:text="${param.username}"></span>
<br/>
<br/>
（10）显示[[${'[[${userinfo.age}]]'}]]的写法
<br/>
[[${'[[${userinfo.age}]]'}]]
<br/>
<span th:text="${'[[${userinfo.age}]]'}"></span>
<br/>
<span th:inline="none">[[${userinfo.age}]]</span><!—原样输出-->
<br/>
<span th:inline="text">[[${userinfo.age}]]</span><!—解析表达式并输出-->
<br/>
<br/>
（11）(加-减-乘-除-取余)的运算
<br/>
[[${0+1}]]
<br/>
[[${3-1}]]
<br/>
[[${1*3}]]
<br/>
[[${20/5}]]
<br/>
[[${105%100}]]
<br/>
<br/>
<span th:text="${0+1}"></span>
<br/>
<span th:text="${3-1}"></span>
```

```
<br/>
<span th:text="${1*3}"></span>
<br/>
<span th:text="${20/5}"></span>
<br/>
<span th:text="${105%100}"></span>
<br/>
<br/>
<br/>
（12）声明变量
<br/>
<th:block th:with="a=${100}">
    [[${a}]]
</th:block>
<br/>
<th:block th:with="a=${100},b=${a+900}">
    [[${a}]]__[[${b}]]
</th:block>
<br/>
<br/>
<br/>
（13）字符串相加
<br/>
[['我的姓名是'+'杰克']]
<br/>
<span th:text="'我的姓名是'+'杰克'"></span>
<br/>
<br/>
[[${'我的姓名是'+'杰克'}]]
<br/>
<span th:text="${'我的姓名是'+'杰克'}"></span>
<br/>
<br/>
[[|我的姓名是${userinfo.username}|]]
<br/>
<span th:text="|我的姓名是${userinfo.username}|"></span>
<br/>
<br/>
[['我的年龄是 40 岁'+'  '+|我的姓名是${userinfo.username}|]]
<br/>
<span th:text="'我的年龄是 40 岁'+'  '+|我的姓名是${userinfo.username}|"></span>
<br/>
<br/>
<br/>
（14）三元运算符
<br/>
[[${userinfo.username=='杰克'}?'我是杰克':'我不是杰克']]
<br/>
<span th:text="${userinfo.username=='杰克'}?'我是杰克':'我不是杰克'"></span>
<br/>
<br/>
<br/>
（15）switch 语句
<br/>
<th:block th:switch="${userinfo.username}">
    <span th:case="'杰克 1'">1 等于杰克 1</span>
    <span th:case="${userinfo.age}">2 等于${userinfo.age}</span>
    <span th:case="*">3 其他值</span>
```

```
        </th:block>
        <br/>
        <th:block th:switch="${requestKey}">
            <th:block th:case="'request 值'">request 值</th:block>
            <th:block th:case="${userinfo.age}">${userinfo.age}</th:block>
            <th:block th:case="*">其他值</th:block>
        </th:block>
        (16) 注释
        <br/>
        <!-- (A)我是注释,我在 HTML 代码中-->
        <br/>
        <!--/*(B)我是注释,我不在 HTML 代码中*/-->
        <br/>
        <!--/*--> (C) 开始-我不在 HTML 代码中<span th:text="${userinfo.username}"></span>结
束-我不在 HTML 代码中<!--*/-->
        <br/>
        <!--/*/
        (D) <br />原型注释开始-我在 HTML 代码中
        <br />
        <span th:text="${userinfo.username}"></span>
        <br />原型注释结束-我在 HTML 代码中
        <br />本注释块的作用是在浏览器中直接打开*.html 文件时，并不显示<span>标签
        <br />但是在结合 Thymeleaf 框架时，解析此标签，并显示出<span>
        /*/-->
        <br/>
        <br/>
        <br/>
        (17) 格式化日期
        <br/>
        [[${#dates.format(nowDate, 'yyyy 年 MM 月 dd 日 hh 时 mm 分 ss 秒')}]]
        <br/>
        <span th:text="${#dates.format(nowDate, 'yyyy 年 MM 月 dd 日 hh 时 mm 分 ss 秒')}"></span>

        templateEngine.process("test110.html", webContext, response.getWriter());
    }

}
```

输入如下网址来运行程序：

```
http://localhost:8080/thymeleafTest/test110?username=urlValue
```

7.2.2 实现循环

创建 Servlet 类，代码如下：

```
public class test120 extends HttpServlet {
    protected void doGet(HttpServletRequest request, HttpServletResponse response)
            throws ServletException, IOException {
        ServletContextTemplateResolver templateResolver = new ServletContextTemplateResolver(
                request.getServletContext());
        templateResolver.setTemplateMode(TemplateMode.HTML);
        templateResolver.setPrefix("/");
        templateResolver.setCharacterEncoding("utf-8");

        TemplateEngine templateEngine = new TemplateEngine();
        templateEngine.setTemplateResolver(templateResolver);
```

```
            response.setContentType("text/html;charset=UTF-8");
            response.setHeader("Pragma", "no-cache");
            response.setHeader("Cache-Control", "no-cache");
            response.setDateHeader("Expires", 0);

            WebContext webContext = new WebContext(request, response, request.getServletContext(),
request.getLocale());
            // (1) 数组
            String[] stringArray = new String[] { "a", "b", "c" };
            Userinfo[] userinfoArray = new Userinfo[] { new Userinfo("账号1", 1), new Userinfo
("账号2", 2),
                    new Userinfo("账号3", 3) };
            webContext.setVariable("stringArray", stringArray);
            webContext.setVariable("userinfoArray", userinfoArray);
            (1) Array
            <th:block th:each="each:${stringArray}">
                [[${each}]]
                <br/>
            </th:block>
            <br/>
            <th:block th:each="each:${userinfoArray}">
                [[${each.username}]]__[[${each.age}]]
                <br/>
            </th:block>

            // (2) List
            ArrayList listString = new ArrayList();
            listString.add("listString1");
            listString.add("listString2");
            listString.add("listString3");
            ArrayList listUserinfo = new ArrayList();
            listUserinfo.add(new Userinfo("list Userinfo1", 1));
            listUserinfo.add(new Userinfo("list Userinfo2", 2));
            listUserinfo.add(new Userinfo("list Userinfo3", 3));
            webContext.setVariable("listString", listString);
            webContext.setVariable("listUserinfo", listUserinfo);
            (2) List
            <br/>
            <th:block th:each="each:${listString}">
                [[${each}]]
                <br/>
            </th:block>
            <br/>
            <th:block th:each="each:${listUserinfo}">
                [[${each.username}]]__[[${each.age}]]
                <br/>
            </th:block>

            // (3) Set
            Set setString = new LinkedHashSet();
            setString.add("setString1");
            setString.add("setString2");
            setString.add("setString3");
            Set setUserinfo = new LinkedHashSet();
            setUserinfo.add(new Userinfo("set Userinfo1", 1));
            setUserinfo.add(new Userinfo("set Userinfo2", 2));
            setUserinfo.add(new Userinfo("set Userinfo3", 3));
```

```
webContext.setVariable("setString", setString);
webContext.setVariable("setUserinfo", setUserinfo);
（3）Set
<br/>
<th:block th:each="each:${setString}">
    [[${each}]]
    <br/>
</th:block>
<br/>
<th:block th:each="each:${setUserinfo}">
    [[${each.username}]]__[[${each.age}]]
    <br/>
</th:block>

// （4）Map
Map mapString = new LinkedHashMap();
mapString.put("key1", "mapString1");
mapString.put("key2", "mapString2");
mapString.put("key3", "mapString3");
Map mapUserinfo = new LinkedHashMap();
mapUserinfo.put("key1", new Userinfo("map Userinfo1", 1));
mapUserinfo.put("key2", new Userinfo("map Userinfo2", 2));
mapUserinfo.put("key3", new Userinfo("map Userinfo3", 3));
webContext.setVariable("mapString", mapString);
webContext.setVariable("mapUserinfo", mapUserinfo);
（4）Map
<br/>
<th:block th:each="each:${mapString}">
    [[${each.key}]]__[[${each.value}]]
    <br/>
</th:block>
<br/>
<th:block th:each="each:${mapUserinfo}">
    [[${each.key}]]__[[${each.value.username}]]__[[${each.value.age}]]
    <br/>
</th:block>

// （5）隔行变色
ArrayList colorListUserinfo = new ArrayList();
colorListUserinfo.add(new Userinfo("list Userinfo1", 1));
colorListUserinfo.add(new Userinfo("list Userinfo2", 2));
colorListUserinfo.add(new Userinfo("list Userinfo3", 3));
colorListUserinfo.add(new Userinfo("list Userinfo4", 4));
colorListUserinfo.add(new Userinfo("list Userinfo5", 5));
colorListUserinfo.add(new Userinfo("list Userinfo6", 6));
colorListUserinfo.add(new Userinfo("list Userinfo7", 7));
colorListUserinfo.add(new Userinfo("list Userinfo8", 8));
webContext.setVariable("colorListUserinfo", colorListUserinfo);
（5）使用&lt;tr th:each&gt;的方式生成&lt;tr&gt;
<table border="1">
    <tr th:each="eachUserinfo : ${listUserinfo}">
        <td th:text="${eachUserinfo.username}">
        </td>
        <td th:text="${eachUserinfo.age}">
        </td>
    </tr>
</table>
<br/>
```

```
（6）使用&lt;tr th:each&gt;的方式生成&lt;input type='text'&gt;
<br/>
<th:block th:each="eachUserinfo : ${listUserinfo}">
    <input type="text" th:value="${eachUserinfo.username}">
    <br/>
</th:block>
<br/>
（7）使用&lt;tr th:each&gt;的方式生成&lt;tr&gt;并获得显式 status 对象
<br/>
<table border="1">
    <tr th:each="eachUserinfo,myStatus: ${colorListUserinfo}" th:styleappend=
"${myStatus.odd==true}?'color:red':'color:blue'">
        <td th:text="${eachUserinfo.username}">
        </td>
        <td th:text="${eachUserinfo.age}">
        </td>
        <td th:text="'odd='+${myStatus.odd}"><!-是否为奇数-->
        </td>
        <td th:text="'even='+${myStatus.even}"><!-是否为偶数-->
        </td>
        <td th:text="'size='+${myStatus.size}"><!-一共循环的次数-->
        </td>
        <td th:text="'count='+${myStatus.count}"><!-现在循环的次数-->
        </td>
        <td th:text="'index='+${myStatus.index}"><!-当前循环的索引-->
        </td>
        <td th:text="'current='+${myStatus.current}"><!-当前循环的对象-->
        </td>
        <td th:text="'first='+${myStatus.first}"><!-是否为第一个-->
        </td>
        <td th:text="'last='+${myStatus.last}"><!-是否为最后一个-->
        </td>
    </tr>
</table>
<br/>
（8）使用&lt;tr th:each&gt;的方式生成&lt;tr&gt;并获得隐式 status 对象，隐式 status 对象
的名称默认为：循环变量名称加上 Stat 后缀
<br/>
<table border="1">
    <tr th:each="eachUserinfo: ${colorListUserinfo}" th:styleappend
="${eachUserinfoStat.odd==true}?'color:red':'color:blue'">
        <td th:text="${eachUserinfo.username}">
        </td>
        <td th:text="${eachUserinfo.age}">
        </td>
        <td th:text="'odd='+${eachUserinfoStat.odd}">
        </td>
        <td th:text="'even='+${eachUserinfoStat.even}">
        </td>
        <td th:text="'size='+${eachUserinfoStat.size}">
        </td>
        <td th:text="'count='+${eachUserinfoStat.count}">
        </td>
        <td th:text="'index='+${eachUserinfoStat.index}">
        </td>
        <td th:text="'current='+${eachUserinfoStat.current}">
        </td>
        <td th:text="'first='+${eachUserinfoStat.first}">
```

```
                </td>
                <td th:text="'last='+${eachUserinfoStat.last}">
                </td>
            </tr>
        </table>
        隐式 status 对象的名称是在 eachUserinfo 后添加 Stat
        <br/>
        即 eachUserinfoStat 就是 status 状态对象的名称

        templateEngine.process("test120.html", webContext, response.getWriter());
    }

}
```

7.2.3 实现国际化与转义

创建属性文件 test130.properties，代码如下：

```
propTextHasParam=prop file has param : {0} and {1} and {2}
showHTMLCode=<a href='http://www.baidu.com'>\u767E\u5EA6</a>
```

创建属性文件 test130_en.properties，代码如下：

```
welcome.text=welcome to thymeleaf world!
```

创建属性文件 test130_zh.properties，代码如下：

```
welcome.text=\u6B22\u8FCE\u6765\u5230 thymeleaf \u4E16\u754C !
```

注意：*.properties 属性文件要和*.html 文件存放在同一个文件夹下，并且*.properties 属性文件名称的前缀是*.html 主文件名。

创建 Servlet 类，代码如下：

```
public class test130 extends HttpServlet {
    protected void doGet(HttpServletRequest request, HttpServletResponse response)
            throws ServletException, IOException {
        ServletContextTemplateResolver templateResolver = new ServletContextTemplateResolver(
                request.getServletContext());
        templateResolver.setTemplateMode(TemplateMode.HTML);
        templateResolver.setPrefix("/");
        templateResolver.setCharacterEncoding("utf-8");

        TemplateEngine templateEngine = new TemplateEngine();
        templateEngine.setTemplateResolver(templateResolver);

        response.setContentType("text/html;charset=UTF-8");
        response.setHeader("Pragma", "no-cache");
        response.setHeader("Cache-Control", "no-cache");
        response.setDateHeader("Expires", 0);

        WebContext webContext = new WebContext(request, response, request.getServletContext(),
request.getLocale());

        webContext.getRequest().setAttribute("param1", "参数 1");
        webContext.getRequest().setAttribute("param2", "参数 2");
```

```
        webContext.getRequest().setAttribute("param3", "参数 3");

        webContext.getRequest().setAttribute("showPropName", "propTextHasParam");

        templateEngine.process("test130.html", webContext, response.getWriter());

    }
}
```

创建模板文件，代码如下：

```
<!DOCTYPE html>
<html xmlns:th="http://www.thymeleaf.org">
    <head>
        <meta http-equiv="Content-Type" content="text/html; charset=UTF-8" />
        <title>Insert title here</title>
    </head>
    <body>
        （1）显示国际化文本：
        <br/>
        [[#{welcome.text}]]
        <br/>
        <span th:text="#{welcome.text}"></span>
        <br/>
        <br/>
        （2）向 property 传递字面参数值：
        <br/>
        [[#{propTextHasParam(${'参数 1'},${'参数 2'},${'参数 3'})}]]
        <br/>
        <span th:text="#{propTextHasParam(${'参数 1'},${'参数 2'},${'参数 3'})}"></span>
        <br/>
        <br/>
        （3）向 property 传递变量参数值：
        <br/>
        [[#{propTextHasParam(${param1},${param2},${param3})}]]
        <br/>
        <span th:text="#{propTextHasParam(${param1},${param2},${param3})}"></span>
        <br/>
        <br/>
        （4）显示属性文件中的 key 值来自于变量：
        <br/>
        [[#{${showPropName}(${param1},${param2},${param3})}]]
        <br/>
        <span th:text="#{${showPropName}(${param1},${param2},${param3})}"></span>
        <br/>
        <br/>
        （5）使用[[${'[[]]'}]]和 th:text 显示信息：
        <br/>
        [[#{showHTMLCode}]]
        <br/>
        <span th:text="#{showHTMLCode}"></span>
        <br/>
        <br/>
        （6）使用[[${'[()]'}]]和 th:utext 显示信息：
        <br/>
        [(#{showHTMLCode})]
        <br/>
        <span th:utext="#{showHTMLCode}"></span>
```

```
    </body>
</html>
```

7.2.4 处理链接

创建 Servlet 类，代码如下：

```
public class test140 extends HttpServlet {
    protected void doGet(HttpServletRequest request, HttpServletResponse response)
            throws ServletException, IOException {
        ServletContextTemplateResolver templateResolver = new ServletContextTemplateResolver(
                request.getServletContext());
        templateResolver.setTemplateMode(TemplateMode.HTML);
        templateResolver.setPrefix("/");
        templateResolver.setCharacterEncoding("utf-8");

        TemplateEngine templateEngine = new TemplateEngine();
        templateEngine.setTemplateResolver(templateResolver);

        response.setContentType("text/html;charset=UTF-8");
        response.setHeader("Pragma", "no-cache");
        response.setHeader("Cache-Control", "no-cache");
        response.setDateHeader("Expires", 0);

        WebContext webContext = new WebContext(request, response, request.getServletContext(),
request.getLocale());

        webContext.setVariable("userinfo", new Userinfo("杰克", 123L));
        webContext.setVariable("url", "findUserinfo3");

        templateEngine.process("test140.html", webContext, response.getWriter());

    }

}
```

创建模板文件，代码如下：

```
<!DOCTYPE html>
<html xmlns:th="http://www.thymeleaf.org">
    <head>
        <meta http-equiv="Content-Type" content="text/html; charset=UTF-8" />
        <title>Insert title here</title>
    </head>
    <body>
        <a th:href="@{http://localhost:8080/aaa/findUserinfo(userId=123)}">aaa</a>
        <br/>
        <a th:href="@{http://localhost:8080/bbb/findUserinfo(userId=123,age=456)}">bbb</a>
        <br/>
        <a th:href="@{http://localhost:8080/ccc/findUserinfo(username=${userinfo.
username})}">ccc</a>
        <br/>
        <a th:href="@{http://localhost:8080/ddd/findUserinfo(username=${userinfo.
username},age=${userinfo.age})}">ddd</a>
        <br/>
        <a th:href="@{/findUserinfo1(username=${userinfo.username})}">findUserinfo1</a>
        <br/>
```

```
            <a th:href="@{/findUserinfo2/{username}(username=${userinfo.username})}">
findUserinfo2</a><!—生成 restFul 格式的 URL-->
            <br/>
            <a th:href="@{${url}(username=${userinfo.username})}">findUserinfo3</a>
            <br/>
            <a th:href="@{${url}+'/'+${userinfo.username}}">newURL</a>
            <a th:href="@{'/findUserinfo4/'+${userinfo.username}(age=${userinfo.age})}">
findUserinfo4</a>
            <br/>
            <a th:href="@{~/otherPorject/findUserinfo5}">findUserinfo5</a>
        </body>
    </html>
```

7.2.5 实现 if 处理

创建 Servlet 类，代码如下：

```
public class test150 extends HttpServlet {
    protected void doGet(HttpServletRequest request, HttpServletResponse response)
            throws ServletException, IOException {
        ServletContextTemplateResolver templateResolver = new ServletContextTemplateResolver(
                request.getServletContext());
        templateResolver.setTemplateMode(TemplateMode.HTML);
        templateResolver.setPrefix("/");
        templateResolver.setCharacterEncoding("utf-8");

        TemplateEngine templateEngine = new TemplateEngine();
        templateEngine.setTemplateResolver(templateResolver);

        response.setContentType("text/html;charset=UTF-8");
        response.setHeader("Pragma", "no-cache");
        response.setHeader("Cache-Control", "no-cache");
        response.setDateHeader("Expires", 0);

        WebContext webContext = new WebContext(request, response, request.getServletContext(),
request.getLocale());

        webContext.setVariable("booleanValue", false);

        templateEngine.process("test150.html", webContext, response.getWriter());

    }

}
```

创建模板文件，代码如下：

```
<!DOCTYPE html>
<html xmlns:th="http://www.thymeleaf.org">
    <head>
        <meta http-equiv="Content-Type" content="text/html; charset=UTF-8" />
        <title>Insert title here</title>
    </head>
    <body>
        <span th:if="${booleanValue}==false">${booleanValue}==false==1</span>
        <br/>
        <span th:if="${booleanValue}==true">${booleanValue}==true==2</span>
        <br/>
```

```
        <span th:if="${booleanValue==false}">${booleanValue==false}==3</span>
        <br/>
        <span th:if="${booleanValue==true}">${booleanValue==true}==4</span>
        <br/>
        <span th:unless="${booleanValue==true}">${booleanValue==true}==5</span>
        <br/>
        <span th:unless="${booleanValue==false}">${booleanValue==false}==6</span>
    </body>
</html>
```

7.2.6 实现比较

创建 Servlet 类，代码如下：

```
public class test160 extends HttpServlet {
    protected void doGet(HttpServletRequest request, HttpServletResponse response)
            throws ServletException, IOException {
        ServletContextTemplateResolver templateResolver = new ServletContextTemplateResolver(
                request.getServletContext());
        templateResolver.setTemplateMode(TemplateMode.HTML);
        templateResolver.setPrefix("/");
        templateResolver.setCharacterEncoding("utf-8");

        TemplateEngine templateEngine = new TemplateEngine();
        templateEngine.setTemplateResolver(templateResolver);

        response.setContentType("text/html;charset=UTF-8");
        response.setHeader("Pragma", "no-cache");
        response.setHeader("Cache-Control", "no-cache");
        response.setDateHeader("Expires", 0);

        WebContext webContext = new WebContext(request, response, request.getServletContext(),
request.getLocale());

        templateEngine.process("test160.html", webContext, response.getWriter());

    }

}
```

创建模板文件，代码如下：

```
<!DOCTYPE html>
<html xmlns:th="http://www.thymeleaf.org">
    <head>
        <meta http-equiv="Content-Type" content="text/html; charset=UTF-8" />
        <title>Insert title here</title>
    </head>
    <body>
        <th:block th:if="100 < 200">
            <span>100 &lt; 200</span>
        </th:block>
        <br/>
        <th:block th:if="100 <= 100">
            <span>100 &le; 100</span>
        </th:block>
```

```
        <br/>
        <th:block th:if="201 > 100">
            <span>201 &gt; 100</span>
        </th:block>
        <br/>
        <th:block th:if="201 >= 201">
            <span>201 &ge; 201</span>
        </th:block>
        <br/>
        <th:block th:if="201 == 201">
            <span>201 == 201</span>
        </th:block>
        <br/>
        <th:block th:if="201 != 202">
            <span>201 != 202</span>
        </th:block>
        <br/>
        <th:block th:if="!false">
            <span>true</span>
        </th:block>
    </body>
</html>
```

7.2.7 处理属性值

创建 test170.properties 文件，代码如下：

```
submitText=\u63D0\u4EA4
```

创建 Servlet 类，代码如下：

```
public class test170 extends HttpServlet {
    protected void doGet(HttpServletRequest request, HttpServletResponse response)
            throws ServletException, IOException {
        ServletContextTemplateResolver templateResolver = new ServletContextTemplateResolver(
                request.getServletContext());
        templateResolver.setTemplateMode(TemplateMode.HTML);
        templateResolver.setPrefix("/");
        templateResolver.setCharacterEncoding("utf-8");

        TemplateEngine templateEngine = new TemplateEngine();
        templateEngine.setTemplateResolver(templateResolver);

        response.setContentType("text/html;charset=UTF-8");
        response.setHeader("Pragma", "no-cache");
        response.setHeader("Cache-Control", "no-cache");
        response.setDateHeader("Expires", 0);

        WebContext webContext = new WebContext(request, response, request.getServletContext(),
request.getLocale());

        webContext.setVariable("userinfo", new Userinfo("杰克", 123L));
        webContext.setVariable("checkboxStatus1", false);
        webContext.setVariable("checkboxStatus2", true);

        templateEngine.process("test170.html", webContext, response.getWriter());
```

```
    }
}
```

创建模板文件，代码如下：

```
<!DOCTYPE html>
<html xmlns:th="http://www.thymeleaf.org">
    <head>
        <meta http-equiv="Content-Type" content="text/html; charset=UTF-8" />
        <title>Insert title here</title>
    </head>
    <body>
        (1) <span th:class="'myClass'">myspan1</span>
        <br/>
        (2) <span th:class="${userinfo.username}">myspan2</span>
        <br/>
        (3) <input type="text" th:value="${userinfo.username}">
        </input>
        <br/>
        (4) <input type="text" th:attr="value=${userinfo.username}">
        </input>
        <br/>
        (5) <a th:attr="href=@{findUsername5(username=${userinfo.username})}">5</a>
        <br/>
        (6)<a th:attr="href=@{findUsername6/{username}(username=${userinfo.username})}">6</a>
        <br/>
        (7) <input type="button" th:attr="value=#{submitText}" />
        <br/>
        (8) <input type="text" th:attr="a=${userinfo.username},b=${userinfo.age}" />
        <br/>
        (9) <span class="a b" th:attrappend="class=${' '}+'c'">span</span>
        <br/>
        (10) <span class="a b" th:attrprepend="class=${'c'}+' '">span</span>
        <br/>
        (11) <input type="checkbox" th:checked="${checkboxStatus1}" />
        <br/>
        (12) <input type="checkbox" th:checked="${checkboxStatus2}" />
        <br/>
        (13) <span th:myProperty="'myvalue'">span</span>
        </body>
    </html>
```

7.3 使用 Spring Boot 开发 Web 项目

本节的主要目的就是使用 Spring Boot 开发基于 MVC 模式的 Web 项目。

7.3.1 创建 Maven Web Project

下面开始创建 Maven Web Project，如图 7-20 所示。

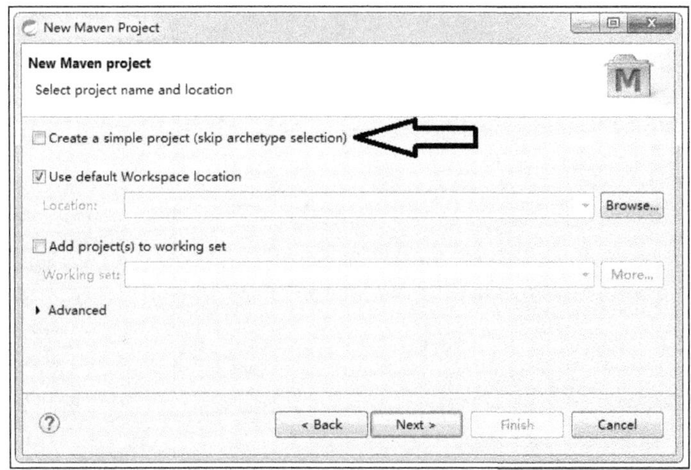

图 7-20　“Select project name and location”界面

在图 7-20 中，取消勾选“Create a simple project（skip archetype selection）”复选框，然后单击“Next”按钮，出现如图 7-21 所示的界面。

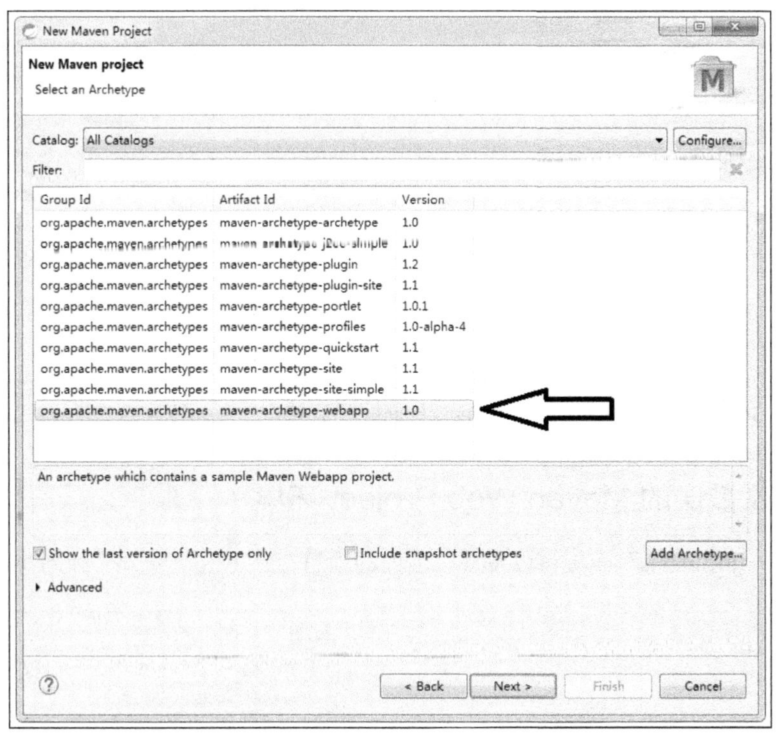

图 7-21　选择“maven-archetype-webapp”模板

在图 7-21 中，选择“maven-archetype-webapp”模板，目的是使创建出来的项目具有 Web 项目的结构，包含 web.xml 文件和 WEB-INF 文件夹等资源。然后单击“Next”按钮，在出现的“Specify Archetype parameters”界面中添加相关配置信息，如图 7-22 所示。

然后，单击“Finish”按钮，开始创建项目，新创建的项目结构如图 7-23 所示。

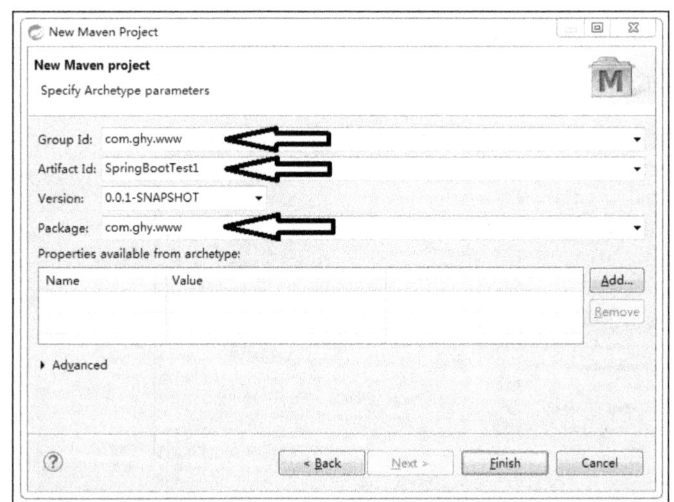

图 7-22 “Specify Archetype parameters”界面配置信息

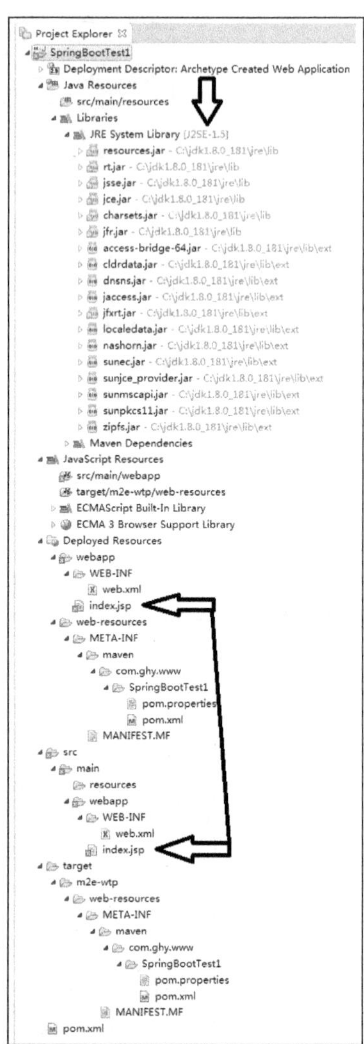

图 7-23 新项目的结构与异常

7.3.2 更改错误的 Maven Web Project 环境

通过查看“Problems”面板可以发现，当前的项目存在 Errors 错误，如图 7-24 所示。

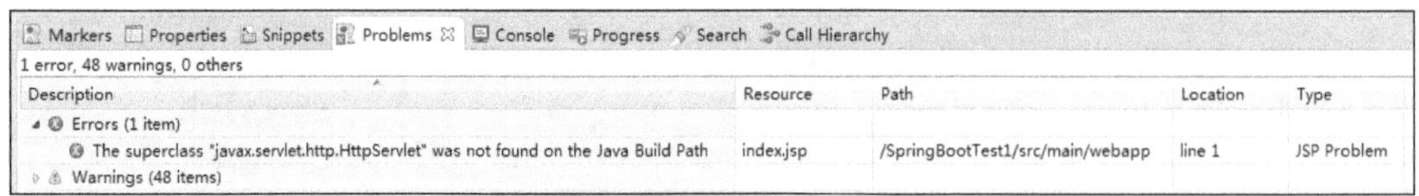

图 7-24 项目存在错误

根据异常信息的内容得知，并没有找到与 Servlet 相关的 JAR 包资源，导致具体的接口或类不存在，更改 pom.xml 配置文件，添加 Web 环境支持，代码如下：

```
<project xmlns="http://maven.apache.org/POM/4.0.0"
    xmlns:xsi="http://www.w3.org/2001/XMLSchema-instance"
```

```
        xsi:schemaLocation="http://maven.apache.org/POM/4.0.0 http://maven.apache.org/
maven-v4_0_0.xsd">
        <modelVersion>4.0.0</modelVersion>
        <groupId>com.ghy.www</groupId>
        <artifactId>SpringBootTest1</artifactId>
        <packaging>war</packaging>
        <version>0.0.1-SNAPSHOT</version>
        <name>SpringBootTest1 Maven Webapp</name>
        <url>http://maven.apache.org</url>

        <parent>
            <!-- 从 Spring Boot 继承默认的配置 -->
            <groupId>org.springframework.boot</groupId>
            <artifactId>spring-boot-starter-parent</artifactId>
            <version>2.0.4.RELEASE</version>
        </parent>

        <dependencies>
            <dependency>
                <groupId>junit</groupId>
                <artifactId>junit</artifactId>
                <version>3.8.1</version>
                <scope>test</scope>
            </dependency>
            <!-- 添加 web 项目典型的依赖 -->
            <dependency>
                <groupId>org.springframework.boot</groupId>
                <artifactId>spring-boot-starter-web</artifactId>
            </dependency>
        </dependencies>
        <build>
            <finalName>SpringBootTest1</finalName>
        </build>
    </project>
```

上面的 pom.xml 文件中使用了 spring-boot-starter-parent，这是一个特殊的启动器，提供很多有用的 Maven 默认配置以及依赖管理。

编辑完 pom.xml 后保存该文件，Eclipse 会自动下载相关的依赖资源，下载成功后的 Maven Dependencies 节点如图 7-25 所示。

这时“Problems”面板中的 Errors 信息发生改变，效果如图 7-26 所示。

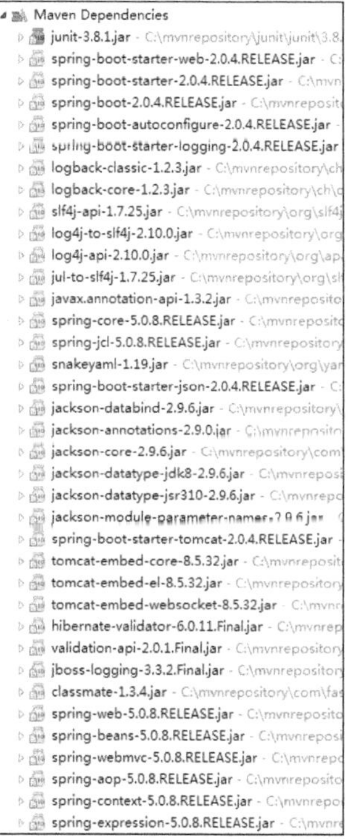

图 7-25　成功下载与 Web 相关的 JAR 包资源

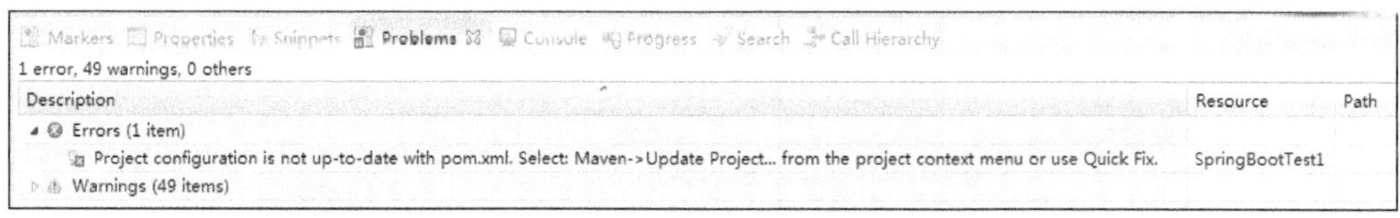

图 7-26　Update Project 提示

在项目上单击右键，然后选择“Update Project...”菜单项，如图 7-27 所示。

在弹出的“Update Maven Project”对话框中，单击“OK”按钮，开始 Update Project 操作，如图 7-28 所示。

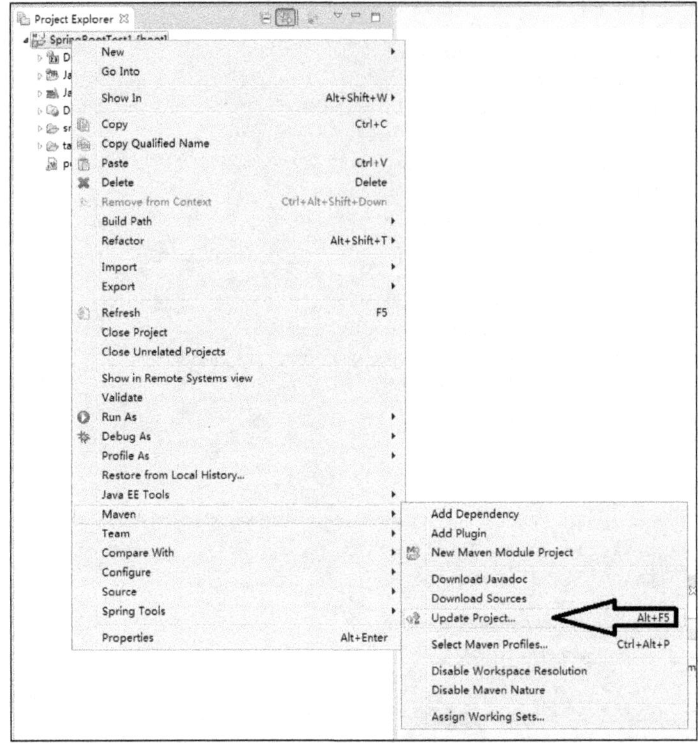

图 7-27 选择“Update Project...”菜单项

项目更新后 Errors 信息没有了，效果如图 7-29 所示。

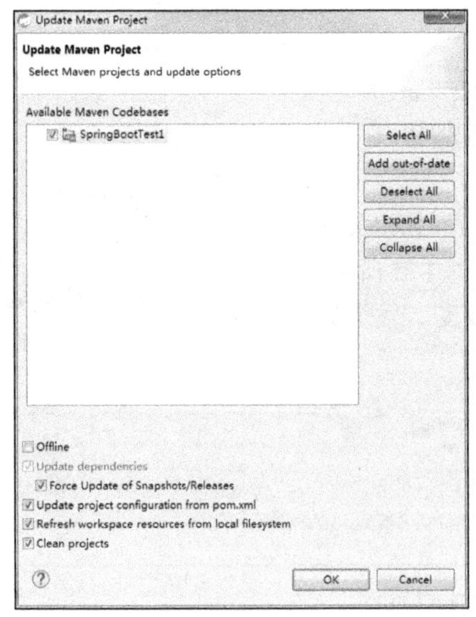

图 7-28 “Update Maven Project”对话框

图 7-29 Errors 信息消失

虽然 Spring Boot 框架没有硬性指定项目结构，但是一个使用 Eclipse 创建 Maven Project 的 Spring Boot 项目结构应包含如下两个资源包：

```
src/main/java
src/main/resources
```

然而，从当前的项目结构来看，src/main/java 路径并不存在，如图 7-30 所示。

右键单击“SpringBootTest1 [boot]”，然后选择“Build Path”→“Configure Build Path...”菜单项，如图 7-31 所示。

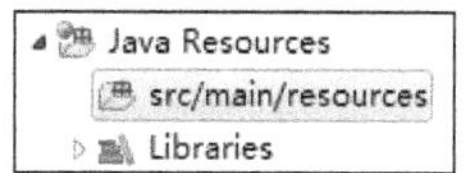

图 7-30　消失不见的 src/main/java 路径

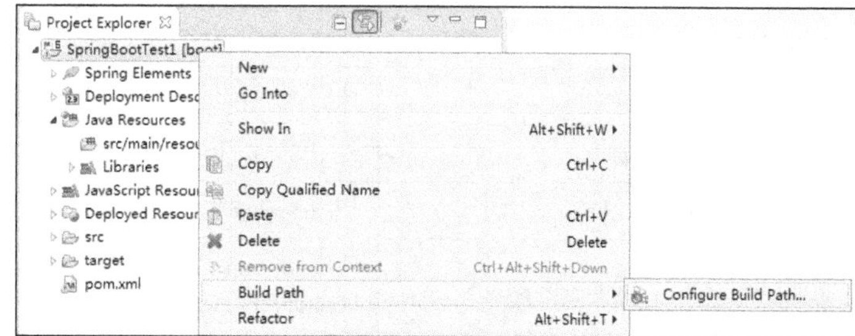

图 7-31　选择“Configure Build Path...”

弹出“Properties for SpringBootTest1”对话框，如图 7-32 所示。

然后，勾选“Maven Dependencies”和“JRE System Library[JavaSE-1.8]”复选框，接着单击“Apply and Close”按钮完成配置，这时在 src 中出现了 java 和 resources 的路径，如图 7-33 所示。

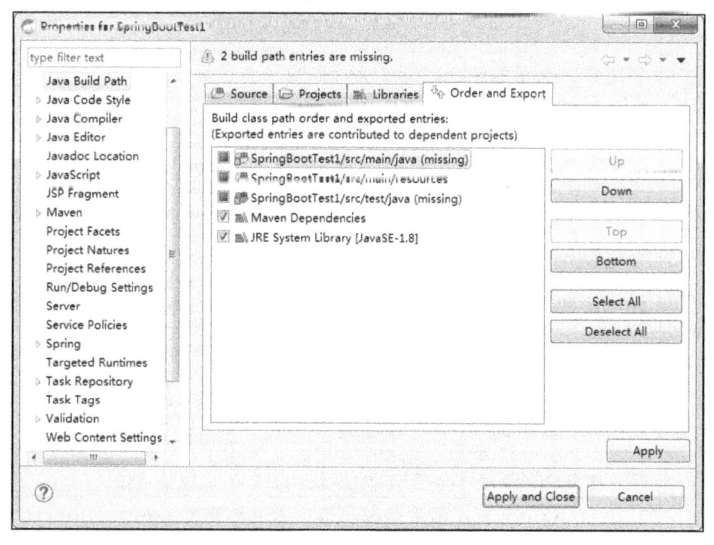

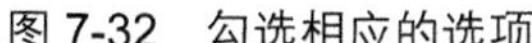

图 7-32　勾选相应的选项

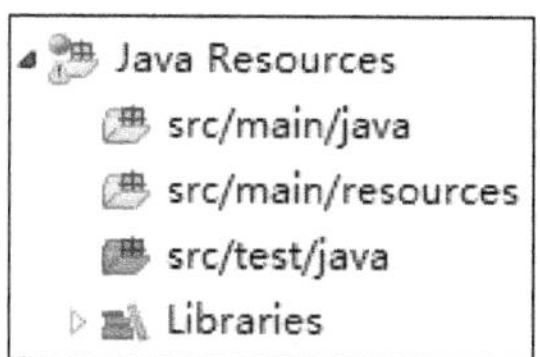

图 7-33　项目的正确结构

上面的步骤将一个错误的 Maven Web Project 环境配置成正确的，相关环境配置正确后就可以编写代码并进行开发了。

7.3.3　常用 Starter 的介绍

Spring Boot 提供了许多 Starter 来将依赖的 JAR 包文件添加到类路径中。Starter 就是将某一个功能依赖的组件进行打包，只需要引用这个 Starter，该 Starter 依赖的组件就会自动下载到项目中。表 7-1 是常见的 Starter 组件。

表 7-1 常见的 Starter 组件

名 称	描 述
spring-boot-starter	核心启动器，包括自动配置支持、日志和 YAML
spring-boot-starter-activemq	使用 Apache ActiveMQ 的 JMS 消息传递启动器
spring-boot-starter-amqp	使用 Spring AMQP 和 Rabbit MQ 的启动器
spring-boot-starter-aop	使用 Spring AOP 和 AspectJ 进行面向切面编程的启动器
spring-boot-starter-artemis	使用 Apache Artemis 的 JMS 消息传递启动器
spring-boot-starter-batch	使用 Spring Batch 的启动器
spring-boot-starter-cache	使用 Spring 框架缓存支持的启动器
spring-boot-starter-cloud-connectors	使用 Spring Cloud 连接器，它简化了连接到云平台的服务，如 Cloud Foundry 和 Heroku
spring-boot-starter-data-cassandra	使用 Cassandra 分布式数据库和针对 Cassandra 的 Spring Data 启动器
spring-boot-starter-data-cassandra-reactive	使用 Cassandra 分布式数据库和 Spring Data Cassandra Reactive 的启动器
spring-boot-starter-data-couchbase	使用 Couchbase 面向文档的数据库和 Spring Data Couchbase 的启动器
spring-boot-starter-data-couchbase-reactive	使用 Couchbase 面向文档的数据库和 Spring Data Couchbase Reactive 的启动器
spring-boot-starter-data-elasticsearch	使用 Elasticsearch 搜索并分析引擎与 Spring Data Elasticsearch 的启动器
spring-boot-starter-data-jpa	使用 Hibernate 的 Spring Data JPA 启动器
spring-boot-starter-data-ldap	使用 Spring Data LDAP 的启动器
spring-boot-starter-data-mongodb	使用 MongoDB 面向文档数据库和 Spring Data MongoDB 的启动器
spring-boot-starter-data-mongodb-reactive	使用 MongoDB 面向文档数据库和 Spring Data MongoDB Reactive 的启动器
spring-boot-starter-data-neo4j	使用 Neo4j 图形数据库和 Spring Data Neo4j 的启动器
spring-boot-starter-data-redis	使用 Redis 键值数据存储以及 Spring Data Redis 和 Lettuce 客户端的启动器
spring-boot-starter-data-redis-reactive	使用 Redis 键值数据存储以及 Spring Data Redis Reactive 和 Lettuce 客户端的启动器
spring-boot-starter-data-rest	使用 Spring Data REST 公开 Spring Data 存储库的启动器
spring-boot-starter-data-solr	使用带有 Spring Data Solr 的 Apache Solr 搜索平台的启动器
spring-boot-starter-freemarker	使用 FreeMarker 视图构建 MVC Web 应用程序的启动器
spring-boot-starter-groovy-templates	使用 Groovy 模板视图构建 MVC Web 应用程序的启动器
spring-boot-starter-hateoas	使用 Spring MVC 和 Spring HATEOAS 构建基于超媒体的 RESTful Web 应用程序的启动器
spring-boot-starter-integration	使用 Spring Integration 的启动器
spring-boot-starter-jdbc	使用 JDBC 与 HikariCP 连接池的启动器
spring-boot-starter-jersey	使用 JAX-RS 和 Jersey 构建基于 RESTful 的 Web 应用程序的启动器，另一种选择是 spring-boot-starter-web
spring-boot-starter-jooq	使用 jOOQ 访问 SQL 数据库的启动器，另一种选择是 spring-boot-starter-data-jpa 或 spring-boot-starter-jdbc

续表

名　称	描　述
spring-boot-starter-json	读和写 JSON 的启动器
spring-boot-starter-jta-atomikos	使用 Atomikos 的 JTA 事务启动器
spring-boot-starter-jta-bitronix	使用 Bitronix 的 JTA 事务启动器
spring-boot-starter-jta-narayana	使用 Narayana 的 JTA 事务启动器
spring-boot-starter-mail	使用 Java 邮件和 Spring 框架的电子邮件发送支持的启动器
spring-boot-starter-mustache	使用 Mustache 视图构建 Web 应用程序的启动器
spring-boot-starter-quartz	使用 Quartz 调度器的启动器
spring-boot-starter-security	使用 Spring Security 的启动器
spring-boot-starter-test	使用包括 JUnit、Hamcrest 和 Mockito 的测试 Spring Boot 应用程序的启动器
spring-boot-starter-thymeleaf	使用 Thymeleaf 视图构建 MVC Web 应用程序的启动器
spring-boot-starter-validation	使用 Hibernate 验证器验证 Java Bean 的启动器
spring-boot-starter-web	构建 Web，包括 RESTful、使用 Spring MVC 的应用程序，使用 Tomcat 作为默认的嵌入式容器的启动器
spring-boot-starter-web-services	使用 Spring Web Services 的启动器
spring-boot-starter-webflux	使用 Spring 框架的 Reactive Web 支持构建 WebFlux 应用程序的启动器
spring-boot-starter-websocket	使用 Spring 框架的 WebSocket 支持构建 WebSocket 应用程序的启动器
spring-boot-starter-actuator	使用 Spring Boot 执行器提供生产就绪特性，帮助用户监视和管理应用程序的启动器
spring-boot-starter-jetty	使用 Jetty 作为嵌入式 Servlet 容器的启动器，替代 spring-boot-starter-tomcat
spring-boot-starter-log4j2	使用 Log4j2 进行日志记录的启动器，替代 spring-boot-starter-logging
spring-boot-starter-logging	使用 Logback 进行日志记录的启动器，默认的日志启动器
spring-boot-starter-reactor-netty	使用 Reactor Netty 作为嵌入式 Reactive HTTP 服务器的启动器
spring-boot-starter-tomcat	使用 Tomcat 作为嵌入式 Servlet 容器的启动器，默认的 Servlet 容器启动器使用 spring-boot-starter-web
spring-boot-starter-undertow	使用 Undertow 作为嵌入式 Servlet 容器的启动器，替代 spring-boot-starter-tomcat

其他的 Starter 提供了在开发特定类型的项目时可能需要的依赖项。因为当前正在开发一个 Web 应用程序，所以又添加了一个 spring-boot-starter-web 依赖。

在 pom.xml 文件中必须指定 spring-boot-starter-parent 版本号。如果导入其他的 Starter，那么可以省略版本号。其他的 Starter 版本号最终由 spring-boot-starter-parent 决定。

7.3.4 创建控制层

创建控制层，代码如下：

```
package controller;

import java.util.ArrayList;
```

```
import java.util.List;

import javax.servlet.http.HttpServletRequest;
import javax.servlet.http.HttpServletResponse;

import org.springframework.stereotype.Controller;
import org.springframework.web.bind.annotation.RequestMapping;
import org.springframework.web.bind.annotation.ResponseBody;

@Controller
public class TestController {
    @RequestMapping("test1")
    public String testMethod1(HttpServletRequest request, HttpServletResponse response) {
        System.out.println("执行了控制层 1");
        List list = new ArrayList();
        list.add("我是中国人 1");
        list.add("我是中国人 2");
        list.add("我是中国人 3");
        request.setAttribute("list", list);
        return "hello.jsp";
    }

    @RequestMapping("test2")
    @ResponseBody
    public String testMethod2() {
        System.out.println("执行了控制层 2");
        return "我就是返回值";
    }
}
```

7.3.5 添加 JSTL 依赖

在 pom.xml 文件中添加 JSTL 依赖，配置代码如下：

```
<!-- 添加 JSTL 依赖 -->
<dependency>
    <groupId>javax.servlet</groupId>
    <artifactId>jstl</artifactId>
</dependency>
```

7.3.6 创建 JSP 视图文件

创建 hello.jsp 文件，位置如图 7-34 所示。

图 7-34 文件 hello.jsp 的存放位置

hello.jsp 中的代码如下：

```
<%@ page language="java" contentType="text/html; charset=utf-8"
    pageEncoding="utf-8"%>
<%@ taglib uri="http://java.sun.com/jsp/jstl/core" prefix="c"%>
<!DOCTYPE html>
<html>

    <head>
        <meta charset="utf-8">
        <title>Insert title here</title>
    </head>

    <body>欢迎来到 JSP 的世界!
```

```
        <br/>
        <c:forEach var="eachString" items="${list}">${eachString}<br/></c:forEach>
    </body>

</html>
```

7.3.7　创建启动类 Application

创建启动类 Application，完整代码如下：

```
package application;

import org.springframework.boot.SpringApplication;
import org.springframework.boot.autoconfigure.EnableAutoConfiguration;
import org.springframework.context.annotation.ComponentScan;

@EnableAutoConfiguration
@ComponentScan(basePackages = { "controller" })
public class Application {
    public static void main(String[] args) {
        SpringApplication.run(Application.class);
    }
}
```

@EnableAutoConfiguration 注解根据添加的 JAR 依赖项来“猜测”希望如何配置 Spring Boot 的环境。例如，spring-boot-starter-web 需要添加 Tomcat 和 Spring MVC，@EnableAutoConfiguration 可以识别出正在开发的 Web 应用程序并相应地设置 Spring Boot 环境。

7.3.8　运行 Application 类

按照普通的“Java Application”来运行 Application 类，如图 7-35 所示。

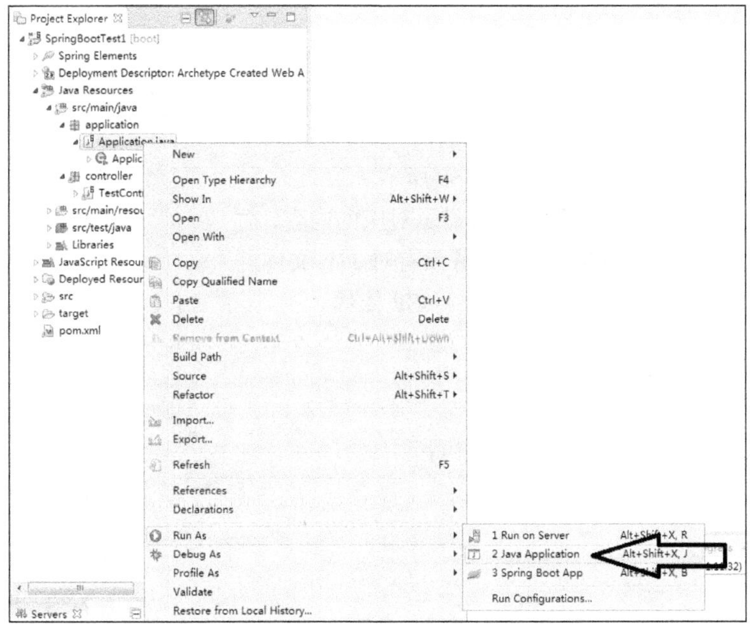

图 7-35　运行 Application.java 类

程序运行后，在控制台输出相关信息，如图 7-36 所示。

```
: Servlet dispatcherServlet mapped to [/]
: Mapping filter: 'characterEncodingFilter' to: [/*]
: Mapping filter: 'hiddenHttpMethodFilter' to: [/*]
: Mapping filter: 'httpPutFormContentFilter' to: [/*]
: Mapping filter: 'requestContextFilter' to: [/*]
: Mapped URL path [/**/favicon.ico] onto handler of type [class
: Looking for @ControllerAdvice: org.springframework.boot.web.s
: Mapped "{[/test2]}" onto public java.lang.String controller.1
: Mapped "{[/test1]}" onto public java.lang.String controller.1
: Mapped "{[/error]}" onto public org.springframework.http.Resp
: Mapped "{[/error],produces=[text/html]}" onto public org.spri
: Mapped URL path [/webjars/**] onto handler of type [class org
: Mapped URL path [/**] onto handler of type [class org.spring
: Registering beans for JMX exposure on startup
: Tomcat started on port(s): 8080 (http) with context path ''
: Started Application in 2.244 seconds (JVM running for 2.618)
```

图 7-36 项目中的两个路径已经被识别

7.3.9 执行 test2 的 URL

输入网址：

```
http://localhost:8080/test2
```

控制台和浏览器显示的内容如图 7-37 所示。

图 7-37 显示效果

7.3.10 执行 test1 的 URL

输入网址：

```
http://localhost:8080/test1
```

控制台和浏览器显示的内容如图 7-38 所示。

执行 test1 路径后，JSP 文件并没有显示对应的内容，而是下载了文件。这说明在当前的环境中并没有对 JSP 文件进行解析，因此，要在 pom.xml 文件中加入 JSP 依赖。

图 7-38 显示效果

7.3.11 添加 JSP 依赖

在 pom.xml 文件中添加如下配置代码：

```
<!-- 添加 JSP 依赖 -->
<dependency>
    <groupId>org.apache.tomcat.embed</groupId>
    <artifactId>tomcat-embed-jasper</artifactId>
    <scope>provided</scope>
</dependency>
```

重启应用后再输入网址：

```
http://localhost:8080/test1
```

浏览器成功显示了 JSP 文件的内容，效果如图 7-39 所示。

图 7-39 成功显示 JSP 文件中的内容

7.3.12 实现项目首页

可以为 Web 项目设置首页，首页的起始点不是 index.jsp，而是默认执行一个控制层，取得数据后再转发到 index.jsp，将数据在 index.jsp 中显示出来。

在与 hello.jsp 文件相同的位置创建 index.jsp 文件，代码如下：

```
<%@ page language="java" contentType="text/html; charset=utf-8"
    pageEncoding="utf-8"%>
<%@ taglib uri="http://java.sun.com/jsp/jstl/core" prefix="c"%>
<!DOCTYPE html>
<html>

    <head>
        <meta charset="utf-8">
        <title>Insert title here</title>
    </head>

    <body>
        这里是 index.jsp 视图层
    </body>

</html>
```

控制层代码如下：

```
@RequestMapping("/")
public String home() {
    System.out.println("执行了首页控制层，现在要转发到 index.jsp 视图层");
    return "index.jsp";
}
```

在浏览器中输入如下网址：

```
http://localhost:8080/
```

控制台和浏览器显示的内容如图 7-40 所示。

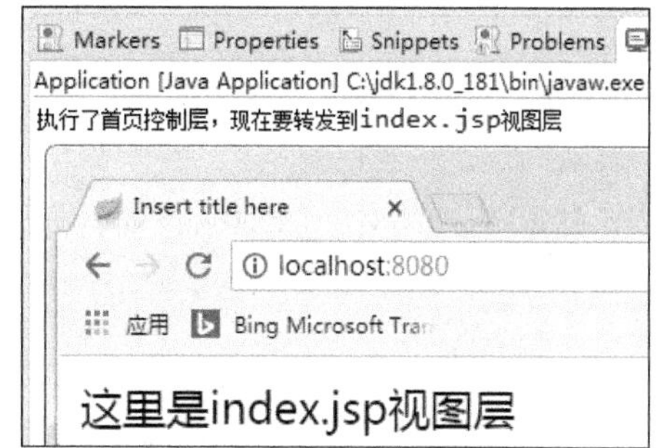

图 7-40 出现首页

7.3.13 在 CMD 中启动项目

在生产环境中，不可能在 Eclipse 中启动项目，而是需要在 CMD 中进行启动。

在 CMD 中，进入项目的根目录，输入命令来启动项目，然后可以正常访问控制层：

```
mvn spring-boot:run
```

注意：使用<Ctrl+C>快捷键可以结束进程。

7.3.14　创建可执行 JAR

可以通过创建一个完全独立的可执行 JAR 文件来运行程序。可执行 JAR 也称为“fat jar”，内部包含已编译的类以及运行项目所需要的所有依赖项。

在 pom.xml 文件中添加配置，代码如下：

```
<build>
    <plugins>
        <plugin>
            <groupId>org.springframework.boot</groupId>
            <artifactId>spring-boot-maven-plugin</artifactId>
        </plugin>
    </plugins>
</build>
```

在 CMD 中进入项目的根目录，输入如下命令：

```
mvn package
```

控制台出现的信息如图 7-41 所示。

```
[INFO]
[INFO] --- maven-war-plugin:3.1.0:war (default-war) @ SpringBootTest1 ---
[INFO] Packaging webapp
[INFO] Assembling webapp [SpringBootTest1] in [C:\spring-tool-suite-3.9.5.RELEASE-workspace\SpringBootTest1\target\SpringBootTest1]
[INFO] Processing war project
[INFO] Copying webapp resources [C:\spring-tool-suite-3.9.5.RELEASE-workspace\SpringBootTest1\src\main\webapp]
[INFO] Webapp assembled in [217 msecs]
[INFO] Building war: C:\spring-tool-suite-3.9.5.RELEASE-workspace\SpringBootTest1\target\SpringBootTest1.war
[INFO]
[INFO] --- spring-boot-maven-plugin:2.0.4.RELEASE:repackage (default) @ SpringBootTest1 ---
[INFO] ------------------------------------------------------------------------
[INFO] BUILD SUCCESS
[INFO] ------------------------------------------------------------------------
```

图 7-41　成功创建 WAR 文件

WAR 文件所处的位置及信息如图 7-42 所示。

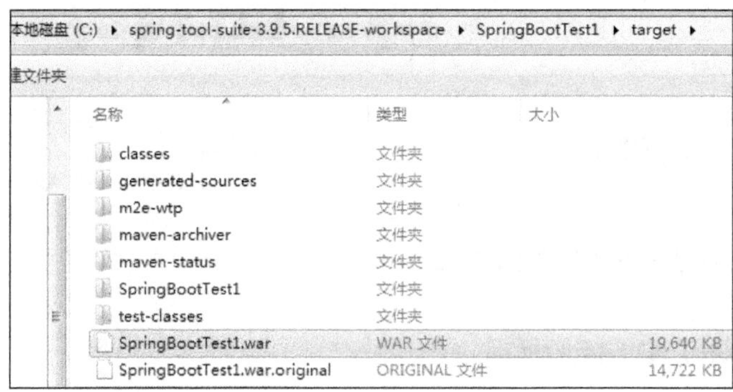

图 7-42　成功创建大小约为 20MB 的 WAR 文件

在 CMD 中使用“java –jar springBootTest1.war”命令启动项目，如图 7-43 所示。

启动后的项目可以正常运行控制层。

注意：使用<Ctrl+C>快捷键结束进程。

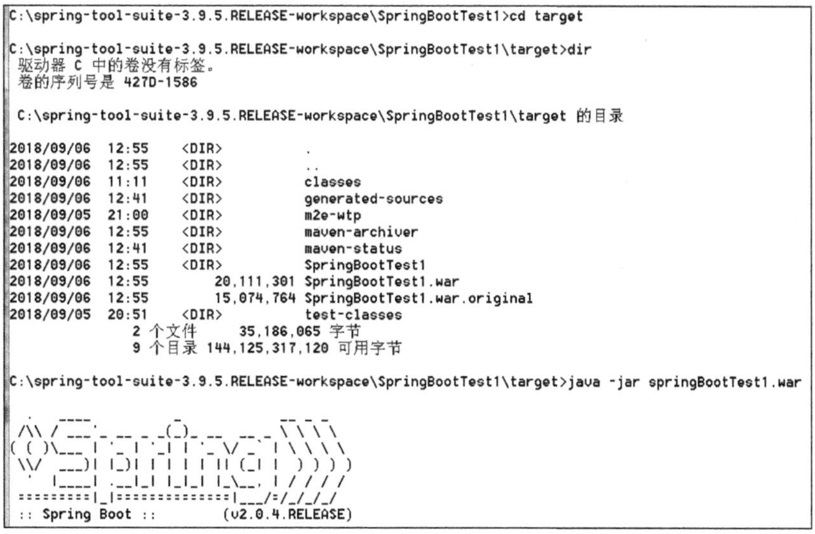

```
C:\spring-tool-suite-3.9.5.RELEASE-workspace\SpringBootTest1>cd target

C:\spring-tool-suite-3.9.5.RELEASE-workspace\SpringBootTest1\target>dir
 驱动器 C 中的卷没有标签。
 卷的序列号是 427D-1586

 C:\spring-tool-suite-3.9.5.RELEASE-workspace\SpringBootTest1\target 的目录

2018/09/06  12:55    <DIR>          .
2018/09/06  12:55    <DIR>          ..
2018/09/06  11:11    <DIR>          classes
2018/09/06  12:41    <DIR>          generated-sources
2018/09/05  21:00    <DIR>          m2e-wtp
2018/09/06  12:55    <DIR>          maven-archiver
2018/09/06  12:41    <DIR>          maven-status
2018/09/06  12:55    <DIR>          SpringBootTest1
2018/09/06  12:55        20,111,301 SpringBootTest1.war
2018/09/06  12:55        15,074,764 SpringBootTest1.war.original
2018/09/05  20:51    <DIR>          test-classes
               2 个文件     35,186,065 字节
               9 个目录 144,125,317,120 可用字节

C:\spring-tool-suite-3.9.5.RELEASE-workspace\SpringBootTest1\target>java -jar springBootTest1.war

  .   ____          _            __ _ _
 /\\ / ___'_ __ _ _(_)_ __  __ _ \ \ \ \
( ( )\___ | '_ | '_| | '_ \/ _` | \ \ \ \
 \\/  ___)| |_)| | | | | || (_| |  ) ) ) )
  '  |____| .__|_| |_|_| |_\__, | / / / /
 =========|_|==============|___/=/_/_/_/
 :: Spring Boot ::        (v2.0.4.RELEASE)
```

图 7-43 项目成功启动

7.3.15 实现注入 IoC

创建业务逻辑层，代码如下：

```java
package service;

import org.springframework.stereotype.Service;

@Service
public class UserinfoService {
    public void insertUserinfoService() {
        System.out.println("public void insertUserinfoService()");
    }
}
```

创建控制层，代码如下：

```java
package controller;

import org.springframework.beans.factory.annotation.Autowired;
import org.springframework.stereotype.Controller;
import org.springframework.web.bind.annotation.RequestMapping;

import service.UserinfoService;

@Controller
public class UserinfoController {

    @Autowired
    private UserinfoService service;

    @RequestMapping("insertUserinfo")
    public String insertUserinfo() {
        service.insertUserinfoService();
        return "index.jsp";
    }
}
```

运行类代码如下：

```
package application;

import org.springframework.boot.SpringApplication;
import org.springframework.boot.autoconfigure.EnableAutoConfiguration;
import org.springframework.context.annotation.ComponentScan;

@EnableAutoConfiguration
@ComponentScan(basePackages = { "controller", "service" })
public class Application {
    public static void main(String[] args) {
        SpringApplication.run(Application.class);
    }
}
```

输入 URL 网址：

```
http://localhost:8080/insertUserinfo
```

控制台输出的结果如下：

```
public void insertUserinfoService()
```

注入成功。

7.3.16 实现切面 AOP

添加 AOP 切面依赖代码：

```
<!-- 添加 AOP 切面依赖 -->
<dependency>
    <groupId>org.springframework.boot</groupId>
    <artifactId>spring-boot-starter-aop</artifactId>
</dependency>
```

创建切面类，代码如下：

```
package aspect;

import org.aspectj.lang.ProceedingJoinPoint;
import org.aspectj.lang.annotation.Around;
import org.aspectj.lang.annotation.Aspect;
import org.springframework.stereotype.Component;

@Aspect
@Component
public class RunTimeAspect {

    @Around(value = "execution(* controller.UserinfoController.insertUserinfo(..))")
    public Object aroundMethod(ProceedingJoinPoint point) throws Throwable {
        Object object = null;
        System.out.println("begin " + System.currentTimeMillis());
        object = point.proceed();
        System.out.println("  end " + System.currentTimeMillis());
        return object;
    }

}
```

运行类代码如下：

```
package application;

import org.springframework.boot.SpringApplication;
import org.springframework.boot.autoconfigure.EnableAutoConfiguration;
import org.springframework.context.annotation.ComponentScan;

@EnableAutoConfiguration
@ComponentScan(basePackages = { "controller", "service", "aspect" })
public class Application {
    public static void main(String[] args) {
        SpringApplication.run(Application.class);
    }
}
```

输入 URL 网址：

```
http://localhost:8080/insertUserinfo
```

控制台输出的结果如下：

```
begin 1536593146457
public void insertUserinfoService()
  end 1536593146457
```

切面成功实现。

7.3.17　官方建议的项目结构

虽然 Spring Boot 官方并没有强制规定 Spring Boot 的项目结构，但是官方有建议使用的项目结构，如图 7-44 所示。

在 com.example.myapplication 包下创建 Application.java 启动类，执行 Application.java 启动类后会自动扫描子孙包中的资源，并不需要显式地指定扫描某个 package 包。

7.3.18　实现 Spring Boot 整合 Thymeleaf 模板

创建新的项目 SpringBootTest2，项目结构如图 7-45 所示。

```
com
 +- example
     +- myapplication
         +- Application.java
         |
         +- customer
         |   +- Customer.java
         |   +- CustomerController.java
         |   +- CustomerService.java
         |   +- CustomerRepository.java
         |
         +- order
             +- Order.java
             +- OrderController.java
             +- OrderService.java
             +- OrderRepository.java
```

图 7-44　官方建议使用的项目结构

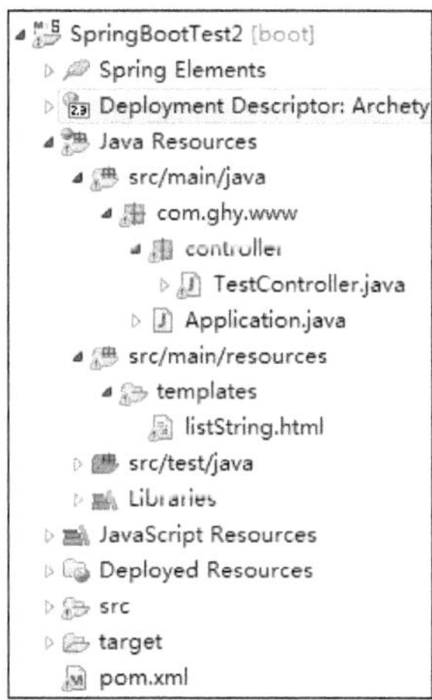

图 7-45　项目结构

文件 pom.xml 代码如下：

```
<project xmlns="http://maven.apache.org/POM/4.0.0"
    xmlns:xsi="http://www.w3.org/2001/XMLSchema-instance"
    xsi:schemaLocation="http://maven.apache.org/POM/4.0.0 http://maven.apache.org/
maven-v4_0_0.xsd">
    <modelVersion>4.0.0</modelVersion>
    <groupId>com.ghy.www</groupId>
    <artifactId>SpringBootTest2</artifactId>
    <packaging>war</packaging>
    <version>0.0.1-SNAPSHOT</version>
    <name>SpringBootTest2 Maven Webapp</name>
    <url>http://maven.apache.org</url>

    <parent>
        <!-- 从 Spring Boot 继承默认的配置 -->
        <groupId>org.springframework.boot</groupId>
        <artifactId>spring-boot-starter-parent</artifactId>
        <version>2.0.4.RELEASE</version>
    </parent>

    <dependencies>
        <dependency>
            <groupId>junit</groupId>
            <artifactId>junit</artifactId>
            <version>3.8.1</version>
            <scope>test</scope>
        </dependency>

        <!-- 添加 Web 项目典型的依赖 -->
        <dependency>
            <groupId>org.springframework.boot</groupId>
            <artifactId>spring-boot-starter-web</artifactId>
        </dependency>

        <!-- 添加 Thymeleaf 依赖 -->
        <dependency>
            <groupId>org.springframework.boot</groupId>
            <artifactId>spring-boot-starter-thymeleaf</artifactId>
        </dependency>
    </dependencies>
    <build>
        <finalName>SpringBootTest1</finalName>
        <plugins>
            <plugin>
                <groupId>org.springframework.boot</groupId>
                <artifactId>spring-boot-maven-plugin</artifactId>
            </plugin>
        </plugins>
    </build>
</project>
```

控制层代码如下：

```
@Controller
public class TestController {
    @RequestMapping("test1")
    public String testMethod1(HttpServletRequest request, HttpServletResponse response) {
```

```
        System.out.println("执行了控制层 test1");
        List list = new ArrayList();
        list.add("我是中国人 1");
        list.add("我是中国人 2");
        list.add("我是中国人 3");
        request.setAttribute("list", list);
        return "listString.html";
    }
}
```

视图文件 listString.html 的代码如下：

```
<!DOCTYPE html>
<html>
<head>
<meta charset="UTF-8">
<title>Insert title here</title>
</head>
<body>
    <th:block th:each="each:${list}">
            [[${each}]]
            <br />
    </th:block>
</body>
</html>
```

注意：要在 src/main/resources 中创建名称为 templates 的文件夹，而不是名称为 templates 的包。在 templates 文件夹中创建 listString.html 文件。

启动类代码如下：

```
@SpringBootApplication
public class Application {
    public static void main(String[] args) {
        SpringApplication.run(Application.class);
    }
}
```

程序运行后，执行控制层的 URL 就可以访问 Thymeleaf 标签文件了。

一个 @SpringbootApplication 相当于 @Configuration、@EnableAuto-Configuration 和@ComponentScan 注解的集合，并具有它们的默认属性值。

7.3.19 使用自定义的 Thymeleaf 模板显示异常信息

创建测试用的项目 SpringBootTest3，项目结构如图 7-46 所示。注意在 pom.xml 文件中添加 Thymeleaf 依赖，配置代码如下：

```
<!-- 添加 Thymeleaf 依赖 -->
<dependency>
    <groupId>org.springframework.boot</groupId>
    <artifactId>spring-boot-starter-thymeleaf</artifactId>
</dependency>
```

```
SpringBootTest3 [boot]
  Spring Elements
  Deployment Descriptor: Archetyp
  Java Resources
    src/main/java
      com.ghy.www
        controller
          TestController.java
        entity
          Userinfo.java
            Userinfo
        errorviews
          ErrorView.java
        exception
          LoginException1.java
          LoginException2.java
        Application.java
    src/main/resources
      templates
        errorPage.html
        loginException1.html
        sqlerror.html
      springMVC-servlet.xml
    src/test/java
    Libraries
  JavaScript Resources
  Deployed Resources
  src
  target
  pom.xml
```

图 7-46 项目结构

创建实体类，代码如下：

```
public class Userinfo {
    private String username;
    private String password;

    public Userinfo() {
    }

    public Userinfo(String username, String password) {
        super();
        this.username = username;
        this.password = password;
    }

    //省略 get 和 set 方法

}
```

控制层代码如下：

```
@RestController
public class TestController {
    @RequestMapping(value = "test1")
    public String test1() {
        System.out.println("public String test1()");
        int i = 10;
        int j = 0;
        int result = i / j;
        return "responseBody1";
    }

    @RequestMapping(value = "test2")
    public String test2() throws Exception {
        System.out.println("public String test2()");
        if (1 == 1) {
            throw new SQLException("SQL 语句错误");
        }
        return "responseBody2";
    }

    @RequestMapping(value = "test3")
    public String test3() throws LoginException1 {
        System.out.println("public String test3()");
        if (1 == 1) {
            throw new LoginException1("账号××登录失败，请重新登录!");
        }
        return "responseBody3";
    }

    @RequestMapping(value = "test4")
    public String test4() throws LoginException2 {
        System.out.println("public String test4()");
        if (1 == 1) {
            throw new LoginException2("账号××登录失败，请重新登录!");
        }
        return "responseBody4";
    }
```

```
    @RequestMapping(value = "test5")
    public Map<String, String> test5() {
        System.out.println("public String test5()");
        Map map = new HashMap();
        map.put("key1", "中国 new");
        map.put("key2", "中国人 new");
        return map;
    }

    @RequestMapping(value = "test6")
    public Userinfo test6() {
        System.out.println("public String test6()");
        Userinfo userinfo = new Userinfo();
        userinfo.setUsername("账号 new");
        userinfo.setPassword("密码 new");
        return userinfo;
    }

}
```

创建自定义异常类，代码如下：

```
public class LoginException1 extends Exception {
    public LoginException1(String exceptionMessage) {
        super(exceptionMessage);
    }
}
```

创建自定义异常类，代码如下：

```
public class LoginException2 extends Exception {
    public LoginException2(String exceptionMessage) {
        super(exceptionMessage);
    }
}
```

创建异常处理类，代码如下：

```
@RestControllerAdvice
public class ErrorView {
    @ExceptionHandler(value = LoginException2.class)
    public Map processLoginException2(LoginException2 ex) {
        Map map = new HashMap();
        map.put("controllerName", "UserinfoController");
        map.put("message", ex.getMessage());
        return map;
    }

    @ExceptionHandler(value = Exception.class)
    public ModelAndView processException(Exception ex) {
        ModelAndView mav = new ModelAndView();
        mav.addObject("ex", ex);
        mav.setViewName("errorPage.html");
        return mav;
    }

    @ExceptionHandler(value = SQLException.class)
    public ModelAndView processSQLException(SQLException ex) {
```

```
        ModelAndView mav = new ModelAndView();
        mav.addObject("ex", ex);
        mav.setViewName("sqlerror.html");
        return mav;
    }

    @ExceptionHandler(value = LoginException1.class)
    public ModelAndView processLoginException1(LoginException1 ex) {
        ModelAndView mav = new ModelAndView();
        mav.addObject("ex", ex);
        mav.setViewName("loginException1.html");
        return mav;
    }

}
```

启动类代码如下：

```
@SpringBootApplication
public class Application {
    public static void main(String[] args) {
        SpringApplication.run(Application.class);
    }
}
```

标签文件在默认的情况下保存在 src/main/resources/templates 路径中。

视图文件 errorPage.html 的代码如下：

```
<!DOCTYPE html>
<html>
<head>
<meta charset="UTF-8">
<title>Insert title here</title>
</head>
<body>
进入了 errorPage.html 文件<br>
错误消息为：[[${ex.getMessage()}]]
</body>
</html>
```

视图文件 loginException1.html 的代码如下：

```
<!DOCTYPE html>
<html>
<head>
<meta charset="UTF-8">
<title>Insert title here</title>
</head>
<body>
进入了 loginException1.html 文件<br>
错误消息为：[[${ex.getMessage()}]]
</body>
</html>
```

视图文件 sqlerror.html 的代码如下：

```
<!DOCTYPE html>
<html>
```

```
<head>
<meta charset="UTF-8">
<title>Insert title here</title>
</head>
<body>
进入了 sqlerror.html 文件<br>
错误消息为：[[${ex.getMessage()}]]
</body>
</html>
```

执行路径 http://localhost:8080/test1 后浏览器显示的内容如图 7-47 所示。

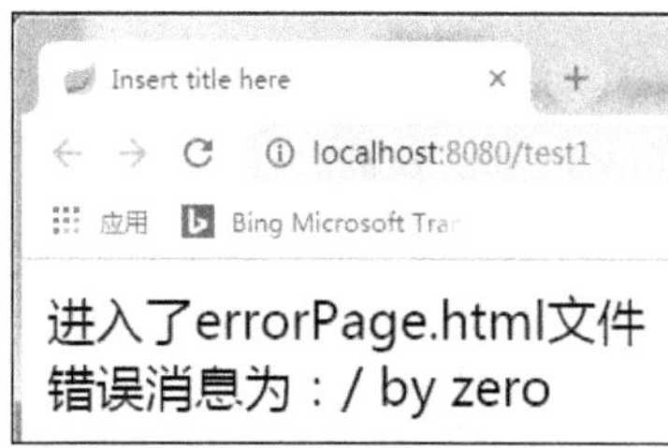

图 7-47　执行效果 1

执行路径 http://localhost:8080/test2 后浏览器显示的内容如图 7-48 所示。

执行路径 http://localhost:8080/test3 后浏览器显示的内容如图 7-49 所示。

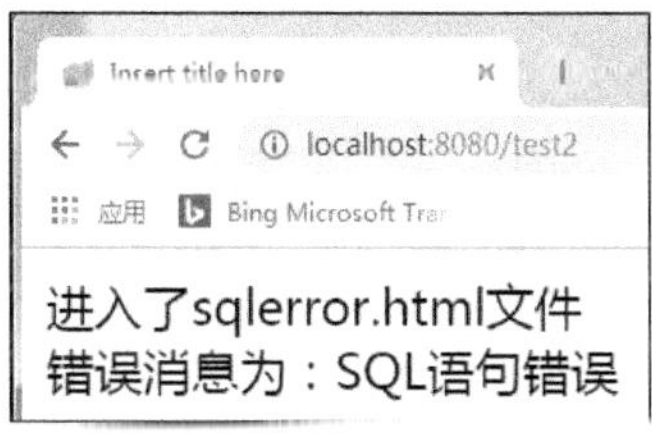

图 7-48　执行效果 2

图 7-49　执行效果 3

执行路径 http://localhost:8080/test4 后浏览器显示的内容如图 7-50 所示。

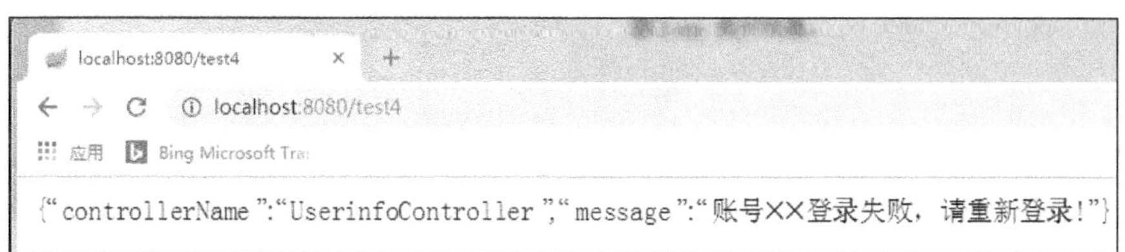

图 7-50　执行效果 4

执行路径 http://localhost:8080/test5 后浏览器显示的内容如图 7-51 所示。

执行路径 http://localhost:8080/test6 后浏览器显示的内容如图 7-52 所示。

图 7-51　执行效果 5

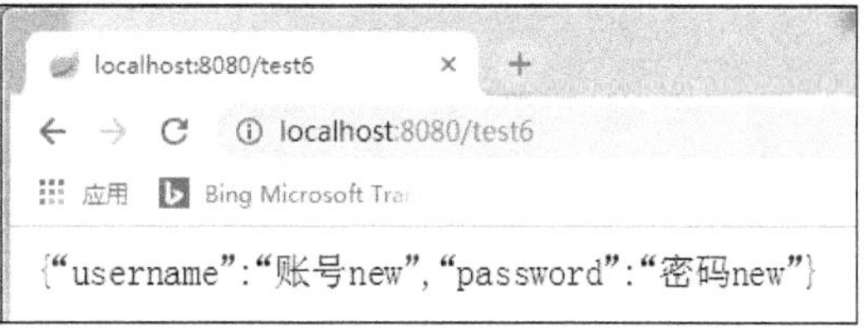

图 7-52　执行效果 6

7.3.20　实现 Spring Boot 整合 MyBatis 框架

创建测试用的项目 SpringBootTest4。

项目结构如图 7-53 所示。

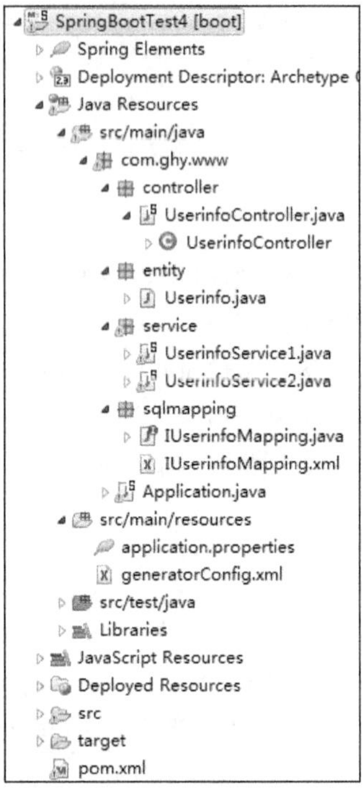

创建实体类，代码如下：

```
package com.ghy.www.entity;

import java.util.Date;

public class Userinfo {
    private Long id;
    private String username;
    private String password;
    private Long age;
    private Date insertdate;

    //省略 set 和 get 方法

}
```

```
SpringBootTest4 [boot]
  Spring Elements
  Deployment Descriptor: Archetype
  Java Resources
    src/main/java
      com.ghy.www
        controller
          UserinfoController.java
            UserinfoController
        entity
          Userinfo.java
        service
          UserinfoService1.java
          UserinfoService2.java
        sqlmapping
          IUserinfoMapping.java
          IUserinfoMapping.xml
        Application.java
    src/main/resources
      application.properties
      generatorConfig.xml
    src/test/java
    Libraries
  JavaScript Resources
  Deployed Resources
  src
  target
  pom.xml
```

图 7-53 项目结构

创建业务层，代码如下：

```
@Service
public class UserinfoService1 {
    @Autowired
    private IUserinfoMapping userinfoMapping;

    public void insertUserinfoService1_OK() {
        Userinfo userinfo1 = new Userinfo();
        userinfo1.setUsername("中国 1");
        userinfo1.setPassword("中国人 1");
        userinfo1.setAge(100L);
        userinfo1.setInsertdate(new Date());
        userinfoMapping.insertUserinfo(userinfo1);
    }
}
```

创建业务层，代码如下：

```
@Service
public class UserinfoService2 {
    @Autowired
    private IUserinfoMapping userinfoMapping;

    public void insertUserinfoService2_OK() {
        Userinfo userinfo1 = new Userinfo();
        userinfo1.setUsername("中国 1");
        userinfo1.setPassword("中国人 1");
        userinfo1.setAge(100L);
        userinfo1.setInsertdate(new Date());
        userinfoMapping.insertUserinfo(userinfo1);
    }

    public void insertUserinfoService2_Error() {
        Userinfo userinfo1 = new Userinfo();
        userinfo1.setUsername("中国 2");
        userinfo1.setPassword(
                    "中国人 2 中国人 2 中国人 2 中国人 2 中国人 2 中国人 2 中国人 2 中国人 2 中国人 2 中国人 2 
中国人 2 中国人 2 中国人 2 中国人 2 中国人 2 中国人 2 中国人 2 中国人 2 中国人 2 中国人 2 中国人 2 中国人 2 中国人 2 中国人 2 中
国人 2 中国人 2 中国人 2 中国人 2 中国人 2 中国人 2 中国人 2 中国人 2");
        userinfo1.setAge(100L);
        userinfo1.setInsertdate(new Date());
```

```
        userinfoMapping.insertUserinfo(userinfo1);
    }

}
```

创建 MyBatis 的 SQL 映射文件 IUserinfoMapping.xml，代码如下：

```
<?xml version="1.0" encoding="UTF-8" ?>
<!DOCTYPE mapper
 PUBLIC "-//mybatis.org//DTD Mapper 3.0//EN"
 "http://mybatis.org/dtd/mybatis-3-mapper.dtd">
<mapper namespace="com.ghy.www.sqlmapping.IUserinfoMapping">
    <insert id="insertUserinfo"
        parameterType="com.ghy.www.entity.Userinfo">
        <selectKey resultType="java.lang.Long" keyProperty="id"
            order="BEFORE">
            select idauto.nextval from dual
        </selectKey>
        insert into userinfo(id,username,password,age,insertdate)
        values(#{id},#{username},#{password},#{age},#{insertdate})
    </insert>
</mapper>
```

创建 IUserinfoMapping.xml 映射的 Java 接口，代码如下：

```
@Mapper
public interface IUserinfoMapping {
    public void insertUserinfo(Userinfo userinfo);
}
```

注意：接口 IUserinfoMapping.java 上方要添加 @Mapper 注解。

创建控制层，代码如下：

```
@Controller
@Transactional
public class UserinfoController {

    @Autowired
    private UserinfoService1 service1;

    @Autowired
    private UserinfoService2 service2;

    @RequestMapping("test1")
    public void test1(HttpServletRequest request, HttpServletResponse response) {
        System.out.println("test1 run !");
        service1.insertUserinfoService1_OK();
        service2.insertUserinfoService2_OK();
    }

    @RequestMapping("test2")
    public void test2(HttpServletRequest request, HttpServletResponse response) {
        System.out.println("test2 run !");
        service1.insertUserinfoService1_OK();
        service2.insertUserinfoService2_Error();
    }

}
```

创建运行类，代码如下：

```
@SpringBootApplication
@EnableTransactionManagement
public class Application {
    public static void main(String[] args) {
        SpringApplication.run(Application.class);
    }
}
```

创建属性文件 application.properties，代码如下：

```
spring.datasource.driver-class-name=oracle.jdbc.OracleDriver
spring.datasource.url=jdbc:oracle:thin:@localhost:1521:orcl
spring.datasource.username=y2
spring.datasource.password=123

spring.datasource.type=com.zaxxer.hikari.HikariDataSource
spring.datasource.hikari.minimum-idle=5
spring.datasource.hikari.maximum-pool-size=15
spring.datasource.hikari.auto-commit=true
spring.datasource.hikari.idle-timeout=30000
spring.datasource.hikari.pool-name=MyHikariCP
spring.datasource.hikari.max-lifetime=1800000
spring.datasource.hikari.connection-timeout=30000
spring.datasource.hikari.connection-test-query=SELECT 1 from dual

spring.http.encoding.charset=utf-8
spring.http.encoding.enabled=true
spring.http.encoding.force=true
```

创建 MyBatis 逆向 Userinfo 实体类的配置文件，代码如下：

```
<?xml version="1.0" encoding="UTF-8"?>
<!DOCTYPE generatorConfiguration PUBLIC "-//mybatis.org//DTD MyBatis Generator
Configuration 1.0//EN" "http://mybatis.org/dtd/mybatis-generator-config_1_0.dtd">
<generatorConfiguration>
    <context id="context1">
        <jdbcConnection
            connectionURL="jdbc:oracle:thin:@localhost:1521:orcl"
            driverClass="oracle.jdbc.OracleDriver" password="123" userId="y2" />
        <javaModelGenerator targetPackage="entity"
            targetProject="SpringBootTest4" />
        <table schema="y2" tableName="userinfo">
            <generatedKey column="id"
                sqlStatement="select idauto.nextval from dual" identity="false" />
        </table>
    </context>
</generatorConfiguration>
```

文件 pom.xml 的配置代码如下：

```
<project xmlns="http://maven.apache.org/POM/4.0.0"
    xmlns:xsi="http://www.w3.org/2001/XMLSchema-instance"
    xsi:schemaLocation="http://maven.apache.org/POM/4.0.0 http://maven.apache.org/
maven-v4_0_0.xsd">
    <modelVersion>4.0.0</modelVersion>
    <groupId>com.ghy.www</groupId>
    <artifactId>SpringBootTest4</artifactId>
```

```
    <packaging>war</packaging>
    <version>0.0.1-SNAPSHOT</version>
    <name>SpringBootTest4 Maven Webapp</name>
    <url>http://maven.apache.org</url>

    <parent>
        <!-- 从 Spring Boot 继承默认的配置 -->
        <groupId>org.springframework.boot</groupId>
        <artifactId>spring-boot-starter-parent</artifactId>
        <version>2.0.4.RELEASE</version>
    </parent>

    <dependencies>
        <dependency>
            <groupId>junit</groupId>
            <artifactId>junit</artifactId>
            <version>3.8.1</version>
            <scope>test</scope>
        </dependency>

        <!-- 添加 Web 项目典型的依赖 -->
        <dependency>
            <groupId>org.springframework.boot</groupId>
            <artifactId>spring-boot-starter-web</artifactId>
        </dependency>

        <!-- 添加 Thymeleaf 依赖 -->
        <dependency>
            <groupId>org.springframework.boot</groupId>
            <artifactId>spring-boot-starter-thymeleaf</artifactId>
        </dependency>

        <!-- 添加 Mybatis 依赖 -->
        <dependency>
            <groupId>org.mybatis.spring.boot</groupId>
            <artifactId>mybatis-spring-boot-starter</artifactId>
            <version>1.3.2</version>
        </dependency>

        <!-- 添加 Oracle 数据库驱动依赖 -->
        <dependency>
            <groupId>com.github.noraui</groupId>
            <artifactId>ojdbc8</artifactId>
            <version>12.2.0.1</version>
        </dependency>

    </dependencies>
    <build>
        <finalName>SpringBootTest4</finalName>
        <plugins>
            <plugin>
                <groupId>org.springframework.boot</groupId>
                <artifactId>spring-boot-maven-plugin</artifactId>
            </plugin>
        </plugins>
    </build>
</project>
```

启动项目后执行路径为 test1 的控制层，成功在数据表中添加了两条记录。

而执行路径为 test2 的控制层后系统出现异常，并未在数据表中添加任何一条记录，事务进行了整体回滚，Spring Boot 和 MyBatis 整合成功。

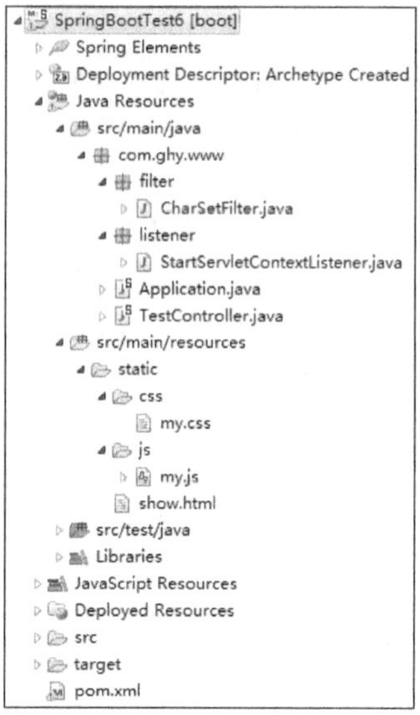

7.3.21 整合 @WebFilter 和 @WebListener 资源

创建测试用的项目 SpringBootTest6。

项目结构如图 7-54 所示。

```
SpringBootTest6 [boot]
  Spring Elements
  Deployment Descriptor: Archetype Created
  Java Resources
    src/main/java
      com.ghy.www
        filter
          CharSetFilter.java
        listener
          StartServletContextListener.java
        Application.java
        TestController.java
    src/main/resources
      static
        css
          my.css
        js
          my.js
        show.html
    src/test/java
    Libraries
  JavaScript Resources
  Deployed Resources
  src
  target
  pom.xml
```

图 7-54 项目结构

创建过滤器，代码如下：

```
@WebFilter(urlPatterns = "/*")
public class CharSetFilter implements Filter {

    @Override
    public void init(FilterConfig filterConfig) throws ServletException {
    }

    @Override
    public void doFilter(ServletRequest request, ServletResponse response, FilterChain chain)
            throws IOException, ServletException {
        System.out.println("CharSetFilter run !");
        request.setCharacterEncoding("utf-8");
        response.setCharacterEncoding("utf-8");
        chain.doFilter(request, response);
    }

    @Override
    public void destroy() {
    }

}
```

创建监听器，代码如下：

```
@WebListener
public class StartServletContextListener implements ServletContextListener {
    @Override
    public void contextInitialized(ServletContextEvent sce) {
        System.out.println("contextInitialized===========");
    }

    @Override
    public void contextDestroyed(ServletContextEvent sce) {
        System.out.println("contextDestroyed===========");
    }
}
```

创建控制层，代码如下：

```
@Controller
public class TestController {
    @RequestMapping("test")
    public String testMethod() {
        System.out.println("test controller run !");
        return "show.html";
    }
}
```

创建启动类，代码如下：

```
@SpringBootApplication
@ServletComponentScan
public class Application {
    public static void main(String[] args) {
        SpringApplication.run(Application.class);
    }
}
```

样式文件 my.css 的代码如下：

```
.colorFont {
    font-size: 50px;
    color: red;
}
```

脚本文件 my.js 的代码如下：

```
function autoRun() {
    alert("autoRunMethod run !");
}
```

视图文件 show.html 的代码如下：

```
<!DOCTYPE html>
<html>
<head>
<meta charset="UTF-8">
<title>Insert title here</title>
<script type="text/javascript" src="js/my.js">

</script>
<link href="css/my.css" rel="stylesheet" type="text/css" />
</head>
<body onload="autoRun()">
    <span class="colorFont">进入了 show.html 文件! </span>
</body>
</html>
```